Macdonald Theory and Beyond

CONTEMPORARY MATHEMATICS

815

Macdonald Theory and Beyond

AMS Special Session
Macdonald Theory and Beyond: Combinatorics, Geometry, and Integrable Systems
March 19–20, 2022
Virtual

Daniel Orr
Joshua Jeishing Wen
Editors

AMERICAN MATHEMATICAL SOCIETY

2020 *Mathematics Subject Classification.* Primary 05E05, 33D52.

Library of Congress Cataloging-in-Publication Data

Names: Virtual AMS Special Session on Macdonald Theory and Beyond : Combinatorics, Geometry, and Integrable Systems (2022 : online), author. | Orr, Daniel, 1986- editor. | Wen, Joshua Jeishing, 1992- editor.

Title: Macdonald theory and beyond : AMS Special Session on Macdonald Theory and Beyond : Combinatorics, Geometry, and Integrable Systems, virtual, March 19-20, 2022 / Daniel Orr, Joshua Jeishing Wen, editor.

Description: Providence, Rhode Island : American Mathematical Society, [2025] | Series: Contemporary mathematics, 0271-4132 ; volume 815 | Includes bibliographical references.

Identifiers: LCCN 2024035041 | ISBN 9781470471897 (paperback) | ISBN 9781470478667 (ebook)

Subjects: LCSH: Polynomials – Congresses. | Combinatorial analysis – Congresses. | Functions, Special – Congresses. : | AMS: Combinatorics – Algebraic combinatorics – Symmetric functions and generalizations. | Special functions – Basic hypergeometric functions – Basic orthogonal polynomials and functions associated with root systems (Macdonald polynomials, etc.).

Classification: LCC QA161.P59 V57 2022 | DDC 512.9/422–dc23/eng/20250121

LC record available at https://lccn.loc.gov/2024035041

Contemporary Mathematics ISSN: 0271-4132 (print); ISSN: 1098-3627 (online)

DOI: https://doi.org/10.1090/conm/815

∞ The paper used in this book is acid-free and falls within the guidelines
established to ensure permanence and durability.
Visit the AMS home page at `https://www.ams.org/`

10 9 8 7 6 5 4 3 2 1 30 29 28 27 26 25

Contents

Preface

The articles in this volume represent a number of recent developments in Macdonald theory while highlighting some of its many connections to other areas of mathematics. An important common thread throughout the volume is the role of combinatorial formulas—for Macdonald polynomials themselves as well as operations on them arising from rich additional structures. The articles of Haglund, Mandelshtam, and Romero concern the type A Macdonald polynomials, which remain a major focus of the subject due to the depth of their combinatorial theory and the power of their specific applications. For arbitrary type Macdonald polynomials, a new combinatorial formula with pseudo-crystal structure is presented in the article of Lenart, Naito, Nomoto, and Sagaki. Finally, the articles of Saied and Wen take up two important new directions in the subject: the SSV polynomials arising from the study of special functions on metaplectic groups, and the wreath Macdonald polynomials associated with certain symplectic resolutions.

Most of the articles in the volume are at least in part expository. It is our hope that this will be useful to those at all levels of experience with the subject.

The idea for this volume originated from our Special Session on "Macdonald Theory and Beyond: Combinatorics, Geometry, and Integrable Systems" at the AMS Spring Eastern Virtual Sectional Meeting on March 19–20, 2022. It is our pleasure to thank Steven H. Weintraub for help with our special session, as well as all those who participated in our session and those who contributed to this volume. Finally, we are especially grateful to Sergei Gelfand, Eriko Hironaka, and Christine Thivierge for their help and guidance in making this volume a reality.

Daniel Orr
Joshua Jeishing Wen

Contemporary Mathematics
Volume **815**, 2025
https://doi.org/10.1090/conm/815/16314

Combinatorial formulas for (type A) Macdonald polynomials

J. Haglund

Abstract. We give an expository overview of the combinatorial formula of Haglund, Haiman and Loehr for the modified Macdonald polynomials $\tilde{H}_\mu(X;q,t)$, as well as versions of the formula which correspond to other variants of Macdonald polynomials. These include Macdonald's original $P_\mu(X;q,t)$ and the nonsymmetric Macdonald polynomials $E_\alpha(X;q,t)$, as well as their integral forms $J_\mu(X;q,t)$ and $\mathcal{E}_\alpha(X;q,t)$. We also discuss more compact versions of these formulas, as well as a quasisymmetric Macdonald polynomial, which grew out of work connecting Macdonald polynomials to the asymmetric simple exclusion process (ASEP).

1. Introduction

In 1988 Macdonald introduced an important family of symmetric functions $P_\mu(X;q,t)$, which depend on a countable set of variables $\{x_1, x_2, \ldots\}$, a partition μ, and two parameters q,t [**27**], [**28**][Ch. 6]. Special cases of the P_μ include the famous Hall Littlewood polynomials (i.e. $P_\mu(X;0,t)$), the Jack polynomials $J^\beta_\mu(X)$ (defined as $\lim_{t\to 1} P_\mu(X;t^\beta,t)$), as well as the classical bases of monomial symmetric functions m_λ, elementary symmetric functions e_λ, and Schur functions s_λ. We refer the reader to [**34**][Ch. 7], [**28**][Ch. 1] for general background on these bases for the ring $\mathrm{Sym}(X)$ of symmetric functions. In order to maintain a certain amount of brevity we will often use notation common to these sources without giving explicit definitions.

In general the coefficient of a given monomial in $P_\mu(X;q,t)$ is a rational function in q,t, but Macdonald found that if he multiplied $P_\mu(X;q,t)$ by a certain product of linear terms in q,t to form the *integral form* $J_\mu(X;q,t)$, the monomial coefficients seemed to always be in $\mathbb{Z}[q,t]$. (We let $\mathbb{Z}, \mathbb{N}, \mathbb{Z}_+$ denote the integers, nonnegative integers, and positive integers, respectively.) More importantly, Macdonald conjectured the much stronger fact that when the J_μ are expanded in a certain variant of the Schur basis, the coefficients are in $\mathbb{N}[q,t]$. Garsia and Haiman found a refinement of this, their famous $n!$ Conjecture, which says that if a modified version of J_μ they denoted $\tilde{H}_\mu(X;q,t)$ is expanded in the Schur basis, the coefficients count the multiplicities of irreducible S_n modules in the decomposition of a certain S_n module V_μ. The $n!$ Conjecture was finally proved by Haiman in 2001 using algebraic geometry. It is still an open problem though to find a combinatorial interpretation for the Schur coefficients of $\tilde{H}_\mu(X;q,t)$.

2020 *Mathematics Subject Classification.* 05-02, 05-06, 05A15, 05A19, 05E05.

In 1996 several groups of researchers independently found proofs of "integrality", i.e. that the monomial coefficients in J_μ are in $\mathbb{Z}[q,t]$. This is equivalent to the same result holding for $\tilde{H}_\mu(X;q,t)$. In 2004 the author found a (conjectural) combinatorial formula for $\tilde{H}_\mu(X;q,t)$ which showed these coefficients were also nonnegative, i.e. in $\mathbb{N}[q,t]$, and this was proved shortly after in joint work with Haiman and Loehr, giving the so-called HHL formula for $\tilde{H}_\mu(X;q,t)$. This eventually led to a second proof of the stronger result of Schur positivity, as well as other nice applications.

This paper is an expository account of the HHL formula and its variants. Section 2 contains a description of the original formula, and outlines some of its main applications. In Section 3 we show how to obtain a corresponding formula for $J_\mu(X;q,t)$. Section 4 describes an extension of this to the nonsymmetric Macdonald polynomial $E_\alpha(X;q,t)$ and its integral form $\mathcal{E}_\alpha(X;q,t)$. We also discuss how this formula for E_α inspires equivalent versions of the HHL formulas for $\tilde{H}_\mu(X;q,t)$ and $J_\mu(X;q,t)$ which involve fillings of any "skyline" diagram obtained by permuting the columns of a partition μ. Section 5 contains somewhat more compact versions of the above formulas, as well as a new quasisymmetric Macdonald polynomial, all of which grew out of the study of the ASEP.

2. The Formula for $\tilde{H}_\mu(X;q,t)$

By the notation $\mu \vdash n$, we mean μ is a partition of n. We let μ refer to both a partition and its Ferrers diagram. We let $\mathrm{dg}(\mu)$ denote the "augmented" diagram of μ, consisting of μ together with a row of squares below μ, referred to as the *basement*, with (x,y) coordinates $(j,0)$, $1 \le j \le \mu_1$. Define a *filling* σ of μ to be an assignment of a positive integer to each square of μ. For $s \in \mu$, we let $\sigma(s)$ denote the integer assigned to s, i.e the integer occupying s. If s has (x,y) coordinates (i,j), we let $\sigma(i,j)$ denote $\sigma(s)$. By the bottom row of μ, also referred to as the first row, or row 1, we mean the set of squares $\{(k,1), 1 \le k \le \mu_1\}$. Define the *reading word* of σ, denoted $\mathrm{read}(\sigma)$, to be the word $\sigma_1\sigma_2\cdots\sigma_n$, obtained be reading the occupants of μ across rows left to right, starting with the top row and working downwards. We say σ is a *standard filling* if $\mathrm{read}(\sigma)$ is a permutation. See Figure 1.

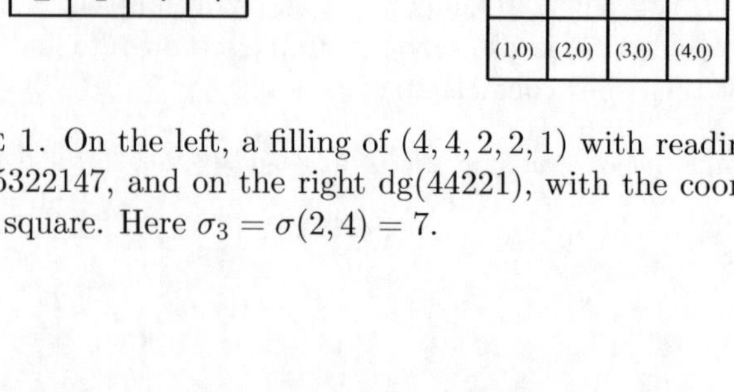

FIGURE 1. On the left, a filling of $(4,4,2,2,1)$ with reading word 1573235322147, and on the right $\mathrm{dg}(44221)$, with the coordinates of each square. Here $\sigma_3 = \sigma(2,4) = 7$.

For each filling σ of μ we associate x, q and t weights. The x weight is defined in a similar fashion to the usual semi-standard Young tableaux (SSYT), namely

$$x^\sigma = \prod_{s\in\mu} x_{\sigma(s)}. \tag{2.1}$$

For $s \in \mu$, let North(s) denote the square of μ right above s (if it exists) in the same column, and South(s) the square of dg(μ) directly below s, in the same column. Let the descent set of σ, denoted Des(σ,μ), be the set of squares s where $\sigma(s) > \sigma(\text{South}(s))$. (We regard the basement as containing virtual infinity symbols, so no square in the bottom row of σ can be in Des(σ,μ).) For $s \in \mu$, let leg(s) denote the number of squares strictly above s in its column, and arm(s) the number of squares strictly to the right of s in its row. Finally set

$$\operatorname{maj}(\sigma,\mu) = \sum_{s\in \operatorname{Des}(\sigma,\mu)} 1 + \operatorname{leg}(s). \tag{2.2}$$

Note that $\operatorname{maj}(\sigma,1^n) = \operatorname{maj}(\sigma)$, where σ is viewed as a word, and maj is the usual major index statistic (i.e., the sum of the elements of the descent set of the word). For the filling σ on the left in Figure 1, we have $\operatorname{maj}(\sigma) = 2+4+1+3 = 10$.

We say a square $u \in \mu$ attacks all other squares $v \in \mu$ in its row and strictly to its right, and all other squares $v \in \mu$ in the row below and strictly to its left. We say u, v attack each other if u attacks v or v attacks u. An *inversion pair* of σ is a pair of squares u, v where u attacks v and $\sigma(u) > \sigma(v)$. Let Invset(σ,μ) denote the set of inversion pairs of σ, $\operatorname{Inv}(\sigma,\mu) = |\operatorname{Invset}(\sigma,\mu)|$ its cardinality and set

$$\operatorname{inv}(\sigma,\mu) = \operatorname{Inv}(\sigma,\mu) - \sum_{s\in \operatorname{Des}(\sigma,\mu)} \operatorname{arm}(s). \tag{2.3}$$

For example, if σ is the filling on the left in Figure 1 then

$$\begin{aligned}
\operatorname{Des}(\sigma) &= \{(1,2),(1,4),(2,2),(2,4)\},\\
\operatorname{maj}(\sigma) &= 4+2+3+1 = 10,\\
\operatorname{Invset}(\sigma) &= \{[(2,4),(1,3)],[(1,3),(2,3)],[(1,2),(4,2)],[(2,2),(3,2)],[(2,2),(4,2)],\\
&[(2,2),(1,1)],[(3,2),(4,2)],[(3,2),(1,1)],[(3,2),(2,1)],[(4,2),(2,1)],[(1,1),(2,1)]\}\\
\operatorname{inv}(\sigma) &= |\operatorname{Invset}(\sigma)| - (3+1+2+0) = 11-6 = 5.
\end{aligned}$$

Note that when μ is a single row, $\operatorname{inv}(\sigma,(n))$ is the usual inversion statistic on words, and when μ is a single column, $\operatorname{maj}(\sigma,(1^n))$ is the usual major index statistic.

Macdonald originally defined $P_\mu(X;q,t)$ as the unique polynomial satisfying certain orthogonality conditions. Later in this section we will describe a translation of these conditions which can be used to define the $\tilde{H}_\mu(X;q,t)$. It is nontrivial to show that there exists an element of Sym(X) satisfying these conditions. We now introduce a polynomial defined combinatorially which one can show satisfies the orthogonality conditions, and hence equals $\tilde{H}_\mu(X;q,t)$.

DEFINITION 2.1. For $\mu \vdash n$, let

$$C_\mu(X;q,t) = \sum_{\sigma:\mu\to\mathbb{Z}_+} x^\sigma t^{\operatorname{maj}(\sigma,\mu)} q^{\operatorname{inv}(\sigma,\mu)}. \tag{2.4}$$

Given a word $\sigma \in \mathbb{Z}_+^n$, we Stan($\sigma$) denote the *standardization* of σ, which is the unique permutation in the symmetric group S_n satisfying $\operatorname{Stan}(\sigma)_i < \operatorname{Stan}(\sigma)_j$ if and only if $\sigma_i \le \sigma_j$ for all $1 \le i < j \le n$. For example, Stan(24254) = 13254.

6	4		
2	4		
3	3	2	
1	4	1	1

11	8		
4	9		
6	7	5	
1	10	2	3

FIGURE 2. On the left, a filling of $(4,3,2,2)$ with reading word 64243321411 and on the right, its standardization.

Furthermore we define the standardization of a filling σ, again denoted $\operatorname{Stan}(\sigma)$, as the standard filling whose reading word is the standardization of $\operatorname{read}(\sigma)$. Figure 2 gives an example of this. Note that the statistics $\operatorname{inv}(\sigma,\mu)$ and $\operatorname{maj}(\sigma,\mu)$ are invariant under standardization. It thus follows easily from Definition 2.1 that

$$C_\mu(X;q,t) = \sum_{\beta\in S_n} t^{\operatorname{maj}(\beta,\mu)} q^{\operatorname{inv}(\beta,\mu)} F_{\operatorname{Des}(\beta^{-1})}(X), \tag{2.5}$$

where we identify a permutation β with the standard filling whose reading word is β. Here for n fixed and any subset γ of $\{1,2,\ldots,n-1\}$, $F_\gamma(X)$ is the Gessel Fundamental Quasisymmetric Function, which can be defined as

$$F_\gamma(X) = \sum_{\substack{1\le a_1\le a_2\le\cdots\le a_n \\ a_i=a_{i+1}\implies i\notin\gamma}} x_{a_1}x_{a_2}\cdots x_{a_n}, \tag{2.6}$$

or equivalently as

$$F_\gamma(X) = \sum_{\substack{\sigma\in Z_+^n \\ \operatorname{Des}(\operatorname{Stan}(\sigma)^{-1})=\gamma}} x_{\sigma_1}x_{\sigma_2}\cdots x_{\sigma_n}. \tag{2.7}$$

REMARK 2.2. There is another way to view $\operatorname{inv}(\sigma,\mu)$ which is makes it clear it is nonnegative, and will be important in the sequel. Call three squares u,v,w, with $u,v\in\mu$, $w=\operatorname{South}(u)$, and with v in the same row as u and strictly to the right of μ, a *triple*. Given a standard filling σ, we define an orientation on such a triple by starting at the square, either u,v or w, with the smallest element of σ in it, and going in a circular motion, towards the next largest element, and ending at the largest element. We say the triple is an inversion triple or a coinversion triple depending on whether this circular motion is counterclockwise or clockwise, respectively. Note that since $\sigma(j,0)=\infty$, if u,v are in the bottom row of σ, they are part of a counterclockwise triple if and only if $\sigma(u)>\sigma(v)$.

Extend this definition to (possibly non-standard) fillings by defining the orientation of a triple to be the orientation of the corresponding triple for the standardized filling $\operatorname{Stan}(\sigma)$. (So for two equal numbers, the one which occurs first in the reading word is regarded as being smaller). It is an easy exercise to show that $\operatorname{inv}(\sigma,\mu)$ is the number of counterclockwise triples. For example, for the filling on the left in Figure 1, the inversion triples are

$$\begin{gathered}[(2,3),(1,3),(1,2)],[(2,1),(3,2),(2,2)],\\ [(2,1),(4,2),(2,2)],[(4,2),(3,2),(3,1)],[(2,1),(1,1)].\end{gathered} \tag{2.8}$$

It is useful to note that if $\sigma(u) = \sigma(v)$ for u, v two entries in the same row, then any triple containing them is always a clockwise triple. Also, if $\sigma(u) = \sigma(w) \neq \sigma(v)$, then u, v, w is always form a counterclockwise triple.

We now introduce the operation of *plethysm*, which is central to the study of Macdonald polynomials. Let $p_k(X) = \sum_i x_i^k$ be the kth *power sum* symmetric function. Given an alphabet of parameters $t_1, t_2, \ldots$ and a function $E(t_1, t_2, \ldots)$ on this alphabet, we define the plethystic substitution of E into p_k, denoted $p_k[E]$, by

$$p_k[E(t_1, t_2, \ldots)] = E(t_1^k, t_2^k, \ldots). \tag{2.9}$$

More generally, for any $G \in \mathrm{Sym}$, we define the plethystic substitution of E into G, denoted $G[E]$, as the result obtained by first expressing G as a polynomial in the power sums (which can be done in a unique way since the power sums generate the ring Sym) and then replacing each p_k by $E(t_1^k, t_2^k, \ldots)$.

REMARK 2.3. Note $p_k[-X] = -p_k(X)$ (since -1 is not a parameter, the plethystic operation doesn't replace it by its kth power.) We define a special symbol ϵ satisfying $p_k[\epsilon X] = (-1)^k p_k(X)$ which gives us a way of distinguishing a negative sign inside plethystic brackets by the operation of negating a set of variables X. If ω is the standard involution on Sym which sends the Schur function s_λ to $s_{\lambda'}$, where λ' is the conjugate partition, then $\omega p_k(X) = (-1)^{k-1} p_k = p_k[-\epsilon X]$, so for any $G \in \mathrm{Sym}$, $\omega G(X) = G[-\epsilon X]$. For a more detailed discussion of plethysm see [**18**][Chap. 1].

Macdonald's Existence Theorem, when translated into a statement about the $\tilde{H}_\mu$, gives the following [**19**], [**14**].

THEOREM 2.4 (Existence Theorem for the modified Macdonald polynomial). *The following three conditions uniquely determine a family $\tilde{H}_\mu(X; q, t)$ which form a basis for the ring of symmetric functions with coefficients in $\mathbb{Q}(q, t)$.*

$$\tilde{H}_\mu[X(q-1); q, t] = \sum_{\rho \leq \mu'} c_{\rho,\mu}(q, t) m_\rho(X), \tag{2.10}$$

$$\tilde{H}_\mu[X(t-1); q, t] = \sum_{\rho \leq \mu} d_{\rho,\mu}(q, t) m_\rho(X), \tag{2.11}$$

$$\text{The coefficient of } x_1^n \text{ in } \tilde{H}_\mu(X; q, t) = 1, \tag{2.12}$$

where in the sums in (2.10) *and* (2.11) *we use the usual dominance partial order, i.e.*

$$\rho \leq \mu \iff \rho_1 + \rho_2 + \ldots + \rho_k \leq \mu_1 + \mu_2 + \ldots + \mu_k \quad \text{for} \quad 1 \leq k. \tag{2.13}$$

The following result was conjectured in [**13**] and proved shortly after in [**14**].

THEOREM 2.5 (HHL Formula). *For any partition μ,*

$$C_\mu[X; q, t] = \tilde{H}_\mu[X; q, t]. \tag{2.14}$$

OPEN PROBLEM 2.6. The $n!$ Theorem implies the well-known symmetry relation

$$\tilde{H}_\mu[Z; q, t] = \tilde{H}_{\mu'}[Z; t, q]. \tag{2.15}$$

(This can also be derived fairly easily from the three axioms in Theorem 2.4 above). Prove that (2.4) satisfies this symmetry bijectively. Note that in the case $\mu = (n)$ this question is equivalent to asking for a bijective proof that maj and inv have the same distribution on arbitrary multisets, which is exactly what Foata's ϕ map [**10**] gives (see also [**18**][Chap.1]).

Before outlining the proof of Theorem 2.5, we discuss a few of its important corollaries.

- The proof of Theorem 2.5 gives a new proof of Macdonald's Existence Theorem.
- Theorem 2.5 not only gives a new proof that the monomial coefficients of $\tilde{H}_\mu$ are in $\mathbb{Z}[q,t]$, but shows they are actually in $\mathbb{N}[q,t]$. Haiman's Schur positivity result mentioned earlier implies this immediately, but (2.14) gave the first proof of monomial positivity that didn't use that.
- With a little work one can show that the HHL formula implies a famous result of Lascoux-Schützenberger [**24**] which says that

$$\langle \tilde{H}_\mu(X;0,t), s_\lambda(X) \rangle = \sum_{T \in \mathrm{SSYT}(\lambda,\mu)} t^{\mathrm{cocharge}(T)}. \tag{2.16}$$

Here $\langle , \rangle$ is the usual Hall scalar product with respect to which the Schur functions are orthonormal, and so (2.16) gives the Schur expansion of the $q = 0$ case of $\tilde{H}_\mu$. The statistic cocharge(T) on SSYT of shape λ and weight μ will be given implicitly by the algorithm outlined below; see [**18**][Ch. 1 and Appendix A] for an explicit definition of cocharge, background on the Hall scalar product, and a full description of how to obtain (2.16) from (2.14).

Say we are given multisets of positive integers R_i of cardinality μ_i for $i \geq 1$, and we want to create a filling σ of μ such that for each i the entries in the ith row of μ are the elements of R_i, and furthermore that there are no inversion triples (so $\mathrm{inv}(\sigma,\mu) = 0$). This can be done in a unique way as follows. Form the bottom row of σ by placing the entries of R_1 in the bottom row in nondecreasing order, read left-to-right. Let w denote the smallest of these entries (i.e. the leftmost). Next, for the first entry in the second row of σ, choose the smallest element of R_2 which is larger than w, or choose the smallest entry in R_2 if none are larger then w. Remove this entry from R_2 to form R_2', and iterate; let w' denote the entry in the second column of the first row; choose the smallest element of R_2' which is larger than w', or choose the smallest entry in R_2' if there are not any larger then w', and let this be the entry in the second column of the second row, and continue filling out the second row in this way by moving one column to the right at a time. To define the entry in the first column, third row of σ, let w denote the entry in the first column, second row of σ, and choose the smallest element of R_3 which is larger than w, or choose the smallest entry in R_3 if there are none larger then w. Remove this entry from R_3 to form R_3', and so on. See Figure 3 for a filling with no inversion triples.

We construct a word cword(σ) by initializing cword to the empty string, then scanning through read(σ), from the beginning to the end, and each time we encounter a 1, adjoin the number of the row containing this 1 to the beginning of cword. After recording the row numbers of all the 1's in this fashion, we go back to the beginning of read(σ), and adjoin the row numbers of squares containing 2's to the beginning of cword. For example, if σ is the filling in Figure 3, then cword$(\sigma) = 11222132341123$.

2				
3	1	2		
2	4	4	1	5
1	1	3	6	7

FIGURE 3. A filling with no inversion triples

We now translate the statistic $\mathrm{maj}(\sigma, \mu)$ into a statistic on $\mathrm{cword}(\sigma)$. Note that $\sigma(1,1)$ corresponds to the rightmost 1 in $\mathrm{cword}(\sigma)$ - denote this 1 by w_{11}. If $\sigma(1,2) > \sigma(1,1)$, $\sigma(1,2)$ corresponds to the rightmost 2 which is left of w_{11}, otherwise it corresponds to the rightmost 2 (in $\mathrm{cword}(\sigma)$). In any case denote this 2 by w_{12}. More generally, the element in $\mathrm{cword}(\sigma)$ corresponding to $\sigma(1,i)$ is the first i encountered when travelling left from $w_{1,i-1}$, looping around and starting at the right end of $\mathrm{cword}(\sigma)$ if necessary. To find the subword $w_{21}w_{22}\cdots w_{2\mu_2'}$ corresponding to the second column of σ, we do the same algorithm on the word $\mathrm{cword}(\sigma)'$ obtained by removing the elements $w_{11}w_{12}\cdots w_{1\mu_1'}$ from $\mathrm{cword}(\sigma)$, then remove $w_{21}w_{22}\cdots w_{2\mu_2'}$ and apply the same process to find $w_{31}w_{32}\cdots w_{3\mu_3'}$ etc..

Clearly $\sigma(i,j) \in \mathrm{Des}(\sigma,\mu)$ if and only if w_{ij} occurs to the left of $w_{i,j-1}$ in $\mathrm{cword}(\sigma)$. The statistic $\mathrm{maj}(\sigma,\mu)$ now becomes transparently equal to the statistic $\mathrm{cocharge}(\mathrm{cword}(\sigma))$ introduced by Lascoux and Schützenberger [**24**]. See also [**18**][pp. 16-17].

Next we associate a two-line array $A(\sigma)$ to a filling σ with no inversions by letting the upper row $A_1(\sigma)$ be nonincreasing with the same weight as σ, and the lower row $A_2(\sigma)$ be $\mathrm{cword}(\sigma)$. For example, to the filling in Figure 3 we associate the two-line array

$$\begin{matrix} 7\,6\,5\,4\,4\,3\,3\,2\,2\,2\,2\,1\,1\,1\,1 \\ 1\,1\,2\,2\,2\,1\,3\,2\,3\,4\,1\,1\,2\,3 \end{matrix} \tag{2.17}$$

By construction, below equal entries in the upper row the entries in the lower row are nondecreasing. Later in this section we will show $C_\mu(X;0,t)$ is a symmetric function, so we can reverse the variables, replacing x_i by x_{n-i+1} for $1 \le i \le n$, without changing the sum. This has the effect of reversing $A_1(\sigma)$, and we end up with an ordered two-line array, of the type occurring in the RSK algorithm. This algorithm associates such an array with a pair of SSYT. We can invert this correspondence since from the two-line array we get the multiset of elements in each row of σ, which uniquely determines σ. Thus

$$C_\mu(x_1,x_2,\ldots,x_n;0,t) = \sum_\sigma x^{\mathrm{weight}(A_1(\sigma))} t^{\mathrm{cocharge}(A_2(\sigma))} \tag{2.18}$$

$$= \sum_{(A_1,A_2)} x^{\mathrm{weight}(A_1)} t^{\mathrm{cocharge}(A_2)}, \tag{2.19}$$

where the sum is over ordered two-line arrays satisfying $\mathrm{weight}(A_2) = \mu$.

It is well known that for any word w of partition weight, $\mathrm{cocharge}(w) = \mathrm{cocharge}(\mathrm{read}(P_w))$, where $\mathrm{read}(P_w)$ is the reading word of the insertion tableau P_w under the RSK algorithm [**30**, pp.48-49], [**34**, p.417]. Hence applying the RSK algorithm to (2.19),

$$C_\mu(x_1,x_2,\dots,x_n;0,t) = \sum_{(P,Q)} x^{\mathrm{weight}(Q)} t^{\mathrm{cocharge}(\mathrm{read}(P))}, \tag{2.20}$$

where the sum is over all pairs (P,Q) of SSYT of the same shape with weight$(P) = \mu$. Since the number of different Q tableau of weight ν matched to a given P tableau of shape λ is the Kostka number $K_{\lambda,\nu}$,

$$\begin{aligned} C_\mu[X;0,t] &= \sum_\nu m_\nu \sum_\lambda \sum_{\substack{P\in SSYT(\lambda,\mu)\\ Q\in SSYT(\lambda,\nu)}} t^{\mathrm{cocharge}(\mathrm{read}(P))} \\ &= \sum_\lambda \sum_{P\in SSYT(\lambda,\mu)} t^{\mathrm{cocharge}(\mathrm{read}(P))} \sum_\nu m_\nu K_{\lambda,\nu} \\ &= \sum_\lambda s_\lambda \sum_{P\in SSYT(\lambda,\mu)} t^{\mathrm{cocharge}(\mathrm{read}(P))}. \qquad (2.21) \end{aligned}$$

This completes the outline of the proof of (2.16).

• A final segment of a word is the last k letters of the word, for some k. We say a word σ is a *Yamanouchi word* if in any final segment of σ, there are at least as many i's as $i+1$'s, for all $i \geq 1$. In [**14**] the following result is derived from Theorem 2.5.

THEOREM 2.7. *For any partition μ with at most two columns,*

$$\langle \tilde{H}_\mu(X;q,t), s_\lambda(X)\rangle = \sum_{Y\in \mathrm{Yam}(\lambda)} q^{\mathrm{inv}(Y,\mu)} t^{\mathrm{maj}(Y,\mu)}. \tag{2.22}$$

Here the sum is over all Yamanouchi words of weight λ, i.e. with λ_i copies of i for $i \geq 1$, and in the summand we identify a word Y with the filling whose reading word is Y.

Other combinatorial formulas for the left-hand-side of (2.22) are known, but the one above is arguably the most elegant. The proof of Theorem 2.7 involves a combinatorial construction which groups together fillings which have the same maj and inv statistics. This is carried out with the aid of *crystal graphs*, which occur in the representation theory of Lie algebras.

• Below we show how C_μ is a positive sum of symmetric functions introduced by Lascoux, Leclerc and Thibon, commonly called LLT polynomials [**25**]. These functions depend on a set of variables X, a tuple of skew shapes, and a parameter q. In an unpublished preprint on Haiman's website titled "Affine Hecke algebras and positivity of LLT and Macdonald polynomials" which is generally accepted by experts to be true, Grojnowski and Haiman prove that LLT polynomials are Schur positive. Using this we have a new proof of the Schur positivity of $\tilde{H}_\mu$.

We now give an overview of the main ideas in the proof of the HHL formula, some of which will be used to derive a corresponding formula for $J_\mu(X;q,t)$ described in the next section. To begin with, we prove C_μ is in Sym (which is not at all obvious) by showing it can be written as a sum of LLT polynomials as described in [**16**], [**18**][Ch. 6].

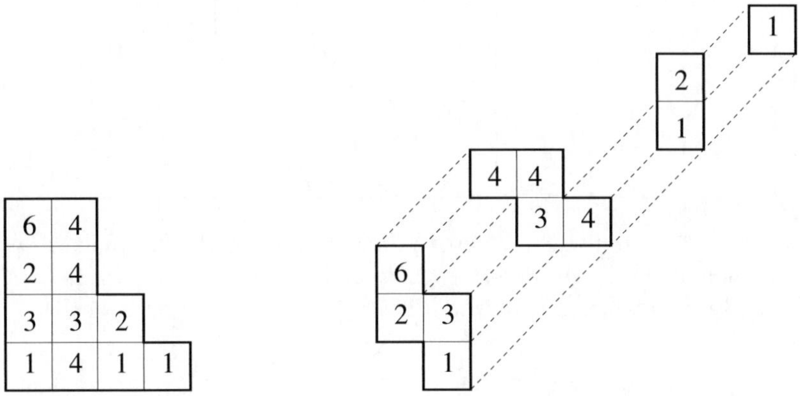

FIGURE 4. On the left, a filling, and on the right, the term in the corresponding LLT product of ribbons

Fix a descent set D, and let

$$R_D(X;q) = \sum_{\substack{\sigma:\mu\to\mathbb{Z}_{>0} \\ \mathrm{Des}(\sigma,\mu)=D}} q^{\mathrm{inv}(\sigma,\mu)} x^\sigma. \tag{2.23}$$

If μ has one column, then R_D is a ribbon Schur function. More generally, $R_D(X;q)$ is an LLT product of ribbons. We illustrate how to transform a filling σ into a term $\gamma(\sigma)$ in the corresponding LLT product in Figure 4. Note that inversion pairs in σ are in direct correspondence with LLT inversion pairs in $\gamma(\sigma)$. Since the shape of the ribbons in $\gamma(\sigma)$ depends only on $\mathrm{Des}(\sigma,\mu)$ we have

$$C_\mu(X;q,t) = \sum_D t^L q^{-A} R_D(X;q), \tag{2.24}$$

where the sum is over all possible descent sets D of fillings of μ, with

$$L = \sum_{s\in D} 1 + \mathrm{leg}(s), \quad A = \sum_{s\in D} \mathrm{arm}(s).$$

Since LLT polynomials are symmetric functions, we have

THEOREM 2.8. *For all μ, $C_\mu(X;q,t)$ is a symmetric function.*

Now that we know $C_\mu \in \mathrm{Sym}$, we proceed to show that C_μ satisfies the three defining conditions of $\tilde{H}_\mu$. We mention that any symmetric function $f(X)$ of homogeneous degree n is uniquely determined by its value on $X_n = \{x_1,\ldots,x_n\}$ so in Definition 2.1 we may as well restrict ourselves to fillings from the set $\{1,2,\ldots,n\}$. First of all, if we consider the filling of μ where all entries equal 1, then there are no descents and no inversion pairs, and the x-weight is x_1^n, so C_μ satisfies (2.12).

The two other conditions are more involved, and depend on a process called the *superization* of an element of $\mathrm{Sym}(X)$. This involves two alphabets, a positive alphabet $\mathcal{A}_+ = \{1,2,\ldots,n\}$ and a negative alphabet $\mathcal{A}_- = \{\bar{1},\bar{2},\ldots,\bar{n}\}$; we denote their union by $\mathcal{A}_\pm$. We define a *super word* $\tilde{\sigma}$ to be a word in the alphabet $\mathcal{A}_\pm$, and let $|\tilde{\sigma}|$ be the word obtained by replacing each negative letter $\bar{i}$ in $\tilde{\sigma}$ by its "absolute

3	$\bar{3}$		
4	$\bar{2}$		
4	5	3	
$\bar{1}$	3	$\bar{1}$	1

5	8		
9	4		
10	11	6	
3	7	2	1

2	8		
5	9		
6	7	3	
11	4	10	1

FIGURE 5. On the left, a super filling. In the middle, its standardization assuming the ordering (2.26) and on the right, its standardization assuming the ordering (2.27).

value" i. Set

$$(xy)^{\tilde{\sigma}} = \prod_{\substack{i \\ \tilde{\sigma}_i \in \mathcal{A}_+}} x_{\tilde{\sigma}_i} \prod_{\substack{i \\ \tilde{\sigma}_i \in \mathcal{A}_-}} y_{|\tilde{\sigma}_i|}. \tag{2.25}$$

So, for example, if $\tilde{\sigma} = 21\bar{1}3\bar{2}$, $(xy)^{\tilde{\sigma}} = x_2 x_1 y_1 x_3 y_2$, and $|\tilde{\sigma}| = 21132$.

We extend the concept of standardization to super words as follows. Fix an arbitrary ordering of the elements of $\mathcal{A}_{\pm}$. Two examples of orderings we will be working with are

$$1 < \bar{1} < 2 < \bar{2} \cdots < n < \bar{n} \tag{2.26}$$

$$1 < 2 < \cdots < n < \bar{n} \cdots < \bar{2} < \bar{1}. \tag{2.27}$$

To standardize a word according to a specific ordering, order two equal positive letters by assuming the leftmost is the smaller of the two, and order two equal negative letters by assuming the leftmost is the larger of the two. For unequal letters order them by the given total ordering on $\mathcal{A}_{\pm}$. For example, if we use (2.26), then $\mathrm{Stan}(31\bar{1}1\bar{1}\bar{2}\bar{2}2) = 81423765$, while if we use (2.27), then $\mathrm{Stan}(31\bar{1}1\bar{1}\bar{2}\bar{2}2) = 41827653$. See Figure 5.

The following lemma, true for any fixed ordering of $\mathcal{A}_{\pm}$, is proved in [**16**]; see also [**18**][pp. 99-100].

LEMMA 2.9. *Say $G \in \mathrm{Sym}(X)$ satisfies*

$$G(X) = \sum_{\beta \in S_n} c(\beta) F_{\mathrm{Des}(\beta^{-1})}(X), \tag{2.28}$$

for some constants $c(\beta)$ independent of X. Then if $Y = \{y_1, y_2, \ldots\}$ is another set of variables,

$$\omega^Y G[X+Y] = \sum_{\beta \in S_n} c(\beta) \tilde{F}_{\mathrm{Des}(\beta^{-1})}(X,Y), \tag{2.29}$$

where ω^Y is the usual involution, acting only on the Y variables, and

$$\tilde{F}_\gamma(X,Y) = \sum_{\substack{a_1 \le a_2 \le \cdots \le a_n \\ a_i = a_{i+1} \in \mathcal{A}_+ \implies i \notin \gamma \\ a_i = a_{i+1} \in \mathcal{A}_- \implies i \in \gamma}} \prod_{a_i \in \mathcal{A}_+} x_{a_i} \prod_{a_i \in \mathcal{A}_-} y_{|a_i|} \tag{2.30}$$

is a superized version of the Gessel Fundamental $F_\gamma(X)$.

We note that for any fixed $\beta \in S_n$ with $\mathrm{Des}(\beta^{-1}) = \gamma$,

$$\tilde{F}_\gamma(X,Y) = \sum_{\substack{\tilde{\sigma}\in\mathcal{A}_\pm^n \\ \mathrm{Des}(\mathrm{Stan}(\tilde{\sigma}))=\beta}} (xy)^{\tilde{\sigma}}. \tag{2.31}$$

By Lemma 2.9 and (2.5) we have

$$\omega^Y C_\mu[X+Y;q,t] = \sum_{\beta\in S_n} t^{\mathrm{maj}(\beta,\mu)} q^{\mathrm{inv}(\beta,\mu)} \tilde{F}_{\mathrm{Des}(\beta^{-1})}(X,Y). \tag{2.32}$$

Now for a real parameter α, $C_\mu[X(\alpha-1);q,t] = C_\mu[X\alpha - X;q,t]$, and by Remark 2.3 can be obtained from $\omega^Y C_\mu[X+Y;q,t] = C_\mu[X-\epsilon Y;q,t]$ by replacing x_i by αx_i and y_i by ϵx_i, for $1 \le i \le n$ (since $\omega f(X) = (-1)^n f[-X]$). Hence (2.32) implies

$$C_\mu[X\alpha - X;q,t] = \sum_{\tilde{\sigma}:\mu\to\mathcal{A}_\pm} x^{|\tilde{\sigma}|} t^{\mathrm{maj}(\tilde{\sigma},\mu)} q^{\mathrm{inv}(\tilde{\sigma},\mu)} \alpha^{\mathrm{pos}(\tilde{\sigma})} (-1)^{\mathrm{neg}(\tilde{\sigma})}. \tag{2.33}$$

Here $\mathrm{pos}(\tilde{\sigma})$ and $\mathrm{neg}(\tilde{\sigma})$ denote the number of positive letters and negative letters in $\tilde{\sigma}$, respectively.

Define the *critical pair* of squares of a filling $\tilde{\sigma}$ as the last pair (in the reading word order) of attacking squares u, v which both contain elements of the set $\{1, \bar{1}\}$, with u say denoting the "critical square", i.e. the earliest of these two squares in the reading order. If $\tilde{\sigma}$ has no such attacking pair, then we define the critical pair as the last pair of attacking squares which both contain elements of the set $\{2, \bar{2}\}$, and if no such pair exists, as the last pair of attacking squares which both contain elements of the set $\{3, \bar{3}\}$, etc. Let $\mathrm{IH}(\tilde{\sigma})$ be the sign-reversing involution obtained by starting with $\tilde{\sigma}$ and switching the sign of the element in the critical square. For example, for the filling on the left in Figure 5, the critical pair is $(3,1),(4,1)$ and IH changes the $\bar{1}$ in $(3,1)$ to a 1.

Assume our alphabets satisfy (2.26), and let $\alpha = q$ in (2.33). Clearly $\mathrm{IH}(\tilde{\sigma})$ fixes the x weight. Let u, v be critical with $|\tilde{\sigma}(u)| = a$. If $|\tilde{\sigma}(\mathrm{North}(u))| = a$ we have $u \in \mathrm{Des}(\tilde{\sigma},\mu)$ if and only if $\tilde{\sigma}(\mathrm{North}(u)) = \bar{a}$. Furthermore if u is not in the bottom row of μ, then $|\tilde{\sigma}(\mathrm{South}(u))| \ne a$, since otherwise v and $\mathrm{South}(u)$ are an attacking pair later in the reading order, and u, v would not be the critical pair. It follows that $\mathrm{IH}(\tilde{\sigma})$ fixes the descent set and hence also the t weight.

Since $\mathrm{IH}(\tilde{\sigma})$ fixes the descent set, it will fix the q-weight if and only if changing a to $\bar{a}$ increases the number of inversion pairs Inv by 1 (since this change decreases pos by 1). Changing a to $\bar{a}$ or vice-versa has no affect on inversion pairs involving numbers not equal to a or $\bar{a}$. The definition of critical implies that changing a to $\bar{a}$ creates a new inversion pair between u and v, but does not affect any other inversion pairs. Similarly, changing $\bar{a}$ to a decreases the number of inversion pairs by 1, hence if $\alpha = q$ in (2.33), then $\mathrm{IH}(\tilde{\sigma})$ fixes the q-weight.

Call a super filling nonattacking if there are no critical pairs of squares. (These are the fixed points of IH). The above reasoning shows

$$C_\mu[X(q-1);q,t] = \sum_{\tilde{\sigma}:\mu\to\mathcal{A}_\pm,\ \text{nonattacking}} x^{|\tilde{\sigma}|} t^{\mathrm{maj}(\tilde{\sigma},\mu)} q^{\mathrm{inv}(\tilde{\sigma},\mu)} q^{\mathrm{pos}(\tilde{\sigma})} (-1)^{\mathrm{neg}(\tilde{\sigma})}, \tag{2.34}$$

where the sum is over all nonattacking superfillings of μ. Since nonattacking implies there is at most one occurrence of a number from the set $\{a, \bar{a}\}$ in any row, the exponent of x_1 in $x^{|\tilde{\sigma}|}$ is at most μ_1', the sum of the exponents of x_1 and x_2 is at most $\mu_1' + \mu_2'$, etc., which shows C_μ satisfies axiom (2.10).

Call the first square w (in the reading word order) such that $|\tilde{\sigma}(w)| = 1$, with w not in the bottom row, the *pivotal square.* If there is no such square, let w denote the first square, not in the bottom *two* rows of μ, with $|\tilde{\sigma}(w)| = 2$, and again if there is no such square, search for the first square not in the bottom three rows satisfying $|\tilde{\sigma}(w)| = 3$, etc. Let $\mathrm{IL}(\tilde{\sigma})$ denote the sign-reversing involution obtained by switching the sign on $\tilde{\sigma}(w)$. For the filling on the left in Figure 5, the pivotal square is $(2,3)$, and IL switches this $\bar{2}$ to a 2.

Let our alphabets satisfy (2.27), with $\alpha = t$ in (2.33). As before, $\mathrm{IL}(\tilde{\sigma})$ trivially fixes the x-weight. Note that by construction, if the pivotal square is in row k, then after standardization it contains either the smallest element occurring anywhere in rows $k-1$ and higher or the largest element occurring in rows $k-1$ and higher. Thus in any triple containing the pivotal square, IL either switches an element smaller than the other two to an element larger than the other two, or vice-versa, and this preserves the orientation of the triple. By Remark 2.2, IL thus fixes the q-weight. We leave it as an exercise for the reader to show that IL also fixes the t-weight.

Call a super filling *primary* if there are no pivotal squares. (These are the fixed points of IL). We have

$$C_\mu[X(t-1);q,t] = \sum_{\tilde{\sigma}:\mu\to\mathcal{A}_\pm,\text{primary}} x^{|\tilde{\sigma}|} t^{\mathrm{maj}(\tilde{\sigma},\mu)} q^{\mathrm{inv}(\tilde{\sigma},\mu)} t^{\mathrm{pos}(\tilde{\sigma})} (-1)^{\mathrm{neg}(\tilde{\sigma})}, \tag{2.35}$$

where the sum is over all primary superfillings of μ. Now primary implies that all 1 or $\bar{1}$'s are in the bottom row, all 2 or $\bar{2}$'s are in the bottom two rows, etc., so in $x^{|\tilde{\sigma}|}$ the power of x_1 is at most μ_1, the sum of the powers of x_1 and x_2 is at most $\mu_1 + \mu_2$, etc. Thus C_μ satisfies axiom (2.11), which completes the proof of Theorem 2.5.

REMARK 2.10. In a private communication to the author in 2007, Haiman noted that the involution IH can be easily extended to any LLT polynomial. This implies that any LLT polynomial, plethystically evaluated at $X(q-1)$, can be expanded as a sum over nonattacking superfillings of the tuple of skew shapes.

3. Formulas for J_μ

By definition, Macdonald's original integral form J_μ is related to Garsia and Haiman's modified Macdonald as follows, where $n(\mu) = \sum_i (i-1)\mu_i$.

$$\begin{aligned} J_\mu(X;q,t) &= t^{n(\mu)} \tilde{H}_\mu[X(1-t);q,1/t] && (3.1)\\ &= t^{n(\mu)} \tilde{H}_\mu[Xt(1/t-1);q,1/t] \\ &= t^{n(\mu)+n} \tilde{H}_{\mu'}[X(1/t-1);1/t,q] && (3.2)\end{aligned}$$

using (2.15). Formula (2.34), with q,t interchanged thus implies

$$J_\mu(X;q,t) = \sum_{\text{nonattacking super fillings } \tilde{\sigma} \text{ of } \mu'} x^{|\tilde{\sigma}|} q^{\mathrm{maj}(\tilde{\sigma},\mu')} t^{\mathrm{coinv}(\tilde{\sigma},\mu')} (-t)^{\mathrm{neg}(\tilde{\sigma})}, \tag{3.3}$$

where coinv $= n(\mu) -$ inv is the number of coinversion triples (i.e. triples with a clockwise orientation), and we use the ordering (2.26).

2	4	
2	3	
1	3	2

FIGURE 6. A nonattacking filling of $(3,3,1)'$

We can derive a more compact form of (3.3) by grouping together all the 2^n super fillings $\tilde{\sigma}$ of μ' whose absolute value equals a fixed positive filling σ of μ'. The basic idea is that if you consider an entry $\tilde{\sigma}(s)$, where $|\tilde{\sigma}(\text{South}(s))| \neq |\tilde{\sigma}(s)|$, then switching the sign of $\tilde{\sigma}(s)$ will not change the descent set, nor will it affect the orientation of any triples. The only thing that changes is the number of negative entries. This observation uses the fact that the filling is nonattacking. On the other hand, if $|\tilde{\sigma}(\text{South}(s))| = |\tilde{\sigma}(s)|$, then switching the sign of $\tilde{\sigma}(s)$ from a to $\bar{a}$ will create a new descent, while switching it from $\bar{a}$ to a will remove a descent there. Putting everything together, from (3.3) we get a nice combinatorial formula for J_μ.

COROLLARY 3.1 ([**14**]).

$$J_\mu(X;q,t) = \sum_{\text{nonattacking fillings } \sigma \text{ of } \mu'} x^\sigma q^{\text{maj}(\sigma,\mu')} t^{\text{coinv}(\sigma,\mu')} \tag{3.4}$$
$$\times \prod_{\substack{u\in\mu' \\ \sigma(u)=\sigma(\text{South}(u))}} (1-q^{\text{leg}(u)+1}t^{\text{arm}(u)+1}) \prod_{\substack{u\in\mu' \\ \sigma(u)\neq\sigma(\text{South}(u))}} (1-t),$$

where each square in the bottom row is included in the last product, and nonattacking means no two squares which attack each other contain the same number.

EXAMPLE 3.2. Let $\mu = (3,3,1)$. Then for the nonattacking filling σ of μ' in Figure 6, maj $= 3$, coinv $= 3$, squares $(1,1),(1,2),(2,1),(2,3)$ and $(3,1)$ each contribute a $(1-t)$, square $(1,3)$ contributes a $(1-qt^2)$, and $(2,2)$ contributes a $(1-q^2t)$. Thus the term in (3.4) corresponding to σ is

$$x_1x_2^3x_3^2x_4q^3t^3(1-qt^2)(1-q^2t)(1-t)^5. \tag{3.5}$$

Let α denote a real parameter. The (integral form) Jack polynomial $J_\mu^{(\alpha)}(X)$ can be obtained from $J_\mu(X;q,t)$ by

$$J_\mu^{(\alpha)}(X) = \lim_{t\to 1} \frac{J_\mu(X;t^\alpha,t)}{(1-t)^{|\mu|}}. \tag{3.6}$$

If we set $q = t^\alpha$ in (3.4) and then divide by $(1-t)^{|\mu|}$ and take the limit as $t \to 1$ we get the following result of Knop and Sahi [**23**].

(3.7)
$$J_\mu^{(\alpha)}(X) = \sum_{\text{nonattacking fillings } \sigma \text{ of } \mu'} x^\sigma \prod_{\substack{u\in\mu' \\ \sigma(u)=\sigma(\text{South}(u))}} (\alpha(\text{leg}(u)+1)+\text{arm}(u)+1).$$

There is another formula for J_μ corresponding to involution IL and (2.27). First note that by applying the $t \to 1/t$, $X \to tX$ case of (2.35) to the right-hand side of (3.1) we get

$$J_\mu(X;q,t) = \sum_{\tilde\sigma:\mu\to\mathcal{A}_\pm,\ \text{primary}} x^{|\tilde\sigma|} t^{\text{nondes}(\tilde\sigma,\mu)} q^{\text{inv}(\tilde\sigma,\mu)} (-t)^{\text{neg}(\tilde\sigma)}, \tag{3.8}$$

where the sum is over all primary super fillings of μ, and nondes $= n(\mu) - \text{maj}$ is the sum of leg $+1$ over all squares of μ which are *not* in $\text{Des}(\tilde\sigma,\mu)$.

For a positive filling σ and $s \in \mu$, let

$$\text{maj}_s(\sigma,\mu) = \begin{cases} \text{leg}(s) \text{ if North}(s) \in \text{Des}(\sigma,\mu) \\ 0 \text{ otherwise} \end{cases} \tag{3.9}$$

$$\text{nondes}_s(\sigma,\mu) = \begin{cases} \text{leg}(s) + 1 \text{ if } s \notin \text{Des}(\sigma,\mu) \\ 0 \text{ otherwise,} \end{cases} \tag{3.10}$$

so $\text{maj} = \sum_s \text{maj}_s$ and $\text{nondes} = \sum_s \text{nondes}_s$. Note that a square s makes a nonzero contribution to $\text{nondes}(\tilde\sigma,\mu)$ if it contains a positive letter and $\text{nondes}_s(|\tilde\sigma|,\mu) > 0$, or it contains a negative letter and $\text{maj}_s(|\tilde\sigma|,\mu) > 0$. Thus by letting $q = 1$ in (3.8) we get

$$J_\mu(X;1,t) = \sum_{\sigma:\mu\to\mathcal{A}_+,\text{primary}} x^\sigma \prod_{s\in\mu} (t^{\text{nondes}_s(\sigma,\mu)} - t^{1+\text{maj}_s(\sigma,\mu)}). \tag{3.11}$$

By a similar analysis we can incorporate the q-parameter into (3.11). Given a triple u, v, w of μ with $w \in \mu$, we define the "middle square" of u, v, w, with respect to $\tilde\sigma$, to be the square containing the middle of the three numbers of the set $\{\text{Stan}(|\tilde\sigma|)(u), \text{Stan}(|\tilde\sigma|)(v), \text{Stan}(|\tilde\sigma|)(w)\}$. In other words, we replace all letters by their absolute value, then standardize, then take the square containing the number which is neither the largest, nor the smallest, of the three. For squares $(1,1),(1,2),(3,2)$ for the super filling on the left in Figure 5, square (3,2) is the middle square. For squares $(1,4),(1,3),(2,4)$ square $(2,4)$ is the middle square. Extend this definition to triples u, v, w of μ with u, v in the bottom row by letting the middle square be u if $|\tilde\sigma|(u) \le |\tilde\sigma|(v)$, otherwise let it be v (here, as usual, v is to the right of u). By checking the eight possibilities for choices of signs of the letters in u, v, w (or four possibilities if u, v are in the bottom row), one finds that if u, v, w form a coinversion triple in $|\tilde\sigma|$, then they form a coinversion triple in $\tilde\sigma$ if the middle square contains a positive letter, otherwise they form an inversion triple in $\tilde\sigma$. Similarly, if u, v, w form an inversion triple in $|\tilde\sigma|$, they form a coinversion triple in $\tilde\sigma$ if the middle square contains a negative letter, otherwise they form an inversion triple. Thus by defining $\text{coinv}_s(\sigma,\mu)$ to be the number of coinversion triples for which s is the middle square, and $\text{inv}_s(\sigma,\mu)$ to be the number of inversion triples for which s is the middle square, we get the following result of the author [**18**][p.133].

COROLLARY 3.3.

$$J_\mu(X;q,t) = \sum_{\substack{\sigma:\mu\to\mathcal{A}_+ \\ primary}} x^\sigma \prod_{s\in\mu} (q^{inv_s(\sigma,\mu)} t^{\text{nondes}_s(\sigma,\mu)} - q^{coinv_s(\sigma,\mu)} t^{1+\text{maj}_s(\sigma,\mu)}). \tag{3.12}$$

Although the products in (3.12) are more complicated than those in (3.4), one advantage this formula has is that the q and t-weights are invariant under standardization, hence we can express it as

(3.13)
$$J_\mu(X;q,t) = \sum_{\substack{\beta\in S_n\\ \text{primary}}} F_{\mathrm{Des}(\beta^{-1})}(X) \prod_{s\in\mu} (q^{\mathrm{inv}_s(\beta,\mu)} t^{\mathrm{nondes}_s(\beta,\mu)} - q^{\mathrm{coinv}_s(\beta,\mu)} t^{1+\mathrm{maj}_s(\beta,\mu)}),$$

which gives an expansion of J_μ into Gessel's fundamental quasisymmetric functions. The sum is over all standard fillings (permutations) which are also primary fillings, although the equation is still true if we extend the sum to all permutations, since if β is a standard filling which is not primary, with pivotal square s, then the term $(q^{\mathrm{inv}_s(\sigma,\mu)} t^{\mathrm{nondes}_s(\sigma,\mu)} - q^{\mathrm{coinv}_s(\sigma,\mu)} t^{1+\mathrm{maj}_s(\sigma,\mu)})$ is zero. M. Yoo has used (3.12) to obtain the expansion of $J_\mu(X;q,t)$ into Schur functions for certain shapes μ [**35**], [**36**].

4. Nonsymmetric Macdonald Polynomials

In 1995 Macdonald [**29**] introduced polynomials $E_\alpha(X_n;q,t)$ which form a basis for the polynomial ring $\mathbb{Q}(q,t)[X_n]$, and in many ways are analogous to the $P_\lambda(X;q,t)$. Here $X_n = \{x_1,\ldots,x_n\}$ as in Section 2. Further development of the theory was made by Cherednik [**5**], Sahi [**33**], Knop [**22**], Ion [**20**], and others. Both the E_α and the P_λ have versions for any affine root system, and are both orthogonal families of polynomials with respect to a certain scalar product.

We will be working with the type A_{n-1} case, where there is a special structure which allows us to assume α is a composition, i.e. $\alpha\in\mathbb{N}^n$. In [**15**] Haiman, Loehr and the author showed how the $E_\alpha(X_n;q,t)$ and their integral forms $\mathcal{E}_\alpha(X_n;q,t)$ also satisfy a combinatorial formula, and that this formula implies formula (3.4) for $J_\mu(X;q,t)$. The proof of the formula for $\mathcal{E}_\alpha(X_n;q,t)$ involves operators T_i typically used to define the type A Hecke algebra. Briefly, the T_i are linear operators on the polynomial ring, which act on monomials x^λ by

$$T_i x^\lambda = t x^{s_i(\lambda)} + (t-1)\frac{x^\lambda - x^{s_i(\lambda)}}{1-x^{\alpha_i}} \quad \text{for all } i = 0,\,1,\,\ldots,\,n-1. \tag{4.1}$$

Here for $1\le i\le n-1$, s_i permutes x_i and x_{i+1}, while for any polynomial $f(x_1,\ldots,x_n)$,

$$s_0 f(x_1,\ldots,x_n) = f(qx_n, x_2,\ldots,x_{n-1}, x_1/q).$$

See [**15**][p. 3] for background on the T_i.

The HHL formula for $E_\alpha(X_n;q,t)$ from [**15**] involves nonattacking fillings of $\mathrm{dg}(\alpha')$, as in the formula (3.4) for J_μ, but with the added feature of having the number $n-i+1$ in the "basement" square with coordinates $(i,0)$, for $1\le i\le n$. Shortly after the publication of this formula, in a private communication to the author Haiman conjectured that for any $\sigma\in S_n$, the polynomial $T_\sigma E_\alpha(X_n;q,t)$ satisfies the same HHL formula, with the basement replaced by σ. Here T_σ is a product of the T_i corresponding to a reduced word for σ. Utilizing some unpublished notes of the author which outlined a proof of the conjecture, Per Alexandersson [**1**] completed the proof and obtained many related results about this more general function. Below we describe Alexandersson's formula for the *permuted basement nonsymmetric Macdonald polynomial.*

We first extend some of the definitions from earlier sections to more general "skyline" diagrams which can be obtained by permuting the columns of a partition.

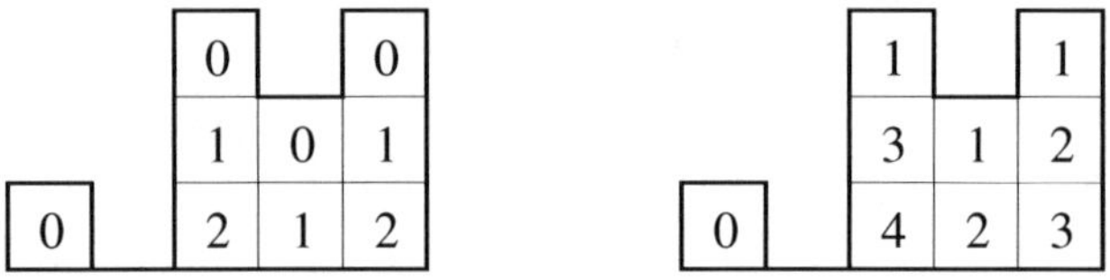

FIGURE 7. The leg lengths (on the left) and the arm lengths (on the right) for $(1,0,3,2,3)'$.

Given $\alpha \in \mathbb{N}^n$, let α' denote the transpose graph of α, consisting of the squares

$$\alpha' = \{(i,j), \quad 1 \le i \le n, \quad 1 \le j \le \alpha_i)\} \tag{4.2}$$

Furthermore let $\mathrm{dg}(\alpha')$ denote the augmented diagram obtained by adjoining the "basement" row of n squares with coordinates $\{(j,0), 1 \le j \le n\}$ below α'. Given $s \in \alpha'$, we let $\mathrm{leg}(s)$ be the number of squares of α' above s and in the same column of s. Define $\mathrm{arm}(s)$ to be the set of squares of $\mathrm{dg}(\alpha')$ which are either to the right and in the same row as s, and also in a column not taller than the column containing s, or to the left and in the row below the row containing s, and in a column strictly shorter than the column containing s. Then set $\mathrm{arm}(s) = |\,\mathrm{arm}(s)|$. For example, for $\alpha = (1,0,3,2,3,0,0,0,0)$, the leg lengths of the squares of $(1,0,3,2,3,0,0,0,0)'$ are listed on the left in Figure 7 and the arm lengths on the right. Note that if α is a partition μ, the leg and arm definitions agree with those previously given for μ'.

A filling σ of α' is an assignment of positive integers to the squares of $\mathrm{dg}(\alpha')$. The reading word of such a σ is obtained by reading the entries iin α' along rows, right-to-left, top-to-bottom, as in the case of partition diagrams. We standardize as before; given two equal letters in σ, the one occurring first in the reading word is viewed as being smaller. We say two squares $u, v \in \mathrm{dg}(\alpha')$ attack each other if they are in the same row, or if $u \in \alpha'$ and $v \in \mathrm{dg}(\alpha')$, v is in the row below u, and in a column strictly to the left of u. A filling (from a positive alphabet) of $\mathrm{dg}(\alpha')$ is nonattacking if no two squares which attack each other contain the same number.

A triple of $\mathrm{dg}(\alpha')$ is three squares with $u \in \alpha'$, $v \in \mathrm{arm}(u)$, and $w = \mathrm{South}(u)$. Note that both v and w could be in the basement. Given a filling σ of α', we say a triple u, v, w is a coinversion triple if either v is in a column to the right of u, and u, v, w has a clockwise orientation, or v is in a column to the left of u, and u, v, w has a counterclockwise orientation. A triple of the first kind, with v to the right of u, is called a Type A triple, while a triple of the second kind, where v is to the left of u and one row below, is called a Type B triple. Let $\mathrm{coinv}(\sigma, \alpha')$ denote the number of coinversion triples of σ. For example, the filling in Figure 8 has 7 coinversion triples:

$$\{[(3,2),(3,1),(4,2)],[(3,2),(3,1),(5,2)],[(3,2),(3,1),(1,1)],[(4,2),(4,1),(1,1)] \tag{4.3}$$

$$[(5,1),(5,0),(1,0)],[(5,1),(5,0),(2,0)],[(5,1),(5,0),(4,0)]\}, \tag{4.4}$$

where the first two are Type A triples and the rest are Type B.

Given $\beta \in S_n$ and $\alpha \in \mathbb{N}^n$, Alexandersson's formulas for the nonsymmetric Macdonald polynomial corresponding to α with basement β, denoted $E_\alpha^\beta(X_n; q, t)$,

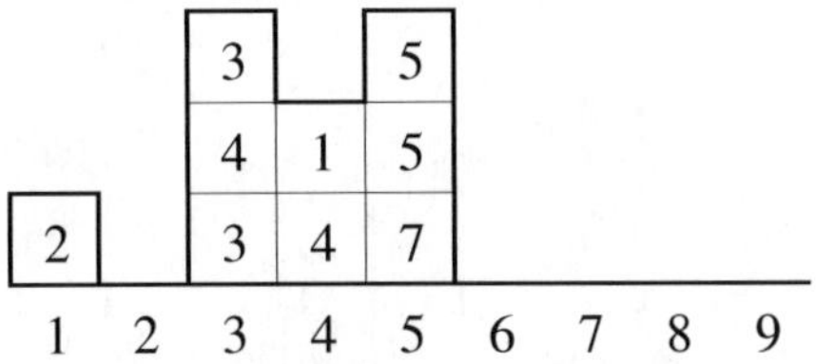

FIGURE 8. A nonattacking filling of $(1, 0, 3, 2, 3, 0, 0, 0, 0)'$.

and its integral form $\mathcal{E}_\alpha^\beta(X_n; q, t)$ are [**1**];

$$E_\alpha^\beta(X_n; q, t) = \sum_{\substack{\sigma \text{ nonattacking} \\ 1 \le \sigma_i \le n}} x^\sigma q^{\mathrm{maj}(\sigma)} t^{\mathrm{coinv}(\sigma)} \prod_{\substack{s \in \alpha' \\ \sigma(s) \neq \sigma(\mathrm{South}(s))}} \frac{1-t}{1 - q^{\mathrm{leg}+1} t^{\mathrm{arm}+1}} \tag{4.5}$$

and

$$\mathcal{E}_\alpha^\beta(X_n; q, t) = \sum_{\substack{\sigma \\ \text{nonattacking}}} x^\sigma q^{\mathrm{maj}(\sigma)} t^{\mathrm{coinv}(\sigma)} \tag{4.6}$$

$$\times \prod_{\substack{s \in \alpha' \\ \sigma(s) \neq \sigma(\mathrm{South}(s))}} (1-t) \prod_{\substack{s \in \alpha' \\ \sigma(s) = \sigma(\mathrm{South}(s))}} (1 - q^{\mathrm{leg}+1} t^{\mathrm{arm}+1}). \tag{4.7}$$

Here the sums are over all nonattacking fillings σ of the diagram $\mathrm{dg}(\alpha')$ with basement β, i.e. $\sigma(j, 0) = \beta_j$ for $1 \le j \le n$. By definition,

$$\mathcal{E}_\alpha^\beta(X_n; q, t) = E_\alpha^\beta(X_n; q, t) \prod_{s \in \alpha'} (1 - q^{\mathrm{leg}+1} t^{\mathrm{arm}+1}), \tag{4.8}$$

and it is easy to see formulas (4.5) and (4.6) are equivalent.

EXAMPLE 4.1. By (4.3) the nonattacking filling in Figure 8 has coinv $= 7$. There are descents at squares $(1, 1)$, $(3, 2)$ and $(5, 1)$, with maj-values 1, 2 and 3, respectively. The squares $(3, 1)$, $(4, 1)$ and $(5, 3)$ satisfy the condition $\sigma(u) = \sigma(\mathrm{South}(u))$ and contribute factors $(1 - q^3 t^5)$, $(1 - q^2 t^3)$ and $(1 - qt^2)$, respectively. Hence the total weight associated to this filling in (4.5) would be

$$x_1 x_2 x_3^2 x_4^2 x_5^2 x_7 \, q^6 t^7 (1 - q^3 t^5)(1 - q^2 t^3)(1 - qt^2)(1 - t)^6. \tag{4.9}$$

If $\mathrm{rev}(\alpha) = (\alpha_n, \ldots, \alpha_1)$, then $E_{\mathrm{rev}(\alpha)}^{(n, n-1, \ldots, 1)}(X_n; q, t)$ is Macdonald's original $E_\alpha(X_n; q, t)$, and $E_\alpha^{(1, 2, \ldots, n)}(X_n; q, t)$ is the variant $E_{\mathrm{rev}(\alpha)}(x_n, \ldots, x_2, x_1; 1/q, 1/t)$ of Macdonald's E studied by Marshall [**31**]. Hence (4.5) gives combinatorial formulae for both these special cases, which are equivalent to ones that first appeared in [**15**].

5. A formula for P_μ and a Quasisymmetric Macdonald Polynomial

Given a vector $\alpha = (\alpha_1, \alpha_2, \ldots, \alpha_n)$ of nonnegative integers, we let $\mathrm{inc}(\alpha)$ be the vector α, sorted so the parts are in nondecreasing order, and $\mathrm{dec}(\alpha)$ the vector α, sorted so the parts are in nonincreasing order. Furthermore let $\beta(\alpha)$ be the permutation in S_n of maximal length with the property that β applied to the vector

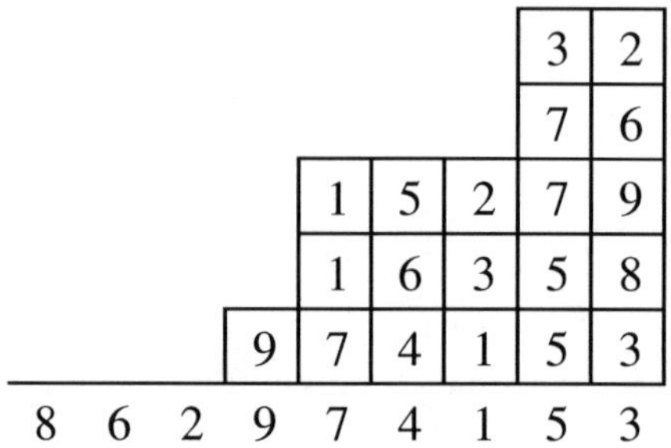

FIGURE 9. A nonattacking filling of $(0,0,0,1,3,3,3,5,5)'$ with an ordered basement.

α yields $\mathrm{inc}(\alpha)$, let α^+ be the (strong) composition consisting of α with the zeros removed, and $\ell(\alpha)$ the number of parts of α^+. For example, if $\alpha = (0,2,0,2,1,3)$ then $\mathrm{inc}(\alpha) = (0,0,1,2,2,3)$, $\mathrm{dec}(\alpha) = (3,2,2,1,0,0)$, $\beta(\alpha) = (3,1,5,4,2,6)$, $\alpha^+ = (2,2,1,3)$, and $\ell(\alpha) = 4$.

In [**4**][Lemma 3], Cantini, de Gier and Wheeler showed that

$$P_\mu(X_n;q,t) = \sum_{\substack{\alpha \\ \mathrm{dec}(\alpha^+)=\mu}} f_\alpha(X_n;q,t), \tag{5.1}$$

for certain polynomials $f_\alpha \in \mathbb{Q}(q,t)[X_n]$. In [**8**], Corteel, Mandelshtam and Williams proved that

$$f_\alpha(X_n;q,t) = E^{\beta(\alpha)}_{\mathrm{inc}(\alpha)}(X_n;q,t), \tag{5.2}$$

and using this in (5.1) we get

$$P_\mu(X_n;q,t) = \sum_{\substack{\alpha \\ \mathrm{dec}(\alpha^+)=\mu}} E^{\beta(\alpha)}_{\mathrm{inc}(\alpha)}(X_n;q,t) \tag{5.3}$$

Note that for all the terms in the right-hand-side of (5.2), the shape $\mathrm{inc}(\alpha)$ being filled is the same, namely the shape $0^{n-\ell(\mu)}\,\mathrm{inc}(\mu)$. Also, for this shape the definition of $\beta(\alpha)$ implies the entries in the basement below columns of $0^{n-\ell(\mu)}\,\mathrm{inc}(\mu)$ of equal height are decreasing left-to-right. Say a basement with this property is *ordered.* There is a bijection between ordered basements and permutations of the form $\beta(\alpha)$, where $\alpha \in \mathbb{N}^n$ and $\mathrm{dec}(\alpha^+) = \mu$. See Figure 9 for an example of a filling contributing to (5.1).

It is a general fact that in any nonattacking filling of a shape α' with a given basement, any entry in the first row of α' must equal the number just below it in the basement. Thus the basement $\beta(\alpha)$ in a filling σ in a given term contributing to (5.3) is uniquely determined by σ. Furthermore, if $\sigma(u) = \sigma(w) \neq \sigma(v)$, then (as in the type A case) u, v, w always from an inversion triple. Thus none of the entries in the basement columns of height 0 in (5.3) are involved in any coinversion triples, so contribute nothing to the power of t. They also contribute nothing to the x or q-weights. Since this part of the basement is determined by the part below columns of positive height, we can erase it. But the rest of the basement is just a copy of the first row, so we can just sum over fillings of $\mathrm{inc}(\mu)$ with the property that in the first row, entries in columns of the same height are decreasing left-to-right; call such a filling ordered. We will keep the concept of the basement here, consisting of a copy of the first row, so our definition of coinv still applies. We get the following final formula for P_μ, which is implicit in [**6**].

(5.4)
$$P_\mu(X;q,t) = \sum_{\substack{\text{fillings } \sigma \text{ of inc}(\mu)' \\ \sigma \text{ nonattacking, ordered}}} x^\sigma q^{\text{maj}(\sigma)} t^{\text{coinv}(\sigma)} \prod_{\substack{s \in \text{inc}(\mu)' \\ s \text{ not in bottom row} \\ \sigma(s) \neq \sigma(\text{South}(s))}} \frac{1-t}{1-q^{\text{leg}+1} t^{\text{arm}+1}}.$$

This formula for $P_\mu(X;q,t)$ is more elegant than previous ones from ([**15**]). For example, one formula there starts with a decomposition of $P_\mu(X;q,t)$ as a sum of the Marshall E_α 's in a way similar to (5.3), but the coefficient of a given E_α there is a ratio of complicated products.

By iterating the intertwiner recurrence for nonsymmetric Macdonald polynomials, Ram and Yip [**32**] obtained a very general formula, parameterized in terms of alcove walks, for symmetric and nonsymmetric Macdonald polynomials associated to an arbitrary affine root system. In the type A (specifically GL_n) case, these formulas have a lot more terms than the HHL formulas, but for a partition μ with distinct parts, Lenart [**26**] has shown how to "compress" the Ram-Yip formula for $P_\mu(X;q,t)$, and the resulting identity is equivalent to (5.4). More recently, Guo and Ram [**12**] have shown how type GL_n permuted basement nonsymmetric Macdonald polynomials (which they have suggested be renamed relative Macdonald polynomials - perhaps a good idea) satisfy a different "box-by-box" recurrence, which upon iteration leads directly to the nonattacking formulas (4.5) and (5.4). Hopefully their method will lead to compression of the Ram-Yip formula for other root systems.

For $\mu \vdash n$ and $\alpha \in \mathbb{N}^n$, let

$$\text{PR}\,1(\mu) = \prod_{s\in\mu}(1-q^{\text{arm}(s)}t^{\text{leg}(s)+1}) = \prod_{s\in\mu'}(1-q^{\text{leg}(s)}t^{\text{arm}(s)+1}) \tag{5.5}$$

$$\text{PR}\,2(\alpha) = \prod_{i\geq 1}(t;t)_{m_i} \prod_{\substack{s\in\text{inc}(\alpha') \\ s \text{ not in the bottom row}}} (1-q^{\text{leg}(s)+1}t^{\text{arm}(s)+1}), \tag{5.6}$$

where for $i \geq 1$, m_i is the number of times i occurs in α. In [**6**] it is noted that $\text{PR}\,1(\mu) = \text{PR}\,2(\text{inc}(\mu))$. Now Macdonald's definition of J_μ in terms of P_μ is

$$J_\mu(X;q,t) = \text{PR1}(\mu) P_\mu(X;q,t), \tag{5.7}$$

and so by multiplying both sides of (5.4) by $\text{PR}\,1(\mu)$ we get an equivalent formula for J_μ;

COROLLARY 5.1.

(5.8)
$$J_\mu(X;q,t) = \prod_i (t;t)_{m_i} \sum_{\textit{ordered, nonattacking fillings } \sigma \textit{ of inc}(\mu)'} x^\sigma q^{\text{maj}(\sigma)} t^{\text{coinv}(\sigma)}$$
$$\times \prod_{\substack{s\in\mu',\ s \textit{ not in row } 1 \\ \sigma(s)=\sigma(\text{South}(s))}} (1-q^{\text{leg}(s)+1}t^{\text{arm}(s)+1}) \prod_{\substack{s\in\mu',\ s \textit{ not in row } 1 \\ \sigma(s)\neq\sigma(\text{South}(s))}} (1-t),$$

Note that (5.8) implies $J_\mu(X_n;q,t)$ is $\prod_i (t;t)_{m_i}$ times an element of $\mathbb{Z}[X_n,q,t]$. By taking limits as in (3.6) we recover a result of Knop and Sahi [**23**], namely that the coefficient of a given monomial in the integral form Jack polynomial $J^\alpha_\mu(X)$ is $\prod_{i\geq 1} m_i(\mu)!$ times an element of $\mathbb{N}[\alpha]$.

Let $\gamma \in \mathbb{Z}_+^k$, where $k \in \mathbb{Z}_+$, be a (strong) composition. Haglund, Luoto, Mason and van Willigenburg [**17**] showed that if we let

$$\mathrm{QS}_\gamma(X_n) = \sum_{\substack{\alpha \in \mathbb{N}^n \\ \alpha^+ = \gamma}} E_\alpha(X_n; 0, 0) \tag{5.9}$$

then the $\{\mathrm{QS}_\gamma\}$ form a basis for the ring QSym of quasisymmetric functions they called the quasisymmetric Schur basis. They noted that

$$\sum_{\substack{\gamma \in \mathbb{Z}_+^{\ell(\mu)} \\ \mathrm{dec}(\gamma) = \mu}} \mathrm{QS}_\gamma(X_n) = s_\mu(X_n), \tag{5.10}$$

so the QS basis decomposes the Schur basis into positive components. Since $P_\mu(X_n; 0, 0) = s_\mu(X_n)$, a natural question to ask is whether or not there is a way to incorporate q, t parameters into $\mathrm{QS}_\gamma(X_n)$ to form a quasisymmetric basis for $\mathrm{QSym}(q, t)$, the ring of quasisymmetric functions with coefficients in $\mathbb{Q}(q, t)$. Identity (5.2) has led to a solution to this question. In [**6**] it is shown if we define

$$\mathrm{G}_\gamma(X_n; q, t) = \sum_{\substack{\alpha \in \mathbb{N}^n \\ \alpha^+ = \gamma}} E^{\beta(\alpha)}_{\mathrm{inc}(\alpha)}(X_n; q, t), \tag{5.11}$$

then the G_γ form a basis for $\mathrm{QSym}(q, t)$, $G_\gamma(X_n; 0, 0) = \mathrm{QS}_\gamma(X_n)$, and (by (5.2)),

$$P_\mu(X_n; q, t) = \sum_{\substack{\gamma \\ \mathrm{dec}(\gamma) = \mu}} G_\gamma(X_n; q, t). \tag{5.12}$$

We mention that Corteel, Mandelshtam and Roberts [**7**] have found the expansion of G_γ into Gessel fundamental quasisymmetric functions.

6. Refinements of the HHL formula for $\tilde{H}_\mu(X; q, t)$

In this section we list several identities for $\tilde{H}_\mu(X; q, t)$ which refine and extend the HHL formula (2.4). First of all, in [**15**] it is noted that if in formula (3.3) we replace the sum over nonattacking fillings of μ' by a sum over nonattacking fillings of γ', where γ is any permutation of the parts of μ (i.e. $\gamma^+ = \mu$), and use the same q, t, x-weights, with type A and type B triples to define coinv in terms of triples (or skyline diagram leg and arm lengths to define coinv via inversion pairs and descents) then the result still equals $J_\mu(X; q, t)$. Furthermore one can show that if we apply the same type of idea to (2.4), the obvious extensions of the involutions IH and IL fix the q, t, x-weights and switch the sign. This implies that for any shape τ obtained by permuting the columns of μ,

$$\tilde{H}_\mu(X; q, t) = \sum_{\sigma : \tau \to \mathbb{Z}_+} x^\sigma t^{\mathrm{maj}(\sigma, \tau)} q^{\mathrm{inv}(\sigma, \tau)}. \tag{6.1}$$

There is also [**6**] a more compact version of Theorem 2.5. We say a filling σ of μ is *sorted* if entries in the first row, in columns of the same height, are weakly increasing left-to-right. Letting Col_j denotes the set of columns of μ of height j for $1 \le j \le n$, for a sorted filling σ of μ set

$$\mathrm{perm}_q(\sigma, \mu) = \prod_{j=1}^{n} \binom{m_1 + m_2 + \dots}{m_1, m_2, \dots}_q, \tag{6.2}$$

4							
6	2	6	2				
7	2	5	7				
1	2	3	5				
3	3	5	5	1	1	2	2

FIGURE 10. A sorted filling of 84441.

where inside the sum m_j is the number of times j occurs in σ in the first row and in columns Col_j, i.e. in squares $\{(p,1) : p \in \mathrm{Col}_j\}$. For the filling in Figure 10,

$$\begin{aligned} \mathrm{perm}_q &= \binom{3}{1,2}_q \binom{4}{2,2}_q \\ &= (1+q+q^2)(1+q+q^2)(1+q^2). \end{aligned} \tag{6.3}$$

THEOREM 6.1 ([**6**]).

$$\tilde{H}_\mu(X;q,t) = \sum_{\substack{\sigma:\mu\to\mathbb{Z}_+ \\ \sigma \text{ sorted}}} x^\sigma t^{\mathrm{maj}(\sigma,\mu)} q^{\mathrm{inv}(\sigma,\mu)} \,\mathrm{perm}_q(\sigma,\mu). \tag{6.4}$$

The proof of (6.4) is fairly long; it involves a detailed analysis of the structure of the statistic inv on fillings.

We now describe another recent result which grew out of the study of the ASEP. Given a filling σ of μ, we call a set of squares u, v, w, where $w = \mathrm{South}(u)$, and v is in the same row as w and to the right of w, a quinv triple of σ. Also, we call two squares w, v satisfying the second of these conditions, with both w, v at the top of their respective columns, a (degenerate) quinv triple. Given a quinv triple with $\sigma(u)$, $\sigma(v)$ and $\sigma(w)$ all distinct, we say they form an inversion triple if they have a counterclockwise orientation; otherwise they form a coinversion triple. (Or, if w, v form a degenerate triple, then they from an inversion triple if $\sigma(w) < \sigma(v)$, otherwise they form a coinversion triple.) If the three entries of $\sigma(u), \sigma(v), \sigma(w)$ are not all distinct, we standardize them, using the reading order that goes across rows from right-to-left, starting with the top row and going downwards one row at a time. For example, the quinv reading word for the filling in Figure 11 is 32243135233214321544575l, some of the inversion triples are

$$\begin{aligned} \{[(2,2),(2,1),(3,1)],[(3,3),(4,3)],[(6,2),(6,1),(7,1)],\\ [(6,2),(6,1),(8,1)][(1,5),(2,5)]\}, \end{aligned} \tag{6.5}$$

while some coinversion triples are

$$\{[(1,2),(1,1),(2,1)],[(6,2),(7,2)],[(6,2),(8,2)]\}. \tag{6.6}$$

THEOREM 6.2 ([**2**]).

$$\tilde{H}_\mu(X;q,t) = \sum_{\sigma:\mu\to\mathbb{Z}_+} x^\sigma t^{\mathrm{maj}(\sigma,\mu)} q^{\mathrm{quinv}(\sigma,\mu)}, \tag{6.7}$$

where quinv(σ,μ) *is the number of* quinv *inversion triples.*

2	3						
4	2						
5	3	1	3				
2	3	4	1	2	3	3	2
1	5	7	5	4	4	5	1

FIGURE 11. A filling of 88422

There have been a number of other exciting combinatorial formulas for Macdonald polynomials that have appeared in recent years, but due to space and time considerations we will have to content ourselves with merely listing a few of them. Several of these formulas have emerged from statistical mechanics and representation theory. For example, Garbali and Wheeler [**11**] have a formula for the monomial expansion of $\tilde{H}_\mu(X; q, t)$ where, when $q = 0$, the coefficients reduce to the *rigged configurations* occurring in the description Kirillov and Reshetikhin gave of the Schur expansion of modified Hall-Littlewood polynomials [**21**]. Also, Borodin and Wheeler [**3**] give a new expansion for the nonsymmetric Macdonald polynomial, and then show how their formula can be bijectively mapped to terms in the HHL formula. In addition, Diaconis and Ram [**9**] have a probabilistic model for the coefficients in the expansion of $P_\mu(X; q, t)$ into the power-sum basis, with impressive applications to the study of statistics and Markov chains.

References

[1] Per Alexandersson, *Non-symmetric Macdonald polynomials and Demazure-Lusztig operators*, Sém. Lothar. Combin. **76** ([2016), Art. B76d, 27. MR4063906

[2] Arvind Ayyer, Olya Mandelshtam, and James B. Martin, *Modified Macdonald polynomials and the multispecies zero-range process: I*, Algebr. Comb. **6** (2023), no. 1, 243–284, DOI 10.5802/alco.248. MR4552715

[3] Alexei Borodin and Michael Wheeler, *Nonsymmetric Macdonald polynomials via integrable vertex models*, Trans. Amer. Math. Soc. **375** (2022), no. 12, 8353–8397, DOI 10.1090/tran/8309. MR4504641

[4] Luigi Cantini, Jan de Gier, and Michael Wheeler, *Matrix product formula for Macdonald polynomials*, J. Phys. A **48** (2015), no. 38, 384001, 25, DOI 10.1088/1751-8113/48/38/384001. MR3400909

[5] Ivan Cherednik, *Nonsymmetric Macdonald polynomials*, Internat. Math. Res. Notices **10** (1995), 483–515, DOI 10.1155/S1073792895000341. MR1358032

[6] Sylvie Corteel, Jim Haglund, Olya Mandelshtam, Sarah Mason, and Lauren Williams, *Compact formulas for Macdonald polynomials and quasisymmetric Macdonald polynomials*, Selecta Math. (N.S.) **28** (2022), no. 2, Paper No. 32, 33, DOI 10.1007/s00029-021-00721-7. MR4359567

[7] Sylvie Corteel, Olya Mandelshtam, and Austin Roberts, *Fundamental expansion of quasisymmetric Macdonald polynomials*, Sém. Lothar. Combin. **85B** (2021), Art. 9, 12. MR4311890

[8] Sylvie Corteel, Olya Mandelshtam, and Lauren Williams, *From multiline queues to Macdonald polynomials via the exclusion process*, Sém. Lothar. Combin. **82B** (2020), Art. 97, 12. MR4098318

[9] Persi Diaconis and Arun Ram, *A probabilistic interpretation of the Macdonald polynomials*, Ann. Probab. **40** (2012), no. 5, 1861–1896, DOI 10.1214/11-AOP674. MR3025704

[10] Dominique Foata, *On the Netto inversion number of a sequence*, Proc. Amer. Math. Soc. **19** (1968), 236–240, DOI 10.2307/2036179. MR223256

[11] Alexandr Garbali and Michael Wheeler, *Modified Macdonald polynomials and integrability*, Comm. Math. Phys. **374** (2020), no. 3, 1809–1876, DOI 10.1007/s00220-020-03680-w. MR4076089

[12] Weiying Guo and Arun Ram, *Comparing formulas for type GL_n Macdonald polynomials*, Algebr. Comb. **5** (2022), no. 5, 849–883. MR4511154

[13] J. Haglund, *A combinatorial model for the Macdonald polynomials*, Proc. Natl. Acad. Sci. USA **101** (2004), no. 46, 16127–16131, DOI 10.1073/pnas.0405567101. MR2114585

[14] J. Haglund, M. Haiman, and N. Loehr, *A combinatorial formula for Macdonald polynomials*, J. Amer. Math. Soc. **18** (2005), no. 3, 735–761, DOI 10.1090/S0894-0347-05-00485-6. MR2138143

[15] J. Haglund, M. Haiman, and N. Loehr, *A combinatorial formula for non-symmetric Macdonald polynomials*, Amer. J. Math. **103** (2008), 359–383.

[16] J. Haglund, M. Haiman, N. Loehr, J. B. Remmel, and A. Ulyanov, *A combinatorial formula for the character of the diagonal coinvariants*, Duke Math. J. **126** (2005), no. 2, 195–232, DOI 10.1215/S0012-7094-04-12621-1. MR2115257

[17] J. Haglund, K. Luoto, S. Mason, and S. van Willigenburg, *Quasisymmetric Schur functions*, J. Combin. Theory Ser. A **118** (2011), no. 2, 463–490, DOI 10.1016/j.jcta.2009.11.002. MR2739497

[18] James Haglund, *The q, t-Catalan numbers and the space of diagonal harmonics*, University Lecture Series, vol. 41, American Mathematical Society, Providence, RI, 2008. With an appendix on the combinatorics of Macdonald polynomials, DOI 10.1007/s10711-008-9270-0. MR2371044. Available online at `math.upenn.edu/~jhaglund/`.

[19] Mark Haiman, *Macdonald polynomials and geometry*, New perspectives in algebraic combinatorics (Berkeley, CA, 1996), Math. Sci. Res. Inst. Publ., vol. 38, Cambridge Univ. Press, Cambridge, 1999, pp. 207–254. MR1731818

[20] Bogdan Ion, *Nonsymmetric Macdonald polynomials and Demazure characters*, Duke Math. J. **116** (2003), no. 2, 299–318, DOI 10.1215/S0012-7094-03-11624-5. MR1953294

[21] S. V. Kerov, A. N. Kirillov, and N. Yu. Reshetikhin, *Combinatorics, the Bethe ansatz and representations of the symmetric group* (Russian, with English summary), Zap. Nauchn. Sem. Leningrad. Otdel. Mat. Inst. Steklov. (LOMI) **155** (1986), no. Differentsial′naya Geometriya, Gruppy Li i Mekh. VIII, 50–64, 193, DOI 10.1007/BF01247087; English transl., J. Soviet Math. **41** (1988), no. 2, 916–924. MR869576

[22] Friedrich Knop, *Integrality of two variable Kostka functions*, J. Reine Angew. Math. **482** (1997), 177–189, DOI 10.1515/crll.1997.482.177. MR1427661

[23] Friedrich Knop and Siddhartha Sahi, *A recursion and a combinatorial formula for Jack polynomials*, Invent. Math. **128** (1997), no. 1, 9–22, DOI 10.1007/s002220050134. MR1437493

[24] Alain Lascoux and Marcel-Paul Schützenberger, *Sur une conjecture de H. O. Foulkes* (French, with English summary), C. R. Acad. Sci. Paris Sér. A-B **286** (1978), no. 7, A323–A324. MR472993

[25] Alain Lascoux, Bernard Leclerc, and Jean-Yves Thibon, *Ribbon tableaux, Hall-Littlewood functions, quantum affine algebras, and unipotent varieties*, J. Math. Phys. **38** (1997), no. 2, 1041–1068, DOI 10.1063/1.531807. MR1434225

[26] Cristian Lenart, *On combinatorial formulas for Macdonald polynomials*, Adv. Math. **220** (2009), no. 1, 324–340, DOI 10.1016/j.aim.2008.09.007. MR2462843

[27] I. G. Macdonald, *A new class of symmetric polynomials*, Actes du 20e Séminaire Lotharingien, Publ. Inst. Rech. Math. Av. **372** (1988), 131–171.

[28] I. G. Macdonald, *Symmetric functions and Hall polynomials*, 2nd ed., Oxford Mathematical Monographs, The Clarendon Press, Oxford University Press, New York, 1995. With contributions by A. Zelevinsky; Oxford Science Publications. MR1354144

[29] I. G. Macdonald, *Affine Hecke algebras and orthogonal polynomials*, Astérisque **237** (1996), Exp. No. 797, 4, 189–207. Séminaire Bourbaki, Vol. 1994/95. MR1423624

[30] Laurent Manivel, *Symmetric functions, Schubert polynomials and degeneracy loci*, SMF/AMS Texts and Monographs, vol. 6, American Mathematical Society, Providence, RI; Société Mathématique de France, Paris, 2001. Translated from the 1998 French original by John R. Swallow; Cours Spécialisés [Specialized Courses], 3. MR1852463

[31] Dan Marshall, *Symmetric and nonsymmetric Macdonald polynomials*, Ann. Comb. **3** (1999), no. 2-4, 385–415, DOI 10.1007/BF01608794. On combinatorics and statistical mechanics. MR1772356

[32] Arun Ram and Martha Yip, *A combinatorial formula for Macdonald polynomials*, Adv. Math. **226** (2011), no. 1, 309–331, DOI 10.1016/j.aim.2010.06.022. MR2735761
[33] Siddhartha Sahi, *The binomial formula for nonsymmetric Macdonald polynomials*, Duke Math. J. **94** (1998), no. 3, 465–477, DOI 10.1215/S0012-7094-98-09419-4. MR1639523
[34] Richard P. Stanley, *Enumerative combinatorics. Vol. 2*, Cambridge Studies in Advanced Mathematics, vol. 62, Cambridge University Press, Cambridge, 1999. With a foreword by Gian-Carlo Rota and appendix 1 by Sergey Fomin, DOI 10.1017/CBO9780511609589. MR1676282
[35] Meesue Yoo, *A combinatorial formula for the Schur coefficients of the integral form of the Macdonald polynomials in the two column and certain hook cases*, Ann. Comb. **16** (2012), no. 2, 389–410, DOI 10.1007/s00026-012-0133-x. MR2927615
[36] Meesue Yoo, *Schur coefficients of the integral form Macdonald polynomials*, Tokyo J. Math. **38** (2015), no. 1, 153–173, DOI 10.3836/tjm/1437506242. MR3374619

Department of Mathematics, University of Pennsylvania, Philadelphia, Pennsylvania 19104-6395

Email address: jhaglund@math.upenn.edu

Contemporary Mathematics
Volume **815**, 2025
https://doi.org/10.1090/conm/815/16315

Symmetric and nonsymmetric Macdonald polynomials via a path model with a pseudo-crystal structure

Cristian Lenart, Satoshi Naito, Fumihiko Nomoto, and Daisuke Sagaki

ABSTRACT. In this paper we derive a counterpart of the well-known Ram-Yip formula for symmetric and nonsymmetric Macdonald polynomials of arbitrary type. Our new formula is in terms of a generalization of Lakshmibai-Seshadri paths (originating in standard monomial theory), which we call pseudo-quantum Lakshmibai-Seshadri (LS) paths. This model carries less information than the alcove walks in the Ram-Yip formula, and it is therefore more efficient. Furthermore, we construct a connected pseudo-crystal structure on the pseudo-quantum LS paths, which is expected to lead to a simple Littlewood-Richardson rule for multiplying Macdonald polynomials. By contrast with the Kashiwara crystals, our pseudo-crystals have edges labeled by arbitrary roots.

1. Introduction

Macdonald polynomials are an important family of polynomials associated with an irreducible affine root system, with rational function coefficients in q, t. They have deep connections with several areas of mathematics: double affine Hecke algebras (DAHAs), the representations of affine Lie algebras, p-adic groups, integrable systems, conformal field theory, statistical mechanics, Hilbert schemes, etc.

There are two versions of Macdonald polynomials: the symmetric ones (under the Weyl group action), denoted $P_\lambda(q,t)$, where λ is a nonaffine dominant weight, and the nonsymmetric ones $E_\mu(q,t)$, where μ is an arbitrary nonaffine weight. They were defined in [**M03**] in the setup of the corresponding double affine Hecke algebras, and are orthogonal with respect to the so-called Macdonald scalar product.

The symmetric and nonsymmetric Macdonald polynomials generalize the irreducible characters and Demazure characters of simple Lie algebras, respectively, which are recovered by setting $q = t = 0$. Other specializations are also important: $P_\lambda(q = 0, t)$ are the Hall-Littlewood polynomials, also known as spherical functions corresponding to a Chevalley group over a p-adic field; $P_\lambda(q, t = q^\gamma)$, with γ a positive real number, become the Jack polynomials by setting $q \to 1$; $E_\mu(q = \infty, t)$ are the Iwahori Whittaker functions in number theory; $E_\mu(q, t = 0)$ and $E_\mu(q, t = \infty)$,

2020 *Mathematics Subject Classification.* Primary 05E05; Secondary 33D52, 20G42.

The first author was partly supported by the NSF grant DMS-1855592.

The second author was partly supported by JSPS Grant-in-Aid for Scientific Research (C) 21K03198.

The fourth author was partly supported by JSPS Grant-in-Aid for Scientific Research (C) 19K03415.

as well as their symmetric versions, are closely related to the representation theory of quantum affine algebras (see below).

Given their prominent role, it is important to have explicit formulas for the monomial expansions of Macdonald polynomials, which turn out to have a combinatorial nature. The first such formulas were given in type A: Macdonald's formula for $P_\lambda(q,t)$ in terms of semistandard Young tableaux [**M95**, Chapter VI, Section 7], and the Haglund-Haiman-Loehr formulas for $P_\lambda(q,t)$ and $E_\mu(q,t)$ in [**HHL05**] and [**HHL08**], respectively, which are in terms of so-called nonattacking fillings of certain diagrams. Other formulas for the type A Macdonald polynomials were recently derived through connections with a statistical mechanics model called the multispecies asymmetric simple exclusion process (ASEP) on a circle [**CHM$^+$22, CMW22**], as well as via integrable lattice models [**GW20**]. On another hand, building on work of Schwer [**Sc06**], Ram and Yip [**RY11**] gave a uniform combinatorial formula for both the symmetric and nonsymmetric Macdonald polynomials of arbitrary type, in terms of so-called alcove walks. Recently, a very interesting generalization of the Ram-Yip formula was derived in [**Sa22**] for the so-called metaplectic Macdonald polynomials in [**SSV21, SSV22**]; they specialize to the metaplectic Whittaker functions in number theory.

Several specializations of the Ram-Yip formula, in some of the cases mentioned above, were worked out in [**Le12, OS18**]. In [**Le09, GR21**] it was shown that the Haglund-Haiman-Loehr formulas can be derived as "compressed" versions of the Ram-Yip formulas. On another hand, several applications of the Ram-Yip formula and its specializations were derived. Among them is the uniform combinatorial model, called the quantum alcove model, for tensor products of so-called (single-column) Kirillov-Reshetikhin (KR) crystals of quantum affine algebras [**LNS$^+$17a, LNS$^+$17b**]; see [**K91**] and below for the theory of Kashiwara crystals. A corollary of this work is the identification of $P_\lambda(q,t=0)$ and $E_\mu(q,t=0)$ with the graded characters of tensor products of KR modules and certain Demazure-type submodules, respectively. The quantum alcove model extends the alcove model in [**LP**], which is a uniform combinatorial model for the characters and crystals corresponding to irreducible integrable highest weight modules for complex symmetrizable Kac-Moody algebras. On another hand, the alcove model can be viewed as a discrete counterpart of the Littelmann path model [**Li94, Li95**].

In this paper we derive new combinatorial formulas for the symmetric and nonsymmetric Macdonald polynomials in terms of so-called pseudo-quantum Lakshmibai-Seshadri (LS) paths; see Theorem 2.10. We also specialize our formulas to similar ones for Jack and Hall-Littlewood polynomials; see Corollaries 2.11 and 2.12, respectively. LS paths are certain piecewise-linear paths which were originally defined in the context of standard monomial theory [**LS91**]. They are special cases of Littelmann paths [**Li94, Li95**]. The generalization of LS paths called quantum LS paths was used in [**LNS$^+$17a, LNS$^+$17b**] as yet another model for tensor products of (single-column) KR crystals and Macdonald polynomials specialized at $t=0$. Here we generalize the quantum LS paths even further, by considering the mentioned pseudo-quantum LS paths. Essentially, the mentioned generalizations are obtained by replacing a certain condition based on the Bruhat order on the Weyl group (for LS paths) with similar conditions based on the so-called quantum Bruhat graph (for quantum LS paths, see [**LNS$^+$15**]) and the double Bruhat graph, considered here (for pseudo-quantum LS paths).

We derive the new formulas for the symmetric and nonsymmetric Macdonald polynomials by translating the corresponding Ram-Yip formulas, namely by bijecting the respective alcove walks and pseudo-quantum LS paths, and by translating the corresponding statistics. A similar bijection was constructed in [**LNS$^+$17a**] in the special case of the quantum alcove model and quantum LS paths, and earlier in [**LP**] for the alcove model and LS paths. The map from some version of the alcove model to the respective version of LS paths is a forgetful map, while the inverse map is non-trivial. Thus, the advantage of LS path-type models is that they carry less information than models based on alcove walks, and therefore are more efficient. We explain this fact in a little more detail below.

The alcove model is based on the choice of any shortest sequence of adjacent alcoves joining the fundamental one to its translation by the negative of a dominant weight λ; this is called an alcove path. Consider $\lambda = 2\rho = (4, 2, 0)$ in type A_2, where ρ is half the sum of the positive roots $\alpha_{ij} := \varepsilon_i - \varepsilon_j$. The first figure below shows a corresponding alcove path obtained by deforming the segment joining 0 to $-\lambda$. It is recorded as the sequence of eight hyperplanes (lines) crossed by the blue segment. An element of the alcove model (called an alcove walk) is the "folding" of the alcove path (across the two highlighted lines), which is shown in the second figure; this is recorded as the sequence of eight lines that are crossed or touched by the blue piecewise-linear path. However, if the initial segment is moved back to the segment from 0 to $-\lambda$, then the mentioned piecewise-linear path deforms to the one in the third figure, which is an LS path. Note that the latter retains considerably less information: just the directions and lengths of its two segments (while the total length is normalized to 1).

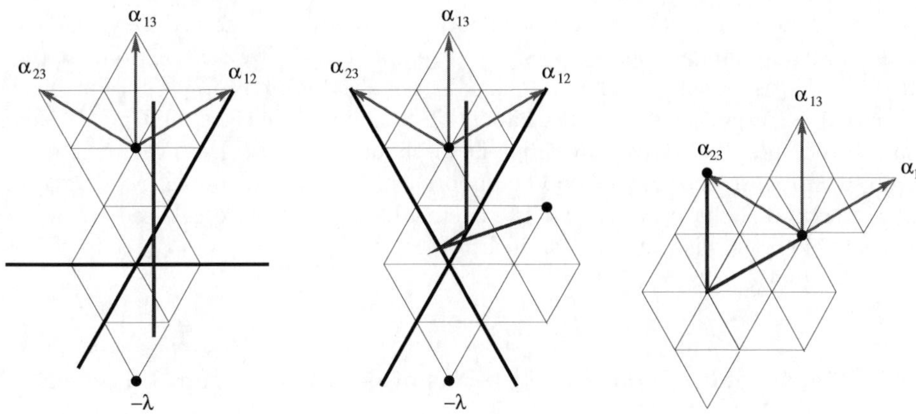

The quantum LS paths were successfully used in studying connections between the specializations of Macdonald polynomials at $t \in \{0, \infty\}$ and the representation theory of quantum affine algebras (in particular, the so-called level-zero extremal weight modules); this was carried out in several recent papers, namely [**NNS18, NNS20, NS05, NS16, NS21, N16, N17**]. Therefore, we expect similar applications of our pseudo-quantum LS paths, in particular to some of the other specializations of Macdonald polynomials mentioned above, which have important roles in representation theory, number theory, and other areas. On another hand, it would be also interesting to develop LS path-type models and similar applications for the metaplectic Macdonald polynomials mentioned above and the metaplectic Whittaker functions (as their specializations).

Finally, in Section 4, we construct a pseudo-crystal structure on pseudo-quantum LS paths. Crystals were defined by Kashiwara [**K91**] as (connected) colored directed graphs encoding certain representations of quantum affine algebras (including the irreducible highest weight ones) in the limit of the quantum parameter going to zero; the edge colors correspond to labeling the edges by simple roots. By contrast, our pseudo-crystals have edges labeled by arbitrary roots, and are also shown to be connected graphs. Therefore, for a fixed dominant weight λ, the pseudo-crystal structure can be used as an alternative way to generate all the pseudo-quantum LS paths of shape λ by starting from the special one $(e; 0, 1)$, which is thought of as the straight-line path joining the origin to λ.

Crystals were successfully used to solve several basic problems in representation theory, such as decomposing tensor products of irreducible representations. Essentially, such a Littlewood-Richardson rule is easily derived as follows: consider the tensor product of the crystals corresponding to the mentioned irreducible representations, which is defined via a specific tensor product rule; this graph has several connected components, each of which is known to be the crystal of some irreducible representation, so it mirrors the decomposition of the original tensor product of representations; record the highest weight vertex in each connected component, which identifies the corresponding irreducible representation. The Littlewood-Richardson rule can be made very explicit by using Littelmann paths [**Li94, Li95**] or the alcove model [**LP**] for the mentioned crystals, as pointed out above.

We expect that a similar procedure can be used to derive a Littlewood-Richardson rule for Macdonald polynomials, which would express the product of two Macdonald polynomials in terms of Macdonald polynomials. More precisely, one can consider the decomposition into connected components of the tensor product (defined via the usual rule) of the two pseudo-crystals (realized via pseudo-quantum LS paths) corresponding to the Macdonald polynomials that are multiplied. Note that such a rule was derived by different methods in [**Y12**]. However, it is quite involved, while we expect the proposed approach to lead to a simpler rule. Indeed, the Littlewood-Richardson rule for the irreducible representations based on tensor products of crystals is the simplest such rule.

2. Macdonald polynomials in terms of pseudo-quantum LS paths

2.1. Pseudo-quantum Lakshmibai-Seshadri paths. Let $\mathfrak{g}$ be a finite-dimensional simple Lie algebra over $\mathbb{C}$, I the vertex set for the Dynkin diagram of $\mathfrak{g}$, and $\{\alpha_i\}_{i\in I}$ (resp., $\{\alpha_i^\vee\}_{i\in I}$) the set of simple roots (resp., coroots) of $\mathfrak{g}$. Then $\mathfrak{h} = \bigoplus_{i\in I}\mathbb{C}\alpha_i^\vee$ is a Cartan subalgebra of $\mathfrak{g}$, with $\mathfrak{h}^* = \bigoplus_{i\in I}\mathbb{C}\alpha_i$ the dual space of $\mathfrak{h}$ and $\mathfrak{h}^*_{\mathbb{R}} = \bigoplus_{i\in I}\mathbb{R}\alpha_i$ its real form; the canonical pairing between $\mathfrak{h}$ and $\mathfrak{h}^*$ is denoted by $\langle\cdot,\cdot\rangle : \mathfrak{h}^*\times\mathfrak{h}\to\mathbb{C}$. Let $Q = \sum_{i\in I}\mathbb{Z}\alpha_i \subset \mathfrak{h}^*_{\mathbb{R}}$ denote the root lattice, $Q^\vee = \sum_{i\in I}\mathbb{Z}\alpha_i^\vee \subset \mathfrak{h}_{\mathbb{R}}$ the coroot lattice, and $P = \sum_{i\in I}\mathbb{Z}\varpi_i \subset \mathfrak{h}^*_{\mathbb{R}}$ the weight lattice of $\mathfrak{g}$, where the ϖ_i, $i\in I$, are the fundamental weights for $\mathfrak{g}$, i.e., $\langle\varpi_i, \alpha_j^\vee\rangle = \delta_{ij}$ for $i, j \in I$; we set $P^+ := \sum_{i\in I}\mathbb{Z}_{\geq 0}\varpi_i$, and call an element λ of P^+ a dominant (integral) weight. Let us denote by Δ the set of roots, and by Δ^+ (resp., Δ^-) the set of positive (resp., negative) roots. Also, let $W := \langle s_i \mid i\in I\rangle$ be the Weyl group of $\mathfrak{g}$, where s_i, $i \in I$, are the simple reflections acting on $\mathfrak{h}^*$ and on $\mathfrak{h}$ as follows:

$$s_i\nu = \nu - \langle \nu, \alpha_i^\vee \rangle \alpha_i \qquad \text{for } \nu \in \mathfrak{h}^*,$$
$$s_i h = h - \langle \alpha_i, h \rangle \alpha_i^\vee \qquad \text{for } h \in \mathfrak{h};$$

we denote the identity element and the longest element of W by e and $w_\circ$, respectively. If $\alpha \in \Delta$ is written as $\alpha = w\alpha_i$ for $w \in W$ and $i \in I$, then its coroot $\alpha^\vee$ is $w\alpha_i^\vee$; we often identify s_α with $s_{\alpha^\vee}$. For $u \in W$, the length of u is denoted by $\ell(u)$, which agrees with the cardinality of the set $\Delta^+ \cap u^{-1}\Delta^-$.

DEFINITION 2.1. The double Bruhat graph, denoted by DBG, is the directed graph with vertex set W and directed edges labeled by arbitrary positive roots: for all $u \in W$ and $\beta \in \Delta^+$, $u \xrightarrow{\beta} us_\beta$ is a directed edge of DBG; a directed edge satisfying $\ell(us_\beta) > \ell(u)$ (resp., $\ell(us_\beta) < \ell(u)$) is called a Bruhat (resp., quantum) edge.

For a subset $S \subset I$, we set $W_S := \langle s_i \mid i \in S \rangle$. We denote the longest element of W_S by $w_\circ(S)$. Also, we set $\Delta_S := Q_S \cap \Delta$, $\Delta_S^+ := \Delta_S \cap \Delta^+$, and $\Delta_S^- := \Delta_S \cap \Delta^-$, where $Q_S := \sum_{i \in S} \mathbb{Z}\alpha_i$. Let $W^S (\subset W)$ denote the set of all minimal-length coset representatives for the cosets in W/W_S. For $w \in W$, we denote the minimal-length coset representative of the coset wW_S by $\lfloor w \rfloor$, and for a subset $T \subset W$, we set $\lfloor T \rfloor := \{\lfloor w \rfloor \mid w \in T\} \subset W^S$.

We consider and fix an arbitrary dominant weight $\lambda \in P^+$, i.e., $\langle \lambda, \alpha_i^\vee \rangle \geq 0$ for all $i \in I$. We set

$$S = S_\lambda := \{i \in I \mid \langle \lambda, \alpha_i^\vee \rangle = 0\} \subset I.$$

DEFINITION 2.2. Let $\lambda \in P^+$ be a dominant weight and $b \in \mathbb{Q} \cap [0,1]$. We denote by $\mathrm{DBG}_{b\lambda}$ the subgraph of DBG with the same vertex set but having only directed edges of the form: $u \xrightarrow{\beta} us_\beta$ with $b\langle \lambda, \beta^\vee \rangle \in \mathbb{Z}$.

REMARK 2.3. Let $\lambda \in P^+$ be a dominant weight and $b \in \mathbb{Q} \cap [0,1]$. Let $u, v \in W$. If there exists a directed path

$$u = x_0 \xrightarrow{\gamma_1} x_1 \xrightarrow{\gamma_2} \cdots \xrightarrow{\gamma_r} x_r = v$$

from u to v in $\mathrm{DBG}_{b\lambda}$, then $b(v\lambda - u\lambda) \in P$ since $b(v\lambda - u\lambda) \in \sum_{k=1}^r \mathbb{Z}\langle \lambda, \gamma_k^\vee \rangle$.

For $\lambda \in P^+$ and $b \in \mathbb{Q} \cap [0,1]$, we define the (b,λ)-degree of an edge $u \xrightarrow{\beta} v$ in $\mathrm{DBG}_{b\lambda}$ by:

$$\deg_{b\lambda}(u \to v) = \begin{cases} \dfrac{t^{-\frac{1}{2}}(1-t)}{1-q^{(1-b)\langle \lambda, \beta^\vee \rangle} t^{\langle \rho, \beta^\vee \rangle}} & \text{if } u \xrightarrow{\beta} v \text{ is a Bruhat edge,} \\ \dfrac{q^{(1-b)\langle \lambda, \beta^\vee \rangle} t^{-\frac{1}{2}+\langle \rho, \beta^\vee \rangle}(1-t)}{1-q^{(1-b)\langle \lambda, \beta^\vee \rangle} t^{\langle \rho, \beta^\vee \rangle}} & \text{if } u \xrightarrow{\beta} v \text{ is a quantum edge,} \end{cases}$$

where q, t are indeterminates, and $\rho := \frac{1}{2}\sum_{\alpha \in \Delta^+} \alpha$.

DEFINITION 2.4. Let $\lambda \in P^+$, and set $S = S_\lambda = \{i \in I \mid \langle \lambda, \alpha_i^\vee \rangle = 0\}$. A pair $\eta = (w_1, w_2, \ldots, w_s; \tau_0, \tau_1, \ldots, \tau_s)$ of a sequence $w_1, \ldots, w_s$ of elements in W^S and an increasing sequence $0 = \tau_0 < \cdots < \tau_s = 1$ of rational numbers, is called a pseudo-quantum Lakshmibai-Seshadri (pQLS) path of shape λ if

(*) for every $1 \leq i \leq s-1$, there exists a directed path from w_{i+1} to w_i in $\mathrm{DBG}_{\tau_i\lambda}$.

Let $\overline{\mathrm{pQLS}}(\lambda)$ denote the set of all pQLS paths of shape λ.

For $\eta = (w_1, w_2, \dots, w_s; \tau_0, \tau_1, \dots, \tau_s) \in \overline{\mathrm{pQLS}}(\lambda)$, we set

$$\mathrm{wt}(\eta) := \sum_{i=0}^{s-1} (\tau_{i+1} - \tau_i) w_{i+1} \lambda.$$

2.2. Relation between $\overline{\mathrm{pQLS}}^{\mu}(\lambda)$ and $\mathrm{pQLS}^{\mu}(\lambda)$. Let $\lambda \in P^+$ be a dominant weight, and $\mu \in W\lambda$. We set $\lambda_- := w_\circ \lambda$. We denote by $v(\mu) \in W^S$ the minimal-length coset representative for the coset $\{w \in W \mid w\lambda = \mu\}$ in W/W_S; recall that $S = S_\lambda = \{i \in I \mid \langle \lambda, \alpha_i^\vee \rangle = 0\}$.We now set $\overline{\mathrm{pQLS}}^{\mu}(\lambda) := \{(w_1, w_2, \dots, w_s; \sigma_0, \dots, \sigma_s) \in \overline{\mathrm{pQLS}}(\lambda) \mid w_1 \leq v(\mu)\}$, $\mu \in W\lambda$, where $\leq$ denotes the Bruhat order on W.

Here we remark that $\ell(v(\mu)w) = \ell(v(\mu)) + \ell(w)$ for all $w \in W_S$. In particular, we have $\ell(v(\mu)w_\circ(S)) = \ell(v(\mu)) + \ell(w_\circ(S))$. When $\mu = \lambda_-$, it is clear that $w_\circ \in \{w \in W \mid w\lambda = \lambda_-\}$. Since $w_\circ$ is the longest element of W, we have $w_\circ = v(\lambda_-)w_\circ(S)$ and $\ell(v(\lambda_-)w_\circ(S)) = \ell(v(\lambda_-)) + \ell(w_\circ(S))$; note that $v(\lambda_-) = w_\circ w_\circ(S) = \lfloor w_\circ \rfloor$. We fix reduced expressions

$$v(\lambda_-)v(\mu)^{-1} = s_{i_1} \cdots s_{i_K},$$
$$v(\mu) = s_{i_{K+1}} \cdots s_{i_M},$$
$$w_\circ(S) = s_{i_{M+1}} \cdots s_{i_N},$$

for $v(\lambda_-)v(\mu)^{-1}$, $v(\mu)$, and $w_\circ(S)$, respectively. Then, $v(\lambda_-) = s_{i_1} \cdots s_{i_M}$ and $w_\circ = s_{i_1} \cdots s_{i_N}$ are reduced expressions for $v(\lambda_-)$ and $w_\circ$, respectively. We set $\beta_j := s_{i_N} \cdots s_{i_{j+1}} \alpha_{i_j}$ for $1 \leq j \leq N$. Then we have

$$\begin{aligned} \Delta^+ \setminus \Delta_S^+ &= \{\beta_1, \dots, \beta_M\}, \\ \Delta_S^+ &= \{\beta_{M+1}, \dots, \beta_N\}, \\ \Delta^+ &= \{\beta_1, \dots, \beta_N\}, \\ w_\circ(S)(\Delta^+ \cap v(\mu)\Delta^-) &= \{\beta_{K+1}, \dots, \beta_M\}. \end{aligned} \tag{2.1}$$

We define a total order $\succ$ on Δ^+ by:

$$\underbrace{\beta_1 \succ \beta_2 \succ \cdots \succ \beta_M}_{\in \Delta^+ \setminus \Delta_S^+} \succ \underbrace{\beta_{M+1} \succ \cdots \succ \beta_N}_{\in \Delta_S^+}. \tag{2.2}$$

REMARK 2.5. The total order $\succ$ above is a reflection order on Δ^+; that is, if $\alpha, \beta, \gamma \in \Delta^+$ with $\gamma^\vee = \alpha^\vee + \beta^\vee$, then $\alpha \prec \gamma \prec \beta$ or $\beta \prec \gamma \prec \alpha$.

For $\alpha \in \Delta$, we set

$$|\alpha| := \begin{cases} \alpha & \text{if } \alpha \in \Delta^+, \\ -\alpha & \text{if } \alpha \in \Delta^-. \end{cases}$$

LEMMA 2.6.

(1) *Let $w \in W$. If $x \xrightarrow{\beta} y$ is an edge of* DBG, *then $wx \xrightarrow{\beta} wy$ and $yw \xrightarrow{|w^{-1}\beta|} xw$ are also edges of* DBG.

(2) *If $x \xrightarrow{\beta} y$ is a Bruhat (resp., quantum) edge of* DBG, *then $yw_\circ \xrightarrow{-w_\circ\beta} xw_\circ$ is also a Bruhat (resp., quantum) edge of* DBG.

PROOF. Part (1) is obvious. Part (2) follows easily from the equality $\ell(y) - \ell(x) = \ell(xw_\circ) - \ell(yw_\circ)$. This proves the lemma. □

LEMMA 2.7. *Let $\lambda \in P^+$ be a dominant weight, and let $b \in \mathbb{Q} \cap [0,1]$. Let $w, v \in W$, and assume that there exists a directed path from w to v in $\mathrm{DBG}_{b\lambda}$; let $q \in \mathbb{Z}_{\geq 0}$ be the length of this path. Then, the following hold.*

(1) *There exists a directed path*

$$w = x_0 \xrightarrow{\gamma_1} x_1 \xrightarrow{\gamma_2} \cdots \xrightarrow{\gamma_p} x_p = v$$

in $\mathrm{DBG}_{b\lambda}$ such that $p \leq q$ and $\gamma_1 \prec \cdots \prec \gamma_p$. Moreover, there exists an element $w' \in wW_S$ such that there exists a label-increasing directed path from w' to v in $\mathrm{DBG}_{b\lambda}$ whose edge labels lie in $\Delta^+ \setminus \Delta_S^+$.

(2) *For each $w'' \in wW_S$ and $v'' \in vW_S$, there exists a directed path from w'' to v'' in $\mathrm{DBG}_{b\lambda}$. In particular, there exists a directed path from $\lfloor w \rfloor$ to $\lfloor v \rfloor$ in $\mathrm{DBG}_{b\lambda}$.*

PROOF. Let

$$w = y_0 \xrightarrow{\xi_1} y_1 \xrightarrow{\xi_2} \cdots \xrightarrow{\xi_q} y_q = v \tag{2.3}$$

be a directed path in $\mathrm{DBG}_{b\lambda}$.

(1) We will prove the assertion of the lemma by induction on q. If $q = 1$, then the assertion is obvious. Assume that $q > 1$. We write $\xi_1 = \beta_j$ for some $1 \leq j \leq N$; we prove the assertion (for the q) by descending induction on j. By the induction hypothesis on q, there exists a directed path

$$y_1 = z_0 \xrightarrow{\zeta_1} z_1 \xrightarrow{\zeta_2} \cdots \xrightarrow{\zeta_r} z_r = v$$

from y_1 to v in $\mathrm{DBG}_{b\lambda}$ such that $r \leq q-1$ and $\zeta_1 \prec \cdots \prec \zeta_r$. Therefore, we obtain the directed path

$$w = y_0 \xrightarrow{\xi_1} y_1 = z_0 \xrightarrow{\zeta_1} z_1 \xrightarrow{\zeta_2} \cdots \xrightarrow{\zeta_r} z_r = v \tag{2.4}$$

in $\mathrm{DBG}_{b\lambda}$. If $r < q-1$, then we can apply the induction hypothesis on q to the directed path above. Hence we may assume that $r = q-1$. Assume that $j = N$, i.e., $\xi_1 = \beta_N$. Then we have $\xi_1 \preceq \zeta_1$ since $\xi_1 = \beta_N$ is the minimal element of Δ^+. If $\xi_1 \prec \zeta_1$, then the directed path (2.4) has the desired property. If $\xi_1 = \zeta_1$, then we see that $w = z_1$, and that the directed path

$$w = z_1 \xrightarrow{\zeta_2} \cdots \xrightarrow{\zeta_r} z_r = v$$

in $\mathrm{DBG}_{b\lambda}$ has the desired property. Assume now that $j < N$. If $\xi_1 \preceq \zeta_1$, then we obtain a directed path having the desired property in exactly the same way as for $j = N$. Assume that $\xi_1 \succ \zeta_1$. Then we see that

$$w = y_0 \xrightarrow{\zeta_1} ws_{\zeta_1} \xrightarrow{|s_{\zeta_1}\xi_1|} z_1 \xrightarrow{\zeta_2} \cdots \xrightarrow{\zeta_r} z_r = v \tag{2.5}$$

is a directed path of length $r+1 = q$ from w to v in $\mathrm{DBG}_{b\lambda}$; here we remark that $b\langle \lambda,\, s_{\zeta_1}\xi_1^\vee \rangle \in \mathbb{Z}$ since $b\langle \lambda,\, \xi_1^\vee \rangle \in \mathbb{Z}$ and $b\langle \lambda,\, \zeta_1^\vee \rangle \in \mathbb{Z}$. Since $\zeta_1 \prec \xi_1$, we can apply the induction hypothesis on j to this path. Thus, we obtain a directed path

$$w = x_0 \xrightarrow{\gamma_1} x_1 \xrightarrow{\gamma_2} \cdots \xrightarrow{\gamma_p} x_p = v$$

in $\mathrm{DBG}_{b\lambda}$ such that $p \leq q$ and $\gamma_1 \prec \cdots \prec \gamma_p$.

Also, by (2.2), there exists $1 \leq k \leq p$ such that $\{\gamma_1, \ldots, \gamma_k\} \subset \Delta_S^+$ and $\{\gamma_{k+1}, \ldots, \gamma_p\} \subset \Delta^+ \setminus \Delta_S^+$. Then the directed path

$$x_k \xrightarrow{\gamma_{k+1}} x_1 \xrightarrow{\gamma_2} \cdots \xrightarrow{\gamma_p} x_p = v$$

is a path from x_k to v in $\mathrm{DBG}_{b\lambda}$ whose edge labels lie in $\Delta^+ \setminus \Delta_S^+$. In addition, we have $x_k \in wW_S$ since $\{\gamma_1, \dots, \gamma_k\} \subset \Delta_S^+$. This proves part (1) of the lemma.

(2) Let

$$w'' = y_{-s} \xrightarrow{\xi_{-s+1}} \cdots \xrightarrow{\xi_0} y_0 = w, \tag{2.6}$$

$$v = y_k \xrightarrow{\xi_{q+1}} \cdots \xrightarrow{\xi_t} y_t = v'' \tag{2.7}$$

be a directed path in $\mathrm{DBG}_{b\lambda}$ from w'' to w and one from v to v'', respectively, whose edge labels lie in Δ_S^+. Since $\langle \lambda, \xi_k^\vee \rangle = 0$ for $-s+1 \le k \le 0$ or $q+1 \le k \le t$, the directed paths (2.6) and (2.7) are paths in $\mathrm{DBG}_{b\lambda}$. Therefore, the path given by concatenating the paths (2.3), (2.6), and (2.7) is a directed path from w'' to v'' in $\mathrm{DBG}_{b\lambda}$. This proves part (2) of the lemma. □

DEFINITION 2.8. Let $\lambda \in P^+$, and set $S = S_\lambda = \{i \in I \mid \langle \lambda, \alpha_i^\vee \rangle = 0\}$. Let

$$\widetilde{\eta} = \left(\left(w_{p,0} \xleftarrow{\beta_{p,1}} \cdots \xleftarrow{\beta_{p,t_p}} w_{p,t_p} \right)_{p=0,\dots,s-1} ; \sigma_0, \dots, \sigma_s \right)$$

be a pair of label-increasing directed paths in DBG whose edge labels lie in $\Delta^+ \setminus \Delta_S^+$ and an increasing sequence $0 = \sigma_0 < \cdots < \sigma_s = 1$ of rational numbers. The pair $\widetilde{\eta}$ is called a pseudo-quantum Lakshmibai-Seshadri (pQLS) pair of shape λ if it satisfies the following conditions:

(i) $\sigma_p \langle \lambda, \beta_{p,m}^\vee \rangle \in \mathbb{Z}$ for $0 \le p \le s-1$, $1 \le m \le t_p$,
(ii) $w_{p,t_p} = w_{p+1,0}$ for $0 \le p \le s-2$.

Let $\mathrm{pQLS}(\lambda)$ denote the set of all pQLS pairs of shape λ. Also, for $\mu \in W\lambda$, let $\mathrm{pQLS}^\mu(\lambda)$ ($\subset \mathrm{pQLS}(\lambda)$) denote the set of pQLS pairs $\widetilde{\eta}$ of shape λ satisfying the following conditions:

(iii) $w_\circ(S)\beta_{0,m} \in \Delta^+ \cap v(\mu)^{-1}\Delta^-$ for $1 \le m \le t_0$,
(iv) $w_{0,0} = v(\mu) w_\circ(S)$.

Recall from [**M03**, (2.2.4) and (2.7.3)] that $\Delta^+ \cap v(\mu)^{-1}\Delta^- \subset \Delta^+ \setminus \Delta_S^+ = w_\circ(S)(\Delta^+ \setminus \Delta_S^+)$. Also, for $w \in W$, we set

$$w\widetilde{\eta} := \left(\left(ww_{p,0} \xleftarrow{\beta_{p,1}} \cdots \xleftarrow{\beta_{p,t_p}} ww_{p,t_p} \right)_{p=0,\dots,s-1} ; \sigma_0, \dots, \sigma_s \right),$$

and $w(\mathrm{pQLS}^\mu(\lambda)) := \{w\widetilde{\eta} \mid \widetilde{\eta} \in \mathrm{pQLS}^\mu(\lambda)\}$; remark that for $\widetilde{\eta} \in \mathrm{pQLS}^\mu(\lambda)$, $w\widetilde{\eta}$ satisfies conditions (i), (ii), and (iii) above.

PROPOSITION 2.9. *We define a map* $\mathrm{pr} : w(\mathrm{pQLS}^\mu(\lambda)) \to \overline{\mathrm{pQLS}}(\lambda)$ *by*

$$\mathrm{pr}\left(\left(\left(w_{p,0} \xleftarrow{\beta_{p,1}} \cdots \xleftarrow{\beta_{p,t_p}} w_{p,t_p} \right)_{p=0,\dots,s-1} ; \sigma_0, \dots, \sigma_s \right) \right)$$
$$:= (\lfloor w_{0,t_0} \rfloor, \dots, \lfloor w_{s-1,t_{s-1}} \rfloor ; \sigma_0, \dots, \sigma_{s-1}, 1).$$

Then the following hold.

(1) $\mathrm{pr}\,(\mathrm{pQLS}^\mu(\lambda)) = \overline{\mathrm{pQLS}}^\mu(\lambda)$.
(2) *If* $\mu = \lambda$, *then* $t_0 = 0$ *and* $\mathrm{pr}\left(\bigsqcup_{w \in W^S} w(\mathrm{pQLS}^\lambda(\lambda)) \right) = \overline{\mathrm{pQLS}}(\lambda)$.

PROOF. (1) We write $(\beta_{0,1}, \dots, \beta_{0,t_0})$ as $(\beta_{j_1}, \dots, \beta_{j_r})$, with $j_1 < \cdots < j_r$. Since $v(\mu) = s_{i_{K+1}} \cdots s_{i_M}$ is a reduced expression, and $\beta_j = s_{i_N} \cdots s_{i_{j+1}} \alpha_{i_j}$ for $1 \le j \le N$, we deduce that $w_{0,t_0} = v(\mu) w_\circ(S) s_{\beta_{0,1}} \cdots s_{\beta_{0,t_0}} = v(\mu) w_\circ(S) s_{\beta_{j_1}} \cdots s_{\beta_{j_r}}$ is less than $v(\mu) w_\circ(S)$ in the Bruhat order. Hence we have $\lfloor w_{0,t_0} \rfloor \le v(\mu)$. Therefore, it follows that $\mathrm{pr}\,(\mathrm{pQLS}^\mu(\lambda)) \subset \overline{\mathrm{pQLS}}^\mu(\lambda)$.

Now, let us consider an arbitrary element $\eta = (w_1, w_2, \dots, w_s; \sigma_0, \dots, \sigma_s)$ in $\overline{\mathrm{pQLS}}^\mu(\lambda)$, and set $\eta_q := (w_1, \dots, w_q; \sigma_0, \dots, \sigma_{q-1}, 1)$ for $1 \le q \le s$. We will prove that $\mathrm{pr}\,(\mathrm{pQLS}^\mu(\lambda)) \supset \overline{\mathrm{pQLS}}^\mu(\lambda)$ by induction on s. It follows from the definition of $\overline{\mathrm{pQLS}}^\mu(\lambda)$ that $w_1 \le v(\mu)$. Therefore, w_1 is obtained as a subexpression of the reduced expression $v(\mu) = s_{i_{K+1}} \cdots s_{i_M}$ for $v(\mu)$, that is, there exists a set $J' := \{k_1 < \cdots < k_q\} \subset \{K+1, \dots, M\}$ such that $w_1 = s_{i_{k_1}} \cdots s_{i_{k_q}}$. We set $J = \{j_1 < \cdots < j_r\} := \{K+1, \dots, M\} \setminus J'$.

CLAIM. The directed path

$$v(\mu) w_\circ(S) \xleftarrow{\beta_{j_1}} \cdots \xleftarrow{\beta_{j_r}} w_1 w_\circ(S)$$

is a label-increasing path in DBG whose labels lie in $w_\circ(S)(\Delta^+ \cap v(\mu)^{-1}\Delta^-)$.

PROOF OF THE CLAIM. By (2.1), we have $\beta_{j_k} \in w_\circ(S)(\Delta^+ \cap v(\mu)^{-1}\Delta^-)$ for $1 \le k \le r$, since $K+1 \le j_k \le M$. Also, by the definition of the total order $\prec$ on Δ^+, we have $\beta_{j_k} \succ \beta_{j_{k+1}}$ for $1 \le k \le r-1$, since $j_k < j_{k+1}$. This proves the claim. ■

From the claim above, we see that the pair

$$\widetilde{\eta}_1 := \left((v(\mu) w_\circ(S) \xleftarrow{\beta_{j_1}} \cdots \xleftarrow{\beta_{j_r}} w_1 w_\circ(S)); 0, 1 \right)$$

satisfies conditions (i)–(iv) in Definition 2.8, and hence that $\widetilde{\eta}_1 \in \mathrm{pQLS}^\mu(\lambda)$ and $\mathrm{pr}(\widetilde{\eta}_1) = \eta_1$.

Let $q < s$. Assume that there exists a pair

$$\widetilde{\eta}_q := \left(\left(w_{p,0} \xleftarrow{\beta_{p,1}} \cdots \xleftarrow{\beta_{p,t_p}} w_{p,t_p} \right)_{p=0,\dots,q-1} ; \sigma_0, \dots, \sigma_{q-1}, 1 \right) \in \mathrm{pQLS}^\mu(\lambda) \tag{2.8}$$

such that $\mathrm{pr}(\widetilde{\eta}_q) = \eta_q$, i.e., such that $\lfloor w_{p,t_p} \rfloor = w_{p+1}$ for $0 \le p \le q-1$. Then, by the definition of $\overline{\mathrm{pQLS}}^\mu(\lambda) \subset \overline{\mathrm{pQLS}}(\lambda)$, there exists a directed path from w_{q+1} to w_q in $\mathrm{DBG}_{\sigma_q \lambda}$. By parts (1) and (2) of Lemma 2.7, there exists a label-increasing directed path

$$w_{q-1,t_{q-1}} =: w_{q,0} \xleftarrow{\beta_{q,1}} \cdots \xleftarrow{\beta_{q,t_q}} w_{q,t_q}$$

from some $w_{q,t_q} \in w_{q+1} W_S$ to $w_{q-1,t_{q-1}}$ in $\mathrm{DBG}_{\sigma_q \lambda}$ whose edge labels lie in $\Delta^+ \setminus \Delta_S^+$. Hence the pair

$$\widetilde{\eta}_{q+1} := \left(\left(w_{p,0} \xleftarrow{\beta_{p,1}} \cdots \xleftarrow{\beta_{p,t_p}} w_{p,t_p} \right)_{p=0,\dots,q} ; \sigma_0, \dots, \sigma_q, 1 \right) \tag{2.9}$$

satisfies the condition that $\lfloor w_{p,t_p} \rfloor = w_{p+1}$ for $0 \le p \le q$, and also conditions (i)-(iv). This proves that $\mathrm{pr}\,(\mathrm{pQLS}^\mu(\lambda)) \supset \overline{\mathrm{pQLS}}^\mu(\lambda)$ by induction on s.

(2) If $\mu = \lambda$, i.e., if $v(\mu) = e$, then $\Delta^+ \cap v(\mu)^{-1}\Delta^- = \emptyset$, and hence $t_0 = 0$ by condition (iii). Also, let $\eta = (w_1, \dots, w_s; \sigma_0, \dots, \sigma_s) \in \overline{\mathrm{pQLS}}(\lambda)$, and set

$w_1^{-1}\eta := (e, \dots, \lfloor w_1^{-1}w_s \rfloor; \sigma_0, \dots, \sigma_s) \in \overline{\mathrm{pQLS}}(\lambda)$. By part (1), there exists $\widetilde{\eta} \in \mathrm{pQLS}^{\lambda}(\lambda)$ such that $\mathrm{pr}(\widetilde{\eta}) = w_1^{-1}\eta$. Hence we deduce that $w_1\widetilde{\eta} \in w_1(\mathrm{pQLS}^{\lambda}(\lambda))$ and $\mathrm{pr}(w_1\widetilde{\eta}) = \eta$. This proves part (2). □

2.3. Description of $E_\mu(q,t)$ and $P_\lambda(q,t)$ in terms of $\mathrm{pQLS}^{\mu}(\lambda)$**.** Let $E_\mu(q,t)$, $\mu \in W\lambda$, denote the nonsymmetric Macdonald polynomial, which is normalized as: $E_\mu(q,t) = e^{\mu} + \sum_{\nu<\mu} f_\nu e^{\nu}$, with $f_\nu \in \mathbb{Q}(q,t)$; here the partial order $<$ on P is the one in [**M03**, (2.7.5)].

For $\lambda \in P^+$, we set $m^{\lambda} := \sum_{\mu \in W\lambda} e^{\mu}$. Let $P_\lambda(q,t)$ denote the symmetric Macdonald polynomial, which is normalized as:

$$P_\lambda(q,t) = m^{\lambda} + \sum_{\substack{\nu \in P^+ \\ \nu<\lambda}} f_\nu m^{\nu}, \quad \text{with } f_\nu \in \mathbb{Q}(q,t);$$

here the partial order $<$ on P^+ is given by:

$$\nu < \lambda \overset{\mathrm{def}}{\Leftrightarrow} \lambda - \nu \in \sum_{i\in I} \mathbb{Z}_{\geq 0}\alpha_i.$$

The following theorem will be proved in Section 3.

THEOREM 2.10. *For*

$$\widetilde{\eta} = \left(\left(w_{p,0} \xleftarrow{\beta_{p,1}} \cdots \xleftarrow{\beta_{p,t_p}} w_{p,t_p} \right)_{p=0,\dots,s-1} ; \sigma_0, \dots, \sigma_s \right) \in w(\mathrm{pQLS}^{\mu}(\lambda)),$$

we set $\mathcal{R}(\widetilde{\eta}) := \prod_{p=0}^{s-1} \prod_{m=1}^{t_p} \deg_{\sigma_p\lambda}(w_{p,m-1} \xleftarrow{\beta_{p,m}} w_{p,m})$. *Then the following hold.*

(1)

$$E_\mu(q,t) = \sum_{\widetilde{\eta} \in \mathrm{pQLS}^{\mu}(\lambda)} t^{\frac{1}{2}(\ell(v(\mu)w_\circ(S)) - \ell(w_{s-1,t_{s-1}}))} e^{\mathrm{wt}(\mathrm{pr}(\widetilde{\eta}))} \mathcal{R}(\widetilde{\eta}).$$

(2)

$$P_\lambda(q,t) = \sum_{\widetilde{\eta} \in \bigsqcup_{w\in W^S} w(\mathrm{pQLS}^{\lambda}(\lambda))} t^{\frac{1}{2}(\ell(ww_\circ(S)) - \ell(w_{s-1,t_{s-1}}))} e^{\mathrm{wt}(\mathrm{pr}(\widetilde{\eta}))} \mathcal{R}(\widetilde{\eta}).$$

Let γ be a positive real number. We consider the specialization of the symmetric Macdonald polynomial $P_\lambda(q,t)$ at $q \to 1$ under the condition that $t = q^{\gamma}$, and denote the resulting polynomial by $J_\lambda^{(\gamma^{-1})}$; namely, we set $J_\lambda^{(\gamma^{-1})} := \lim_{q\to 1} P_\lambda(q, q^{\gamma})$. The polynomial $J_\lambda^{(\gamma^{-1})}$ is called a Jack polynomial. By noting that $(1-q^{\gamma})/(1-q) \to \gamma$ as $q \to 1$, we immediately obtain the following corollary from Theorem 2.10.

COROLLARY 2.11. *Let λ be a dominant weight. Then the following holds:*

$$J_\lambda^{(\gamma^{-1})} = \sum_{\widetilde{\eta} \in \bigsqcup_{w\in W^S} w(\mathrm{pQLS}^{\lambda}(\lambda))} e^{\mathrm{wt}(\mathrm{pr}(\widetilde{\eta}))} \prod_{p=0}^{s-1} \prod_{m=1}^{t_p} \frac{1}{\gamma^{-1}\sigma_p \langle \lambda, \beta_{p,m}^{\vee} \rangle + \langle \rho, \beta_{p,m}^{\vee} \rangle}.$$

Also, we consider the specialization at $q = 0$ of the symmetric Macdonald polynomial $P_\lambda(q,t)$; the resulting polynomial $P_\lambda(0,t)$ is called a Hall-Littlewood polynomial.

Let $\widetilde{\eta} = \left(\left(w_{p,0} \xleftarrow{\beta_{p,1}} \cdots \xleftarrow{\beta_{p,t_p}} w_{p,t_p} \right)_{p=0,\dots,s-1} ; \sigma_0, \dots, \sigma_s \right) \in \mathrm{pQLS}(\lambda)$ be a pQLS pair of shape λ. The pair $\widetilde{\eta}$ is called a pseudo-LS pair of shape λ if

$w_{p,m-1} \xleftarrow{\beta_{p,m}} w_{p,m}$ is a Bruhat edge for all $0 \le p \le s-1$, $1 \le m \le t_p$. Let $\mathrm{pLS}(\lambda)$ ($\subset \mathrm{pQLS}(\lambda)$) denote the set of all pseudo-LS pairs of shape λ. Also, let $\mathrm{pLS}^{\lambda}(\lambda)$ ($\subset \mathrm{pLS}(\lambda)$) denote the set of pseudo-LS pairs $\widetilde{\eta}$ of shape λ lying in $\mathrm{pQLS}^{\lambda}(\lambda)$, that is, the set of pseudo-LS pairs $\widetilde{\eta}$ of shape λ satisfying conditions (iii) and (iv) (with μ replaced by λ) in Definition 2.8. When $q = 0$, we have

$$\deg_{b\lambda}(u \to v) = \begin{cases} t^{-\frac{1}{2}}(1-t) & \text{if } u \xrightarrow{\beta} v \text{ is a Bruhat edge,} \\ 0 & \text{if } u \xrightarrow{\beta} v \text{ is a quantum edge.} \end{cases}$$

Therefore, we obtain the following corollary from Theorem 2.10.

COROLLARY 2.12. *Let λ be a dominant weight. Then the following holds*:

$$P_\lambda(0,t) = \sum_{\widetilde{\eta} \in \bigsqcup_{w \in W^S} w(\mathrm{pLS}^{\lambda}(\lambda))} e^{\mathrm{wt}(\mathrm{pr}(\widetilde{\eta}))} \prod_{p=0}^{s-1} \left(t^{-\frac{1}{2}}(1-t)\right)^{t_p}.$$

EXAMPLE 2.13. Let $\mathfrak{g}$ be of type A_2, $\lambda = \varpi_1 + \varpi_2$, and $\mu = w_\circ\lambda$. We fix a reduced expression $w_\circ = s_1 s_2 s_1$ for $w_\circ$; the total order $\prec$ on Δ^+ is: $\alpha_2 \succ \alpha_1 + \alpha_2 \succ \alpha_1$. Then, $S = S_\lambda = \emptyset$, and the elements $\widetilde{\eta}$ of $\mathrm{pQLS}^{\mu}(\lambda)$ are given as follows:

$\widetilde{\eta}_1 := ((w_\circ); 0, 1)$
$\widetilde{\eta}_2 := \left((w_\circ \xleftarrow{\alpha_1} s_1 s_2); 0, 1\right)$
$\widetilde{\eta}_3 := \left((w_\circ \xleftarrow{\alpha_2} s_2 s_1); 0, 1\right)$
$\widetilde{\eta}_4 := \left((w_\circ \xleftarrow{\alpha_1+\alpha_2} e); 0, 1\right)$
$\widetilde{\eta}_5 := \left((w_\circ \xleftarrow{\alpha_2} s_2 s_1 \xleftarrow{\alpha_1+\alpha_2} s_1); 0, 1\right)$
$\widetilde{\eta}_6 := \left((w_\circ \xleftarrow{\alpha_1+\alpha_2} e \xleftarrow{\alpha_1} s_1); 0, 1\right)$
$\widetilde{\eta}_7 := \left((w_\circ \xleftarrow{\alpha_2} s_2 s_1 \xleftarrow{\alpha_1} s_2); 0, 1\right)$
$\widetilde{\eta}_8 := \left((w_\circ \xleftarrow{\alpha_2} s_2 s_1 \xleftarrow{\alpha_1+\alpha_2} s_1 \xleftarrow{\alpha_1} e); 0, 1\right)$
$\widetilde{\eta}_9 := \left((w_\circ), (w_\circ \xleftarrow{\alpha_1+\alpha_2} e); 0, \frac{1}{2}, 1\right)$
$\widetilde{\eta}_{10} := \left((w_\circ \xleftarrow{\alpha_1} s_1 s_2), (s_1 s_2 \xleftarrow{\alpha_1+\alpha_2} s_2); 0, \frac{1}{2}, 1\right)$
$\widetilde{\eta}_{11} := \left((w_\circ \xleftarrow{\alpha_2} s_2 s_1), (s_2 s_1 \xleftarrow{\alpha_1+\alpha_2} s_1); 0, \frac{1}{2}, 1\right)$
$\widetilde{\eta}_{12} := \left((w_\circ \xleftarrow{\alpha_1+\alpha_2} e), (e \xleftarrow{\alpha_1+\alpha_2} w_\circ); 0, \frac{1}{2}, 1\right)$
$\widetilde{\eta}_{13} := \left((w_\circ \xleftarrow{\alpha_2} s_2 s_1 \xleftarrow{\alpha_1+\alpha_2} s_1), (s_1 \xleftarrow{\alpha_1+\alpha_2} s_2 s_1); 0, \frac{1}{2}, 1\right)$
$\widetilde{\eta}_{14} := \left((w_\circ \xleftarrow{\alpha_1+\alpha_2} e \xleftarrow{\alpha_1} s_1), (s_1 \xleftarrow{\alpha_1+\alpha_2} s_2 s_1); 0, \frac{1}{2}, 1\right)$
$\widetilde{\eta}_{15} := \left((w_\circ \xleftarrow{\alpha_2} s_2 s_1 \xleftarrow{\alpha_1} s_2), (s_2 \xleftarrow{\alpha_1+\alpha_2} s_1 s_2); 0, \frac{1}{2}, 1\right)$
$\widetilde{\eta}_{16} := \left((w_\circ \xleftarrow{\alpha_2} s_2 s_1 \xleftarrow{\alpha_1+\alpha_2} s_1 \xleftarrow{\alpha_1} e), (e \xleftarrow{\alpha_1+\alpha_2} w_\circ); 0, \frac{1}{2}, 1\right)$

Hence we have

$$
\begin{aligned}
&E_\mu(q,t)\\
&= e^{w_\circ\lambda} + e^{s_1s_2\lambda}\times t^{\frac{1}{2}}\times\frac{t^{-\frac{1}{2}}(1-t)}{1-qt} + e^{s_2s_1\lambda}\times t^{\frac{1}{2}}\times\frac{t^{-\frac{1}{2}}(1-t)}{1-qt}\\
&+e^{\lambda}\times t^{\frac{3}{2}}\times\frac{t^{-\frac{1}{2}}(1-t)}{1-q^2t^2} + e^{s_1\lambda}\times t\times\frac{t^{-\frac{1}{2}}(1-t)}{1-qt}\frac{t^{-\frac{1}{2}}(1-t)}{1-q^2t^2}\\
&+e^{s_1\lambda}\times t\times\frac{t^{-\frac{1}{2}}(1-t)}{1-q^2t^2}\frac{qt^{\frac{1}{2}}(1-t)}{1-qt} + e^{s_2\lambda}\times t\times\frac{t^{-\frac{1}{2}}(1-t)}{1-qt}\frac{t^{-\frac{1}{2}}(1-t)}{1-qt}\\
&+e^{\lambda}\times t^{\frac{3}{2}}\times\frac{t^{-\frac{1}{2}}(1-t)}{1-qt}\frac{t^{-\frac{1}{2}}(1-t)}{1-q^2t^2}\frac{t^{-\frac{1}{2}}(1-t)}{1-qt} + e^{0}\times t^{\frac{3}{2}}\times\frac{t^{-\frac{1}{2}}(1-t)}{1-qt^2}\\
&+e^{0}\times t\times\frac{t^{-\frac{1}{2}}(1-t)}{1-qt}\frac{t^{-\frac{1}{2}}(1-t)}{1-qt^2} + e^{0}\times t\times\frac{t^{-\frac{1}{2}}(1-t)}{1-qt}\frac{t^{-\frac{1}{2}}(1-t)}{1-qt^2}\\
&+e^{0}\times 1\times\frac{t^{-\frac{1}{2}}(1-t)}{1-q^2t^2}\frac{qt^{\frac{3}{2}}(1-t)}{1-qt^2} + e^{0}\times t^{\frac{1}{2}}\times\frac{t^{-\frac{1}{2}}(1-t)}{1-qt}\frac{t^{-\frac{1}{2}}(1-t)}{1-q^2t^2}\frac{qt^{\frac{3}{2}}(1-t)}{1-qt^2}\\
&+e^{0}\times t^{\frac{1}{2}}\times\frac{t^{-\frac{1}{2}}(1-t)}{1-q^2t^2}\frac{qt^{\frac{1}{2}}(1-t)}{1-qt}\frac{qt^{\frac{3}{2}}(1-t)}{1-qt^2}\\
&+e^{0}\times t^{\frac{1}{2}}\times\frac{t^{-\frac{1}{2}}(1-t)}{1-qt}\frac{t^{-\frac{1}{2}}(1-t)}{1-qt}\frac{qt^{\frac{3}{2}}(1-t)}{1-qt^2}\\
&+e^{0}\times 1\times\frac{t^{-\frac{1}{2}}(1-t)}{1-qt}\frac{t^{-\frac{1}{2}}(1-t)}{1-q^2t^2}\frac{t^{-\frac{1}{2}}(1-t)}{1-qt}\frac{qt^{\frac{3}{2}}(1-t)}{1-qt^2},
\end{aligned}
$$

where the i-th term on the right-hand side corresponds to $\widetilde{\eta}_i$ in the list above for $1\le i\le 16$.

EXAMPLE 2.14. Let $\mathfrak{g}$ be of type A_2, $\lambda=\varpi_1+\varpi_2$. Then, $S=S_\lambda=\emptyset$, and the elements $\widetilde{\eta}$ of $\bigsqcup_{w\in W} w\left(\mathrm{pQLS}^\lambda(\lambda)\right)$, together with $\mathrm{wt}(\widetilde{\eta})$ and $\mathcal{R}(\widetilde{\eta})$, are given as follows:

$\widetilde{\eta}$	$\mathrm{wt}(\widetilde{\eta})$	$\mathcal{R}(\widetilde{\eta})$
$\widetilde{\eta}_e := ((e);0,1)$	λ	1
$\widetilde{\eta}_{s_1} := ((s_1);0,1)$	$s_1\lambda$	1
$\widetilde{\eta}_{s_2} := ((s_2);0,1)$	$s_2\lambda$	1
$\widetilde{\eta}_{s_1s_2} := ((s_1s_2);0,1)$	$s_1s_2\lambda$	1
$\widetilde{\eta}_{s_2s_1} := ((s_2s_1);0,1)$	$s_2s_1\lambda$	1
$\widetilde{\eta}_{w_\circ} := ((w_\circ);0,1)$	$w_\circ\lambda$	1
$\widetilde{\eta}_1 := ((e),(e \xleftarrow{\alpha_1+\alpha_2} w_\circ);0,\frac{1}{2},1)$	0	$\frac{qt^{\frac{3}{2}}(1-t)}{1-qt^2}$
$\widetilde{\eta}_2 := ((s_1),(s_1 \xleftarrow{\alpha_1+\alpha_2} s_2s_1);0,\frac{1}{2},1)$	0	$\frac{qt^{\frac{3}{2}}(1-t)}{1-qt^2}$
$\widetilde{\eta}_3 := ((s_2),(s_2 \xleftarrow{\alpha_1+\alpha_2} s_1s_2);0,\frac{1}{2},1)$	0	$\frac{qt^{\frac{3}{2}}(1-t)}{1-qt^2}$
$\widetilde{\eta}_4 := ((s_1s_2),(s_1s_2 \xleftarrow{\alpha_1+\alpha_2} s_2);0,\frac{1}{2},1)$	0	$\frac{t^{-\frac{1}{2}}(1-t)}{1-qt^2}$
$\widetilde{\eta}_5 := ((s_2s_1),(s_2s_1 \xleftarrow{\alpha_1+\alpha_2} s_1);0,\frac{1}{2},1)$	0	$\frac{t^{-\frac{1}{2}}(1-t)}{1-qt^2}$
$\widetilde{\eta}_6 := ((w_\circ),(w_\circ \xleftarrow{\alpha_1+\alpha_2} e);0,\frac{1}{2},1)$	0	$\frac{t^{-\frac{1}{2}}(1-t)}{1-qt^2}$

Here

$$\bigsqcup_{w\in W} w\left(\mathrm{pLS}^{\lambda}(\lambda)\right) = \{\widetilde{\eta}_e, \widetilde{\eta}_{s_1}, \widetilde{\eta}_{s_2}, \widetilde{\eta}_{s_1s_2}, \widetilde{\eta}_{s_2s_1}, \widetilde{\eta}_{w_\circ}, \widetilde{\eta}_4, \widetilde{\eta}_5, \widetilde{\eta}_6\}.$$

Hence we have

$$\begin{aligned}P_\lambda(q,t) &= e^{\lambda} + e^{s_1\lambda} + e^{s_2\lambda} + e^{s_1s_2\lambda} + e^{s_2s_1\lambda} + e^{w_\circ\lambda} \\ &\quad + e^0 \times t^{-\frac{3}{2}} \times \frac{qt^{\frac{3}{2}}(1-t)}{1-qt^2} + e^0 \times t^{-\frac{1}{2}} \times \frac{qt^{\frac{3}{2}}(1-t)}{1-qt^2} + e^0 \times t^{-\frac{1}{2}} \times \frac{qt^{\frac{3}{2}}(1-t)}{1-qt^2} \\ &\quad + e^0 \times t^{\frac{1}{2}} \times \frac{t^{-\frac{1}{2}}(1-t)}{1-qt^2} + e^0 \times t^{\frac{1}{2}} \times \frac{t^{-\frac{1}{2}}(1-t)}{1-qt^2} + e^0 \times t^{\frac{3}{2}} \times \frac{t^{-\frac{1}{2}}(1-t)}{1-qt^2}, \\ P_\lambda(0,t) &= e^{\lambda} + e^{s_1\lambda} + e^{s_2\lambda} + e^{s_1s_2\lambda} + e^{s_2s_1\lambda} + e^{w_\circ\lambda} + (t+2)(1-t)e^0,\end{aligned}$$

and

$$J_\lambda^{(\gamma^{-1})} = e^{\lambda} + e^{s_1\lambda} + e^{s_2\lambda} + e^{s_1s_2\lambda} + e^{s_2s_1\lambda} + e^{w_\circ\lambda} + 6 \times \frac{1}{\gamma^{-1}+2} e^0.$$

3. Ram-Yip formulas in terms of $\mathrm{pQLS}^{\mu}(\lambda)$

3.1. Ram-Yip formulas. In this subsection, we review the formulas [**RY11**, Theorems 3.1 and 3.4] for nonsymmetric and symmetric Macdonald polynomials.

Let $\widetilde{\mathfrak{g}}$ denote the finite-dimensional simple Lie algebra whose simple roots are $\{\alpha_i^\vee\}_{i\in I}$, and whose simple coroots are $\{\alpha_i\}_{i\in I}$; the Cartan subalgebra of $\widetilde{\mathfrak{g}}$ is $\mathfrak{h}^*$, and its dual space is $\mathfrak{h}$. We denote the set of all roots of $\widetilde{\mathfrak{g}}$ by $\widetilde{\Delta} = \{\alpha^\vee \mid \alpha \in \Delta\}$, and the set of all positive (resp., negative) roots of $\widetilde{\mathfrak{g}}$ by $\widetilde{\Delta}^+$ (resp., $\widetilde{\Delta}^-$). Also, for a subset $S \subset I$, we set $\widetilde{Q}_S := \sum_{i\in S} \mathbb{Z}\alpha_i^\vee$, $\widetilde{\Delta}_S := \widetilde{\Delta} \cap \widetilde{Q}_S$, $\widetilde{\Delta}_S^+ = \widetilde{\Delta}_S \cap \widetilde{\Delta}^+$, and $\widetilde{\Delta}_S^- = \widetilde{\Delta}_S \cap \widetilde{\Delta}^-$.

Let $\widetilde{\mathfrak{g}}_{\mathrm{aff}}$ denote the untwisted affinization of $\widetilde{\mathfrak{g}}$. We denote by $\widetilde{\Delta}_{\mathrm{aff}}$ the set of all real roots, and by $\widetilde{\Delta}_{\mathrm{aff}}^+$ (resp., $\widetilde{\Delta}_{\mathrm{aff}}^-$) the set of all positive (resp., negative) real roots. Let $\widetilde{\delta}$ be the (primitive) null root of $\widetilde{\mathfrak{g}}_{\mathrm{aff}}$. Then we have $\widetilde{\Delta}_{\mathrm{aff}} = \{\alpha^\vee + a\widetilde{\delta} \mid \alpha \in \Delta, a \in \mathbb{Z}\}$. We set $\alpha_0^\vee := \widetilde{\delta} - \varphi^\vee$, where $\varphi \in \Delta$ denotes the highest short root, and set $I_{\mathrm{aff}} := I \sqcup \{0\}$. Then, $\{\alpha_i^\vee\}_{i\in I_{\mathrm{aff}}}$ is the set of all simple roots of $\widetilde{\mathfrak{g}}_{\mathrm{aff}}$. Also, for $\beta \in \mathfrak{h} \oplus \mathbb{C}\widetilde{\delta}$, we define $\deg(\beta) \in \mathbb{C}$ and $\overline{\beta} \in \mathfrak{h}$ by $\beta = \deg(\beta)\widetilde{\delta} + \overline{\beta}$.

We denote the Weyl group of $\widetilde{\mathfrak{g}}$ by $\widetilde{W}$; since we have $\widetilde{W} \cong W$, we identify these two groups. For $x \in \mathfrak{h}^*$, let $t(x)$ denote the translation in $\mathfrak{h}^*$: $t(x)y = x + y$ for $y \in \mathfrak{h}^*$. The affine Weyl group and the extended affine Weyl group (of $\widetilde{\mathfrak{g}}_{\mathrm{aff}}$) are defined by: $\widetilde{W}_{\mathrm{aff}} := t(Q) \rtimes W$ and $\widetilde{W}_{\mathrm{ext}} := t(P) \rtimes W$, respectively. Also, we define $s_0 : \mathfrak{h}^* \to \mathfrak{h}^*$ by $x \mapsto x - (\langle x, \varphi^\vee\rangle - 1)\varphi$. Then, $\widetilde{W}_{\mathrm{aff}} = \langle s_i \mid i \in I_{\mathrm{aff}}\rangle$.

The extended affine Weyl group $\widetilde{W}_{\mathrm{ext}}$ acts on $\mathfrak{h} \oplus \mathbb{C}\widetilde{\delta}$ as linear transformations, and on $\mathfrak{h}^*$ as affine transformations: for $v \in W$ and $t(x) \in t(P)$,

$$\begin{aligned} vt(x)(\overline{\beta} + r\widetilde{\delta}) &= v\overline{\beta} + (r - \langle x, \overline{\beta}\rangle)\widetilde{\delta} \quad \text{for } \overline{\beta} \in \mathfrak{h},\ r \in \mathbb{C}, \\ vt(x)y &= vx + vy \quad \text{for } y \in \mathfrak{h}^*. \end{aligned}$$

An element $u \in \widetilde{W}_{\mathrm{ext}}$ can be written as $u = t(\mathrm{wt}(u))\,\mathrm{dir}(u)$, where $\mathrm{wt}(u) \in P$ and $\mathrm{dir}(u) \in W$, according to the decomposition $\widetilde{W}_{\mathrm{ext}} = t(P) \rtimes W$. For $w \in \widetilde{W}_{\mathrm{ext}}$, we denote the length of w by $\ell(w)$, which agrees with the cardinality of the set $\widetilde{\Delta}_{\mathrm{aff}}^+ \cap w^{-1}\widetilde{\Delta}_{\mathrm{aff}}^-$. Also, we set $\Omega := \{w \in \widetilde{W}_{\mathrm{ext}} \mid \ell(w) = 0\}$.

For $\mu \in P$, we denote the shortest element in the coset $t(\mu)W$ by $m_\mu \in \widetilde{W}_{\text{ext}}$. In what follows, we fix $\mu \in P$, and consider a reduced expression $m_\mu = us_{\ell_1} \cdots s_{\ell_L} \in \widetilde{W}_{\text{ext}} = \Omega \ltimes \widetilde{W}_{\text{aff}}$, where $u \in \Omega$ and $\ell_1, \dots, \ell_L \in I_{\text{aff}}$. We know the following from [**M03**, Chap. 2].

LEMMA 3.1.

(1) $\text{dir}(m_\mu) = v(\mu)v(\lambda_-)^{-1}$, *and* $\ell(\text{dir}(m_\mu)) + \ell(v(\mu)) = \ell(v(\lambda_-))$. *Moreover,*

$$m_\mu = t(\mu)v(\mu)v(\lambda_-)^{-1}. \tag{3.1}$$

(2) $v(\mu)v(\lambda_-)^{-1}w_\circ = v(\mu)w_\circ(S)$.

(3) $\left(v(\lambda_-)v(\mu)^{-1}\right) m_\mu = m_{\lambda_-}$, *and* $\ell(v(\lambda_-)v(\mu)^{-1}) + \ell(m_\mu) = \ell(m_{\lambda_-})$.

For each $J = \{j_1 < j_2 < j_3 < \cdots < j_r\} \subset \{1, \dots, L\}$ and $w \in W$, we define an alcove path $p_J^{\text{OS}} = \left(wm_\mu = z_0^{\text{OS}}, z_1^{\text{OS}}, \dots, z_r^{\text{OS}}; \beta_{j_1}^{\text{OS}}, \dots, \beta_{j_r}^{\text{OS}}\right)$ as follows: we set $\beta_k^{\text{OS}} := s_{\ell_L} \cdots s_{\ell_{k+1}} \alpha_{\ell_k}^\vee \in \widetilde{\Delta}_{\text{aff}}^+$ for $1 \le k \le L$, and set

$$\begin{aligned}
z_0^{\text{OS}} &= wm_\mu, \\
z_1^{\text{OS}} &= wm_\mu s_{\beta_{j_1}^{\text{OS}}}, \\
z_2^{\text{OS}} &= wm_\mu s_{\beta_{j_1}^{\text{OS}}} s_{\beta_{j_2}^{\text{OS}}}, \\
&\ \ \vdots \\
z_r^{\text{OS}} &= wm_\mu s_{\beta_{j_1}^{\text{OS}}} \cdots s_{\beta_{j_r}^{\text{OS}}}.
\end{aligned}$$

Following [**RY11**, §2.3], we set $\mathcal{B}(w; m_\mu) := \left\{p_J^{\text{OS}} \mid J \subset \{1, \dots, L\}\right\}$ and $\text{end}(p_J^{\text{OS}}) := z_r^{\text{OS}} \in \widetilde{W}_{\text{ext}}$.

REMARK 3.2 ([**M03**, (2.4.7)]). If $j \in J$, then $-\overline{\beta_j^{\text{OS}}}^\vee \in \Delta^+$.

For $p_J^{\text{OS}} \in \mathcal{B}(e; m_\mu)$, let $J^+ \subset J$ (resp., $J^- \subset J$) denote the set of indices $j_i \in J$ for which $\text{dir}(z_{i-1}^{\text{OS}}) \xrightarrow{-\overline{\beta_{j_i}^{\text{OS}}}^\vee} \text{dir}(z_i^{\text{OS}})$ is a Bruhat (resp., quantum) edge of DBG. Also, for $p_J^{\text{OS}} \in \mathcal{B}(w; m_\mu)$, we set

$$\mathcal{L}(p_J^{\text{OS}}) := t^{-\frac{1}{2}\#J}(1-t)^{\#J} \prod_{j \in J^+} \frac{1}{1 - q^{\deg(\beta_j^{\text{OS}})} t^{\langle \rho, -\overline{\beta_j^{\text{OS}}} \rangle}} \prod_{j \in J^-} \frac{q^{\deg(\beta_j^{\text{OS}})} t^{\langle \rho, -\overline{\beta_j^{\text{OS}}} \rangle}}{1 - q^{\deg(\beta_j^{\text{OS}})} t^{\langle \rho, -\overline{\beta_j^{\text{OS}}} \rangle}}.$$

PROPOSITION 3.3 ([**RY11**, Theorem 3.1]). *Let $\mu \in P$. Then, the following holds:*

$$E_\mu(q,t) = \sum_{p_J^{\text{OS}} \in \mathcal{B}(e; m_\mu)} e^{\text{wt}(\text{end}(p_J^{\text{OS}}))} t^{\frac{1}{2}(\ell(\text{dir}(\text{end}(p_J^{\text{OS}}))) - \ell(\text{dir}(m_\mu)))} \mathcal{L}(p_J^{\text{OS}}).$$

PROPOSITION 3.4 ([**RY11**, Theorem 3.4]). *Let $\lambda \in P^+$. Then, the following holds:*

$$P_\lambda(q,t) = \sum_{w \in W^S} \sum_{p_J^{\text{OS}} \in \mathcal{B}(w; m_\lambda)} e^{\text{wt}(\text{end}(p_J^{\text{OS}}))} t^{\frac{1}{2}(\ell(\text{dir}(\text{end}(p_J^{\text{OS}}))) - \ell(w\, \text{dir}(m_\lambda)))} \mathcal{L}(p_J^{\text{OS}}).$$

3.2. A specific total order on $\widetilde{\Delta}^+_{\mathrm{aff}} \cap m_{\lambda_-}^{-1}\widetilde{\Delta}^-_{\mathrm{aff}}$. Let $\lambda \in P^+$ be a dominant weight, and set $\lambda_- = w_\circ\lambda$, $S = \{i \in I \mid \langle \lambda, \alpha_i^\vee \rangle = 0\}$. We fix reduced expressions

$$v(\lambda_-) = s_{i_1} \cdots s_{i_M},$$
$$w_\circ(S) = s_{i_{M+1}} \cdots s_{i_N},$$

for $v(\lambda_-)$ and $w_\circ(S)$, respectively. We set $\beta_k = s_{i_N} \cdots s_{i_{k+1}} \alpha_{i_k}$ for $1 \le k \le N$, and use the total order $\prec$ on Δ^+ defined at the beginning of Section 2.2.

We define an injective map Φ by:

$$\Phi : \widetilde{\Delta}^+_{\mathrm{aff}} \cap m_{\lambda_-}^{-1}\widetilde{\Delta}^-_{\mathrm{aff}} \to \mathbb{Q}_{\ge 0} \times (\Delta^+ \setminus \Delta^+_S),$$
$$\beta = \overline{\beta} + \deg(\beta)\widetilde{\delta} \mapsto \left(\frac{\langle \lambda_-, \overline{\beta} \rangle - \deg(\beta)}{\langle \lambda_-, \overline{\beta} \rangle}, w_\circ \overline{\beta}^\vee \right);$$

note that $\langle \lambda_-, \overline{\beta} \rangle > 0$, $\langle \lambda_-, \overline{\beta} \rangle - \deg(\beta) \ge 0$, and $w_\circ \overline{\beta}^\vee \in \Delta^+ \setminus \Delta^+_S$, since $\langle \lambda_-, \overline{\beta} \rangle = \langle \lambda, w_\circ \overline{\beta} \rangle > 0$ and we know from [**M03**, (2.4.7) (i)] that

$$\widetilde{\Delta}^+_{\mathrm{aff}} \cap m_{\lambda_-}^{-1}\widetilde{\Delta}^-_{\mathrm{aff}} = \{\alpha^\vee + a\widetilde{\delta} \mid \alpha \in \Delta^-, 0 < a \le \langle \lambda_-, \alpha^\vee \rangle\}. \tag{3.2}$$

We now consider the lexicographic order $<$ on $\mathbb{Q}_{\ge 0} \times (\Delta^+ \setminus \Delta^+_S)$ induced by the usual total order on $\mathbb{Q}_{\ge 0}$ and the reverse of the restriction to $\Delta^+ \setminus \Delta^+_S$ of the total order $\prec$ on Δ^+; that is, for $(a, \alpha), (b, \beta) \in \mathbb{Q}_{\ge 0} \times (\Delta^+ \setminus \Delta^+_S)$,

$$(a, \alpha) < (b, \beta) \text{ if and only if } a < b, \text{ or } a = b \text{ and } \alpha \succ \beta.$$

Then we denote by $\prec'$ the total order on $\widetilde{\Delta}^+_{\mathrm{aff}} \cap m_{\lambda_-}^{-1}\widetilde{\Delta}^-_{\mathrm{aff}}$ induced by the lexicographic order on $\mathbb{Q}_{\ge 0} \times (\Delta^+ \setminus \Delta^+_S)$ through the injective map Φ, and write $\widetilde{\Delta}^+_{\mathrm{aff}} \cap m_{\lambda_-}^{-1}\widetilde{\Delta}^-_{\mathrm{aff}}$ as $\{\gamma_1 \prec' \cdots \prec' \gamma_L\}$.

PROPOSITION 3.5 ([**NNS18**, Proposition 3.1.8]). *Keep the notation and setting above. Then, there exists a reduced expression $m_{\lambda_-} = u s_{\ell_1} \cdots s_{\ell_L}$ for m_{λ_-}, with $u \in \Omega$, $\{\ell_1, \dots, \ell_L\} \subset I_{\mathrm{aff}}$, such that $\beta_j^{\mathrm{OS}} \left(= s_{\ell_L} \cdots s_{\ell_{j+1}} \alpha_{\ell_j}^\vee \right) = \gamma_j$ for all $1 \le j \le L$.*

We set $a_k := \deg(\beta_k^{\mathrm{OS}}) \in \mathbb{Z}_{>0}$; since $\widetilde{\Delta}^+_{\mathrm{aff}} \cap m_{\lambda_-}^{-1}\widetilde{\Delta}^-_{\mathrm{aff}} = \{\beta_1^{\mathrm{OS}}, \dots, \beta_L^{\mathrm{OS}}\}$, we see by (3.2) that $0 < a_k \le \langle \lambda_-, \overline{\beta_k^{\mathrm{OS}}} \rangle$.

LEMMA 3.6 ([**NNS18**, Lemma 3.1.10]). *Keep the notation and setting above. Let us rewrite the reduced expression $u s_{\ell_1} \cdots s_{\ell_L}$ for m_{λ_-} as $s_{i'_1} \cdots s_{i'_M} u s_{\ell_{M+1}} \cdots s_{\ell_L}$ by setting $u s_{\ell_k} u^{-1} = s_{i'_k}$ with $i'_k \in I_{\mathrm{aff}}$, for $1 \le k \le M$. Then, $s_{i'_1} \cdots s_{i'_M}$ is a reduced expression for $v(\lambda_-)$, and $u s_{\ell_{M+1}} \cdots s_{\ell_L}$ is a reduced expression for m_λ. Moreover, $i_k = i'_k$ for $1 \le k \le M$.*

REMARK 3.7 ([**LNS$^+$17a**, Theorem 8.3], [**NNS18**, proof of Lemma 3.1.10]).

(1) For $1 \le k \le L$, we set

$$d_k := \frac{\langle \lambda_-, \overline{\beta_k^{\mathrm{OS}}} \rangle - a_k}{\langle \lambda_-, \overline{\beta_k^{\mathrm{OS}}} \rangle}.$$

Here d_k is just the first component of $\Phi(\beta_k^{\mathrm{OS}}) \in \mathbb{Q}_{\ge 0} \times (\Delta^+ \setminus \Delta^+_S)$. For $1 \le k, j \le L$, $\Phi(\beta_k^{\mathrm{OS}}) < \Phi(\beta_j^{\mathrm{OS}})$ if and only if $k < j$, and hence we have

$$0 \le d_1 \le \cdots \le d_L \lneqq 1. \tag{3.3}$$

Also, we have

$$d_i = 0 \Leftrightarrow i \le M. \tag{3.4}$$

(2) For $1 \le k \le M$,

$$w_\circ \overline{\beta_k^{\mathrm{OS}}}^{\vee} = \beta_k. \tag{3.5}$$

LEMMA 3.8 ([**NNS18**, Lemma 3.1.12]). *If $1 \le k < j \le L$ and $d_k = d_j$, then $w_\circ \overline{\beta_k^{\mathrm{OS}}}^{\vee} \succ w_\circ \overline{\beta_j^{\mathrm{OS}}}^{\vee}$.*

3.3. Proof of Theorem 2.10. Let $\lambda \in P^+$ be a dominant weight and $\mu \in W\lambda$; recall that $S = \{i \in I \mid \langle \lambda, \alpha_i^\vee \rangle = 0\}$ and $\lambda_- = w_\circ \lambda$.

We fix reduced expressions for $v(\lambda_-)v(\mu)^{-1}$, $v(\mu)$, and $w_\circ(S)$, and the total order $\prec$ on Δ^+ as at the beginning of Section 2.2. In addition, we use the total order $\prec'$ on $\widetilde{\Delta}_{\mathrm{aff}}^+ \cap m_{\lambda_-}^{-1} \widetilde{\Delta}_{\mathrm{aff}}^-$ defined just before Proposition 3.5, and take the reduced expression $m_{\lambda_-} = u s_{\ell_1} \cdots s_{\ell_L}$ for m_{λ_-} given by Proposition 3.5; recall that $u s_{\ell_k} = s_{i_k} u$ for $1 \le k \le M$. It follows from Lemma 3.1 (3) that $\bigl(v(\mu)v(\lambda_-)^{-1}\bigr) m_{\lambda_-} = m_\mu$ and $-\ell(v(\mu)v(\lambda_-)^{-1}) + \ell(m_{\lambda_-}) = \ell(m_\mu)$. Furthermore, we see that

$$\begin{aligned}\bigl(v(\mu)v(\lambda_-)^{-1}\bigr) m_{\lambda_-} &= (s_{i_K} \cdots s_{i_1})\, u s_{\ell_1} \cdots s_{\ell_L} \\ &= u s_{\ell_K} \cdots s_{\ell_1} s_{\ell_1} \cdots s_{\ell_L} = u s_{\ell_{K+1}} \cdots s_{\ell_L} \quad \text{(by Lemma 3.6)},\end{aligned}$$

and hence that $m_\mu = u s_{\ell_{K+1}} \cdots s_{\ell_L}$ is a reduced expression for m_μ. In particular, when $\mu = \lambda$ (note that $v(\lambda) = e$), $m_\lambda = u s_{\ell_{M+1}} \cdots s_{\ell_L}$ is a reduced expression for m_λ.

Also, recall from Sections 2.2 and 3.1 that $\beta_k = s_{i_N} \cdots s_{i_{k+1}} \alpha_{i_k}$ for $1 \le k \le N$, and $\beta_k^{\mathrm{OS}} = s_{\ell_L} \cdots s_{\ell_{k+1}} \alpha_{\ell_k}^\vee$ for $1 \le k \le L$.

REMARK 3.9. Keep the notation above. We have

$$\begin{aligned}\widetilde{\Delta}_{\mathrm{aff}}^+ \cap m_{\lambda_-}^{-1} \widetilde{\Delta}_{\mathrm{aff}}^- &= \{\beta_1^{\mathrm{OS}}, \dots, \beta_L^{\mathrm{OS}}\}, \\ \widetilde{\Delta}_{\mathrm{aff}}^+ \cap m_{\mu}^{-1} \widetilde{\Delta}_{\mathrm{aff}}^- &= \{\beta_{K+1}^{\mathrm{OS}}, \dots, \beta_L^{\mathrm{OS}}\}, \\ \widetilde{\Delta}_{\mathrm{aff}}^+ \cap m_{\lambda}^{-1} \widetilde{\Delta}_{\mathrm{aff}}^- &= \{\beta_{M+1}^{\mathrm{OS}}, \dots, \beta_L^{\mathrm{OS}}\}, \\ w_\circ(S)\,\bigl(\Delta^+ \cap v(\mu)\Delta^-\bigr) &= \{\beta_{K+1}, \dots, \beta_M\}.\end{aligned}$$

In particular, we have $\widetilde{\Delta}_{\mathrm{aff}}^+ \cap m_{\lambda}^{-1} \widetilde{\Delta}_{\mathrm{aff}}^- \subset \widetilde{\Delta}_{\mathrm{aff}}^+ \cap m_{\mu}^{-1} \widetilde{\Delta}_{\mathrm{aff}}^- \subset \widetilde{\Delta}_{\mathrm{aff}}^+ \cap m_{\lambda_-}^{-1} \widetilde{\Delta}_{\mathrm{aff}}^-$.

LEMMA 3.10 ([**M03**, (2.4.7) (i)]). *If we denote by ϕ the characteristic function of Δ^-, i.e.,*

$$\phi(\gamma) := \begin{cases} 0 & \text{if } \gamma \in \Delta^+, \\ 1 & \text{if } \gamma \in \Delta^-, \end{cases}$$

then we have

$$\widetilde{\Delta}_{\mathrm{aff}}^+ \cap m_{\mu}^{-1} \widetilde{\Delta}_{\mathrm{aff}}^- = \{\alpha^\vee + a\widetilde{\delta} \mid \alpha \in \Delta^-, 0 < a < \phi(v(\mu)v(\lambda_-)^{-1}\alpha) + \langle \lambda_-, \alpha^\vee \rangle\}.$$

REMARK 3.11. Let $\gamma_1, \gamma_2, \dots, \gamma_r \in \widetilde{\Delta}_{\mathrm{aff}}^+ \cap m_{\mu}^{-1} \widetilde{\Delta}_{\mathrm{aff}}^-$, and consider the sequence $(y_0, y_1, \dots, y_r; \gamma_1, \gamma_2, \dots, \gamma_r)$ given by $y_0 = m_\mu$, and $y_i = y_{i-1} s_{\gamma_i}$ for $1 \le i \le r$. Then, the sequence $(y_0, y_1, \dots, y_r; \gamma_1, \gamma_2, \dots, \gamma_r)$ is an element of $\mathcal{B}(w; m_\mu)$ if and only if the following conditions hold:

(1) $\gamma_1 \prec' \gamma_2 \prec' \cdots \prec' \gamma_r$, where the order $\prec'$ is the total order on $\widetilde{\Delta}_{\mathrm{aff}}^+ \cap m_{\mu}^{-1} \widetilde{\Delta}_{\mathrm{aff}}^-$ introduced at the beginning of Section 3.2;

(2) $\mathrm{dir}(y_{i-1}) \xleftarrow{-\overline{\gamma_i}^\vee} \mathrm{dir}(y_i)$ is an edge of DBG for $1 \le i \le r$.

In the following, we will define a map $\Xi : \mathcal{B}(w; m_\mu) \to w(\mathrm{pQLS}^\mu(\lambda))$ for $w \in W$. Let p_J^{OS} be an arbitrary element of $\mathcal{B}(w; m_\mu)$ of the form:

$$p_J^{\mathrm{OS}} = \left(wm_\mu = z_0^{\mathrm{OS}}, z_1^{\mathrm{OS}}, \dots, z_r^{\mathrm{OS}}; \beta_{j_1}^{\mathrm{OS}}, \beta_{j_2}^{\mathrm{OS}}, \dots, \beta_{j_r}^{\mathrm{OS}}\right) \in \mathcal{B}(w; m_\mu),$$

with $J = \{j_1 < \cdots < j_r\} \subset \{K+1, \dots, L\}$. We set $x_k := \mathrm{dir}(z_k^{\mathrm{OS}})$ for $0 \le k \le r$. Then, by the definition of $\mathcal{B}(w; m_\mu)$,

$$wv(\mu)v(\lambda_-)^{-1} = \mathrm{dir}(wm_\mu) = x_0 \xrightarrow{-\overline{\beta_{j_1}^{\mathrm{OS}}}^\vee} x_1 \xrightarrow{-\overline{\beta_{j_2}^{\mathrm{OS}}}^\vee} \cdots \xrightarrow{-\overline{\beta_{j_r}^{\mathrm{OS}}}^\vee} x_r \tag{3.6}$$

is a directed path in DBG; here the first equality follows from Lemma 3.1. We take $0 = u_0 \le u_1 < \cdots < u_{s-1} < u_s = r$ and $0 \le \sigma_1 < \cdots < \sigma_{s-1} < 1 = \sigma_s$ in such a way that (see (3.3))

$$\underbrace{0 = d_{j_1} = \cdots = d_{j_{u_1}}}_{=\sigma_0} < \underbrace{d_{j_{u_1+1}} = \cdots = d_{j_{u_2}}}_{=\sigma_1} < \cdots < \underbrace{d_{j_{u_{s-1}+1}} = \cdots = d_{j_r}}_{=\sigma_{s-1}} < 1 = \sigma_s; \tag{3.7}$$

note that $d_{j_1} > 0$ if and only if $u_1 = 0$. We set $w'_{p,m} := x_{u_p+m}$ for $0 \le p \le s-1$, $0 \le m \le t_p := u_{p+1} - u_p$. Then, by taking a subsequence of (3.6), we obtain the following directed path in DBG for each $0 \le p \le s-1$:

$$w'_{p,0} \xrightarrow{-\overline{\beta_{j_{u_p+1}}^{\mathrm{OS}}}^\vee} w'_{p,1} \xrightarrow{-\overline{\beta_{j_{u_p+2}}^{\mathrm{OS}}}^\vee} \cdots \xrightarrow{-\overline{\beta_{j_{u_{p+1}}}^{\mathrm{OS}}}^\vee} w'_{p,t_p};$$

note that $w'_{p,t_p} = w'_{p+1,0}$ for $0 \le p \le s-2$. We set $w_{p,m} = w'_{p,m}w_\circ$ for $0 \le p \le s-1, 0 \le m \le t_p$. Multiplying this directed path by $w_\circ$ on the right, we obtain the following directed path in DBG for each $0 \le p \le s-1$ (see Lemma 2.6 (2)):

$$w_{p,0} := w'_{p,0}w_\circ \xleftarrow{w_\circ\overline{\beta_{j_{u_p+1}}^{\mathrm{OS}}}^\vee} \cdots \xleftarrow{w_\circ\overline{\beta_{j_{u_{p+1}}}^{\mathrm{OS}}}^\vee} w'_{p+1}w_\circ := w_{p,t_p}. \tag{3.8}$$

Note that the edge labels of this directed path are increasing with respect to the total order $\prec$ on Δ^+ introduced at the beginning of Section 2.2 (see Lemma 3.8), and that they lie in $\Delta^+ \setminus \Delta_S^+$; this property will be used to prove Proposition 3.12. Because

$$\sigma_p\langle \lambda, w_\circ\overline{\beta_{j_u}^{\mathrm{OS}}}\rangle = d_{j_u}\langle \lambda, w_\circ\overline{\beta_{j_u}^{\mathrm{OS}}}\rangle = \frac{\langle \lambda_-, \overline{\beta_{j_u}^{\mathrm{OS}}}\rangle - a_{j_u}}{\langle \lambda_-, \overline{\beta_{j_u}^{\mathrm{OS}}}\rangle}\langle \lambda_-, \overline{\beta_{j_u}^{\mathrm{OS}}}\rangle = \langle \lambda_-, \overline{\beta_{j_u}^{\mathrm{OS}}}\rangle - a_{j_u} \in \mathbb{Z}$$

for $u_p + 1 \le u \le u_{p+1}$, $0 \le p \le s-1$, we find that (3.8) is a directed path in $\mathrm{DBG}_{\sigma_p\lambda}$ for $0 \le p \le s-1$. Thus we obtain

$$\widetilde{\eta} := \left(\left(w_{p,0} \xleftarrow{\beta_{p,1}} \cdots \xleftarrow{\beta_{p,t_p}} w_{p,t_p}\right)_{p=0,\dots,s-1}; \sigma_0, \dots, \sigma_s\right) \in w(\mathrm{pQLS}^\mu(\lambda)); \tag{3.9}$$

here we set $\beta_{p,m} := w_\circ\overline{\beta_{u_p+m}^{\mathrm{OS}}}^\vee$ for $0 \le p \le s-1$, $1 \le m \le t_p$. Indeed, condition (i) in Definition 2.8 has been checked above; condition (ii) is obvious since $w'_{p,t_p} = w'_{p+1,0}$ for $0 \le p \le s-2$. Also, for each $1 \le m \le t_p$, we see that $\beta_{0,m} = w_\circ\overline{\beta_j^{\mathrm{OS}}}^\vee$ for some $1 \le j \le M$ and $w_\circ\overline{\beta_j^{\mathrm{OS}}}^\vee = \beta_j$ by Remark 3.7 (2) and (3.4); here,

$\beta_j \in w_\circ(S)(\Delta^+ \cap v(\mu)^{-1}\Delta^-)$ by Remark 3.9. Hence it follows that $w_\circ(S)\beta_{0,m} \in \Delta^+ \cap v(\mu)^{-1}\Delta^-$, which implies that $\widetilde{\eta}$ satisfies condition (iii). Also, since $x_0 = wv(\mu)v(\lambda_-)^{-1}$, we have

$$w_{0,0} = x_0 w_\circ = wv(\mu)v(\lambda_-)^{-1}w_\circ = wv(\mu)w_\circ(S)$$

by Lemma 3.1(2). Therefore, the pair

$$w^{-1}\widetilde{\eta} = \left(\left(w^{-1}w_{k,0} \xleftarrow{\beta_{k,1}} \cdots \xleftarrow{\beta_{k,t_k}} w^{-1}w_{k,t_k}\right)_{k=0,\dots,s-1} ; \sigma_0, \dots, \sigma_s\right)$$

satisfies conditions (i)-(iv), and hence $w^{-1}\widetilde{\eta} \in \mathrm{pQLS}^\mu(\lambda)$.

We now define $\Xi(p_J^{\mathrm{OS}}) := \widetilde{\eta}$.

PROPOSITION 3.12. *The map* $\Xi : \mathcal{B}(w; m_\mu) \to w(\mathrm{pQLS}^\mu(\lambda))$ *is bijective.*

PROOF. It is clear that the map $\Xi : \mathcal{B}(w; m_\mu) \to w(\mathrm{pQLS}^\mu(\lambda))$ is injective. We will prove that the map $\Xi : \mathcal{B}(w; m_\mu) \to w(\mathrm{pQLS}^\mu(\lambda))$ is surjective. Let us consider an arbitrary element

$$\widetilde{\eta} = \left(\left(v_{p,0} \xleftarrow{\beta_{p,1}} \cdots \xleftarrow{\beta_{p,t_p}} v_{p,t_p}\right)_{p=0,\dots,s-1} ; \tau_0, \dots, \tau_s\right) \in w(\mathrm{pQLS}^\mu(\lambda)). \tag{3.10}$$

For $0 \le p \le s-1$, $1 \le m \le t_p$, we set $a_{p,m} := (\tau_p - 1)\langle \lambda_-, -w_\circ\beta^\vee_{p,m}\rangle$ and $\widetilde{\beta}_{p,m} := w_\circ\beta^\vee_{p,m} + a_{p,m}\widetilde{\delta}$.

CLAIM 1. $\widetilde{\beta}_{p,m} \in \widetilde{\Delta}^+_{\mathrm{aff}} \cap m_\mu^{-1}\widetilde{\Delta}^-_{\mathrm{aff}}$.

PROOF OF CLAIM 1. If $p \neq 0$, then $a_{p,m} = (\tau_p - 1)\langle \lambda_-, -w_\circ\beta^\vee_{p,m}\rangle < \langle \lambda_-, w_\circ\beta^\vee_{p,m}\rangle$ since $\tau_p \neq 0$, and hence $a_{p,m} < \phi(v(\mu)v(\lambda_-)^{-1}w_\circ\beta_{p,m}) + \langle \lambda_-, w_\circ\beta^\vee_{p,m}\rangle$. If $p = 0$, then $w_\circ(S)\beta_{p,m} \in \Delta^+ \cap v(\mu)^{-1}\Delta^-$ by the definition of $\mathrm{pQLS}^\mu(\lambda)$. Thus, $v(\mu)v(\lambda_-)^{-1}(w_\circ\beta_{p,m}) = v(\mu)w_\circ(S)\beta_{p,m} \in \Delta^-$, and hence $\phi(v(\mu)v(\lambda_-)^{-1}(w_\circ\beta_{p,m})) = 1$. Therefore, in both cases, we deduce that $a_{p,m} = (\tau_p - 1)\langle \lambda_-, -w_\circ\beta^\vee_{p,m}\rangle < \phi(v(\mu)v(\lambda_-)^{-1}(w_\circ\beta_{p,m})) + \langle \lambda_-, w_\circ\beta^\vee_{p,m}\rangle$, and hence that $\widetilde{\beta}_{p,m} \in \widetilde{\Delta}^+_{\mathrm{aff}} \cap m_\mu^{-1}\widetilde{\Delta}^-_{\mathrm{aff}}$ by Lemma 3.10. This proves the claim. ∎

CLAIM 2. We have

$$\widetilde{\beta}_{0,1} \prec' \cdots \prec' \widetilde{\beta}_{0,t_0} \prec' \widetilde{\beta}_{1,1} \prec' \cdots \prec' \widetilde{\beta}_{s-1,t_{s-1}},$$

where $\prec'$ denotes the total order on $\widetilde{\Delta}^+_{\mathrm{aff}} \cap m_{\lambda_-}^{-1}\widetilde{\Delta}^-_{\mathrm{aff}}$ introduced at the beginning of Section 3.2; we choose $J' = \{j'_1, \dots, j'_{r'}\} \subset \{K+1, \dots, L\}$ in such way that

$$\left(\beta^{\mathrm{OS}}_{j'_1}, \cdots, \beta^{\mathrm{OS}}_{j'_{r'}}\right) = \left(\widetilde{\beta}_{0,1}, \cdots, \widetilde{\beta}_{0,t_0}, \widetilde{\beta}_{1,1}, \cdots, \widetilde{\beta}_{s-1,t_{s-1}}\right).$$

PROOF OF CLAIM 2. It suffices to show the following:

(i) for $0 \le p \le s-1$ and $1 \le m < t_p$, we have $\widetilde{\beta}_{p,m} \prec' \widetilde{\beta}_{p,m+1}$;

(ii) for $0 \le p \le s-2$, we have $\widetilde{\beta}_{p,t_p} \prec' \widetilde{\beta}_{p+1,1}$.

(i) Since $\frac{\langle \lambda_-, w_\circ\beta^\vee_{p,m}\rangle - a_{p,m}}{\langle \lambda_-, w_\circ\beta^\vee_{p,m}\rangle} = \tau_p$ and $\frac{\langle \lambda_-, w_\circ\beta^\vee_{p,m+1}\rangle - a_{p,m+1}}{\langle \lambda_-, w_\circ\beta^\vee_{p,m+1}\rangle} = \tau_p$, it follows that

$$\Phi(\widetilde{\beta}_{p,m}) = (\tau_p, \beta_{p,m}),$$
$$\Phi(\widetilde{\beta}_{p,m+1}) = (\tau_p, \beta_{p,m+1}).$$

Therefore, the first component of $\Phi(\widetilde{\beta}_{p,m})$ is equal to that of $\Phi(\widetilde{\beta}_{p,m+1})$. Also, since $\beta_{p,m} \succ \beta_{p,m+1}$, we have $\Phi(\widetilde{\beta}_{p,m}) < \Phi(\widetilde{\beta}_{p,m+1})$. This implies that $\widetilde{\beta}_{p,m} \prec' \widetilde{\beta}_{p,m+1}$.

(ii) The proof of (ii) is similar to that of (i). The first components of $\Phi(\widetilde{\beta}_{p,t_p})$ and $\Phi(\widetilde{\beta}_{p+1,1})$ are τ_p and τ_{p+1}, respectively. Since $\tau_p < \tau_{p+1}$, we have $\Phi(\widetilde{\beta}_{p,t_p}) < \Phi(\widetilde{\beta}_{p+1,1})$. This implies that $\widetilde{\beta}_{p,t_p} \prec' \widetilde{\beta}_{p+1,1}$. This proves the claim. ■

Since $J' = \{j'_1, \dots, j'_{r'}\} \subset \{K+1, \dots, L\}$, we can define

$$p_{J'}^{\mathrm{OS}} = \left(wm_\mu = z_0^{\mathrm{OS}}, z_1^{\mathrm{OS}}, \dots, z_{r'}^{\mathrm{OS}}; \beta_{j'_1}^{\mathrm{OS}}, \beta_{j'_2}^{\mathrm{OS}}, \dots, \beta_{j'_{r'}}^{\mathrm{OS}} \right)$$

by: $z_0^{\mathrm{OS}} = wm_\mu$, $z_k^{\mathrm{OS}} = z_{k-1}^{\mathrm{OS}} s_{\beta_{j'_k}^{\mathrm{OS}}}$ for $1 \le k \le r'$; it follows from Remark 3.11 and Claim 2 that $p_{J'}^{\mathrm{OS}} \in \mathcal{B}(w; m_\mu)$.

CLAIM 3. $\Xi(p_{J'}^{\mathrm{OS}}) = \widetilde{\eta}$.

PROOF OF CLAIM 3. In the following description of $p_{J'}^{\mathrm{OS}}$, we employ the notation u_p, σ_p, and $w_{p,m}$ used in the definition of $\Xi(p_J^{\mathrm{OS}})$.

For $1 \le k \le r'$, we see that

$$d_{j'_k} = 1 + \frac{\deg(\beta_{j'_k}^{\mathrm{OS}})}{\langle \lambda_-, -\overline{\beta_{j'_k}^{\mathrm{OS}}} \rangle} = 1 + \frac{\deg(\widetilde{\beta}_{p,m})}{\langle \lambda_-, -\overline{\widetilde{\beta}_{p,m}} \rangle} = 1 + \frac{a_{p,m}}{\langle \lambda_-, \beta_{p,m}^\vee \rangle} = d_{p,m}.$$

Therefore, the sequence (3.7) determined by $p_{J'}^{\mathrm{OS}}$ is

$$\begin{aligned} 0 = \underbrace{d_{0,1} = \cdots d_{0,t_0}}_{=\tau_0} &< \underbrace{d_{1,1} = \cdots = d_{1,t_1}}_{=\tau_1} < \cdots \\ &< \underbrace{d_{s-1,1} = \cdots = d_{s-1,t_{s-1}}}_{=\tau_{s-1}} < 1 = \tau_s = \sigma_s. \end{aligned} \tag{3.11}$$

Because the sequence (3.11) of rational numbers is identical to the sequence (3.7) for $p_{J'}^{\mathrm{OS}}$, we deduce that $u_{p+1} - u_p = t_p$ for $0 \le p \le s-1$, $\beta_{j'_{u_p+m}}^{\mathrm{OS}} = \widetilde{\beta}_{p,m}$ for $0 \le p \le s-1$, $1 \le m \le u_{p+1} - u_p$, and $\sigma_p = \tau_p$ for $0 \le p \le s$. Hence it follows that $w_\circ \overline{\beta_{j'_{u_p+m}}^{\mathrm{OS}}}^\vee = \beta_{p,m}$ and $w_{p,m} = v_{p,m}$. Therefore, we conclude that $\Xi(p_{J'}^{\mathrm{OS}}) = \widetilde{\eta}$. This proves the claim. ■

Thus, we have proved that the map $\Xi : \mathcal{B}(w; m_\mu) \to w(\mathrm{pQLS}^\mu(\lambda))$ is surjective. This completes the proof of Proposition 3.12. □

In order to prove Theorem 2.10, we need the following lemmas.

LEMMA 3.13. *Let $p_J^{\mathrm{OS}} \in \mathcal{B}(w; m_\mu)$, and set*

$$\Xi(p_J^{\mathrm{OS}}) = \left(\left(w_{k,0} \xleftarrow{\beta_{k,1}} \cdots \xleftarrow{\beta_{k,t_k}} w_{k,t_k} \right)_{k=0,\dots,s-1}; \sigma_0, \dots, \sigma_s \right).$$

Then, we have $\ell(\mathrm{dir}(\mathrm{end}(p_J^{\mathrm{OS}}))) - \ell(\mathrm{dir}(wm_\mu)) = \ell(wv(\mu)w_\circ(S)) - \ell(w_{s-1,t_{s-1}})$.

PROOF. Recall from Lemma 3.1 (1), (2) that $\mathrm{dir}(m_\mu) = v(\mu)v(\lambda_-)^{-1} = v(\mu)w_\circ(S)w_\circ$. Also, by the definition of Ξ, we have $\mathrm{dir}(\mathrm{end}(p_J^{\mathrm{OS}})) = w_{s-1,t_{s-1}} w_\circ$.

Therefore, we see that

$$\begin{aligned}\ell(\mathrm{dir}(\mathrm{end}(p_J^{\mathrm{OS}}))) - \ell(\mathrm{dir}(wm_\mu)) &= \ell(w_{s-1,t_{s-1}}w_\circ) - \ell(wv(\mu)w_\circ(S)w_\circ)\\ &= \ell(wv(\mu)w_\circ(S)) - \ell(w_{s-1,t_{s-1}}).\end{aligned}$$

This proves the lemma. □

LEMMA 3.14. *Let* $p_J^{\mathrm{OS}} \in \mathcal{B}(w; m_\mu)$. *Then, the following hold:*

(1) $\mathrm{wt}(\mathrm{end}(p_J^{\mathrm{OS}})) = \mathrm{wt}(\mathrm{pr}(\Xi(p_J^{\mathrm{OS}})))$;
(2) $\mathcal{R}(\Xi(p_J^{\mathrm{OS}}))) = \mathcal{L}(p_J^{\mathrm{OS}})$.

PROOF. We proceed by induction on $\#J$. If $J = \emptyset$, then it is obvious that $\mathcal{L}(p_J^{\mathrm{OS}}) = \mathcal{R}(\Xi(p_J^{\mathrm{OS}})) = 1$ and $\mathrm{wt}(\mathrm{end}(p_J^{\mathrm{OS}})) = \mathrm{wt}(\mathrm{pr}(\Xi(p_J^{\mathrm{OS}}))) = w\mu$, since $\Xi(p_J^{\mathrm{OS}}) = (wv(\mu)w_\circ(S); 0, 1)$.

Let $J = \{j_1 < j_2 < \cdots < j_r\}$ with $r \geq 1$, and set $K := J \setminus \{j_r\}$. Assume that $\Xi(p_K^{\mathrm{OS}})$ is of the form:

$$\Xi(p_K^{\mathrm{OS}}) = \widetilde{\eta} = \left(\left(w_{p,0} \xleftarrow{\beta_{p,1}} \cdots \xleftarrow{\beta_{p,t_p}} w_{p,t_p}\right)_{p=0,\dots,s-1}; \sigma_0, \dots, \sigma_s\right);$$

note that $\mathrm{dir}(\mathrm{end}(p_K^{\mathrm{OS}})) = w_{s-1,t_{s-1}}w_\circ$ and $w_{0,0} = v(\mu)w_\circ(S)$ by the definition of Ξ.

If $d_{j_r} = d_{j_{r-1}} = \sigma_{s-1}$, then $\{d_{j_1} \leq \cdots \leq d_{j_{r-1}} \leq d_{j_r}\} = \{d_{j_1} \leq \cdots \leq d_{j_{r-1}}\}$, and hence

$$\Xi(p_J^{\mathrm{OS}}) = \left(\left(w'_{p,0} \xleftarrow{\beta_{p,1}} \cdots \xleftarrow{\beta_{p,t'_p}} w'_{p,t'_p}\right)_{p=0,\dots,s-1}; \sigma_0, \dots, \sigma_s\right),$$

where

$$\begin{aligned}
t'_p &= t_p && \text{for } 0 \leq p \leq s-2,\\
t'_{s-1} &= t_{s-1} + 1, &&\\
w'_{p,m} &= w_{p,m} && \text{for } 0 \leq p \leq s-1, 1 \leq m \leq t_{p-1},\\
w'_{s-1,t'_{s-1}} &= w_{s-1,t_{s-1}} s_{w_\circ \overline{\beta_{j_r}^{\mathrm{OS}}}}, &&\\
\beta_{s-1,t'_{s-1}} &= w_\circ \overline{\beta_{j_r}^{\mathrm{OS}}}^{\vee}. &&
\end{aligned}$$

If $d_{j_r} > d_{j_{r-1}} = \sigma_{s-1}$, then $\{d_{j_1} \leq \cdots \leq d_{j_{r-1}} \leq d_{j_r}\} = \{d_{j_1} \leq \cdots \leq d_{j_{r-1}} < d_{j_r}\}$, and hence

$$\Xi(p_J^{\mathrm{OS}}) = \left(\left(w'_{p,0} \xleftarrow{\beta_{p,1}} \cdots \xleftarrow{\beta_{p,t'_p}} w'_{p,t'_p}\right)_{p=0,\dots,s}; \sigma_0, \dots, \sigma_{s-1}, d_{j_r}, \sigma_s\right),$$

where

$$\begin{aligned}
t'_p &= t_p && \text{for } 0 \leq p \leq s-1,\\
t'_s &= 1, &&\\
w'_{p,m} &= w_{p,m} && \text{for } 0 \leq p \leq s-1, 1 \leq m \leq t_{p-1},\\
w'_{s,0} &= w_{s-1,t_{s-1}}, &&\\
w'_{s,t'_s} &= w_{s-1,t_{s-1}} s_{w_\circ \overline{\beta_{j_r}^{\mathrm{OS}}}}, &&\\
\beta_{s,t'_s} &= \beta_{s,1} = w_\circ (\overline{\beta_{j_r}^{\mathrm{OS}}})^{\vee}. &&
\end{aligned}$$

For the induction step, it suffices to prove the following claims.

CLAIM 1.

(1) We have

$$\mathrm{wt}(\mathrm{pr}(\Xi(p_J^{\mathrm{OS}}))) = \mathrm{wt}(\mathrm{pr}(\Xi(p_K^{\mathrm{OS}}))) - a_{j_r} w_{s-1,t_{s-1}} w_\circ \overline{\beta_{j_r}^{\mathrm{OS}}}^\vee.$$

(2) We have

$$\mathcal{R}(\Xi(p_J^{\mathrm{OS}})) = \begin{cases} \mathcal{R}(\Xi(p_K^{\mathrm{OS}})) \dfrac{t^{-\frac{1}{2}}(1-t)}{1-q^{\deg(\beta_{j_r}^{\mathrm{OS}})} t^{\langle \rho, -\overline{\beta_{j_r}^{\mathrm{OS}}} \rangle}} \\ \qquad \text{if } \mathrm{dir}(\mathrm{end}(p_K^{\mathrm{OS}})) \xrightarrow{-\overline{\beta_{j_r}^{\mathrm{OS}}}^\vee} \mathrm{dir}(\mathrm{end}(p_J^{\mathrm{OS}})) \text{ is a Bruhat edge,} \\ \mathcal{R}(\Xi(p_K^{\mathrm{OS}})) \dfrac{q^{\deg(\beta_{j_r}^{\mathrm{OS}})} t^{\langle \rho, -\overline{\beta_{j_r}^{\mathrm{OS}}} \rangle - \frac{1}{2}}(1-t)}{1-q^{\deg(\beta_{j_r}^{\mathrm{OS}})} t^{\langle \rho, -\overline{\beta_{j_r}^{\mathrm{OS}}} \rangle}} \\ \qquad \text{if } \mathrm{dir}(\mathrm{end}(p_K^{\mathrm{OS}})) \xrightarrow{-\overline{\beta_{j_r}^{\mathrm{OS}}}^\vee} \mathrm{dir}(\mathrm{end}(p_J^{\mathrm{OS}})) \text{ is a quantum edge;} \end{cases}$$

note that $\mathrm{dir}(\mathrm{end}(p_J^{\mathrm{OS}})) = \mathrm{dir}(\mathrm{end}(p_K^{\mathrm{OS}})) s_{\overline{\beta_{j_r}^{\mathrm{OS}}}}$.

CLAIM 2.

(1) We have

$$\mathrm{wt}(\mathrm{end}(p_J^{\mathrm{OS}})) = \mathrm{wt}(\mathrm{end}(p_K^{\mathrm{OS}})) - a_{j_r} w_{s-1,t_{s-1}} w_\circ \overline{\beta_{j_r}^{\mathrm{OS}}}^\vee.$$

(2) We have

$$\mathcal{L}(\Xi(p_J^{\mathrm{OS}})) = \begin{cases} \mathcal{L}(\Xi(p_K^{\mathrm{OS}})) \dfrac{t^{-\frac{1}{2}}(1-t)}{1-q^{\deg(\beta_{j_r}^{\mathrm{OS}})} t^{\langle \rho, -\overline{\beta_{j_r}^{\mathrm{OS}}} \rangle}} \\ \qquad \text{if } \mathrm{dir}(\mathrm{end}(p_K^{\mathrm{OS}})) \xrightarrow{-\overline{\beta_{j_r}^{\mathrm{OS}}}^\vee} \mathrm{dir}(\mathrm{end}(p_J^{\mathrm{OS}})) \text{ is a Bruhat edge,} \\ \mathcal{L}(\Xi(p_K^{\mathrm{OS}})) \dfrac{q^{\deg(\beta_{j_r}^{\mathrm{OS}})} t^{\langle \rho, -\overline{\beta_{j_r}^{\mathrm{OS}}} \rangle - \frac{1}{2}}(1-t)}{1-q^{\deg(\beta_{j_r}^{\mathrm{OS}})} t^{\langle \rho, -\overline{\beta_{j_r}^{\mathrm{OS}}} \rangle}} \\ \qquad \text{if } \mathrm{dir}(\mathrm{end}(p_K^{\mathrm{OS}})) \xrightarrow{-\overline{\beta_{j_r}^{\mathrm{OS}}}^\vee} \mathrm{dir}(\mathrm{end}(p_J^{\mathrm{OS}})) \text{ is a quantum edge.} \end{cases}$$

PROOF OF CLAIM 1. (1) If $d_{j_r} = d_{j_{r-1}} = \sigma_{s-1}$, then we compute:

$$\begin{aligned}
&\mathrm{wt}(\mathrm{pr}(\Xi(p_J^{\mathrm{OS}}))) \\
&= \sum_{p=1}^{s} (\sigma_p - \sigma_{p-1}) w'_{p-1,t'_{p-1}} \lambda \\
&= \sum_{p=1}^{s-1} (\sigma_p - \sigma_{p-1}) w_{p-1,t_{p-1}} \lambda + (\sigma_s - \sigma_{s-1}) w_{s-1,t_{s-1}} s_{w_\circ \overline{\beta_{j_r}^{\mathrm{OS}}}} \lambda \\
&= \sum_{p=1}^{s} (\sigma_p - \sigma_{p-1}) w_{p-1,t_{p-1}} \lambda + (\sigma_s - \sigma_{s-1}) w_{s-1,t_{s-1}} s_{w_\circ \overline{\beta_{j_r}^{\mathrm{OS}}}} \lambda - (\sigma_s - \sigma_{s-1}) w_{s-1,t_{s-1}} \lambda \\
&= \sum_{p=1}^{s} (\sigma_p - \sigma_{p-1}) w_{p-1,t_{p-1}} \lambda + (1 - d_{j_r}) w_{s-1,t_{s-1}} s_{w_\circ \overline{\beta_{j_r}^{\mathrm{OS}}}} \lambda - (1 - d_{j_r}) w_{s-1,t_{s-1}} \lambda \\
&\qquad\qquad\qquad\qquad\qquad\qquad (\text{since } \sigma_s = 1,\ d_{j_r} = \sigma_{s-1})
\end{aligned}$$

If $d_{j_r} > d_{j_{r-1}} = \sigma_{s-1}$, then we compute:

$$\begin{aligned}
&\mathrm{wt}(\mathrm{pr}(\Xi(p_J^{\mathrm{OS}}))) \\
&= \sum_{p=1}^{s-1}(\sigma_p - \sigma_{p-1})w_{p-1,t_{p-1}}\lambda + (d_{j_r} - \sigma_{s-1})w_{s-1,t_{s-1}}\lambda + (\sigma_s - d_{j_r})w_{s-1,t_{s-1}}s_{w_\circ\overline{\beta_{j_r}^{\mathrm{OS}}}}\lambda \\
&= \sum_{p=1}^{s}(\sigma_p - \sigma_{p-1})w_{p-1,t_{p-1}}\lambda + (d_{j_r} - \sigma_{s-1})w_{s-1,t_{s-1}}\lambda \\
&\qquad + (\sigma_s - d_{j_r})w_{s-1,t_{s-1}}s_{w_\circ\overline{\beta_{j_r}^{\mathrm{OS}}}}\lambda - (\sigma_s - \sigma_{s-1})w_{s-1,t_{s-1}}\lambda \\
&= \sum_{p=1}^{s}(\sigma_p - \sigma_{p-1})w_{p-1,t_{p-1}}\lambda + (\sigma_s - d_{j_r})w_{s-1,t_{s-1}}s_{w_\circ\overline{\beta_{j_r}^{\mathrm{OS}}}}\lambda - (\sigma_s - d_{j_r})w_{s-1,t_{s-1}}\lambda \\
&\overset{\sigma_s=1}{=} \sum_{p=1}^{s}(\sigma_p - \sigma_{p-1})w_{p-1,t_{p-1}}\lambda + (1 - d_{j_r})w_{s-1,t_{s-1}}s_{w_\circ\overline{\beta_{j_r}^{\mathrm{OS}}}}\lambda - (1 - d_{j_r})w_{s-1,t_{s-1}}\lambda.
\end{aligned}$$

In both cases above, since $\mathrm{wt}(\mathrm{pr}(\Xi(p_K^{\mathrm{OS}}))) = \sum_{p=1}^{s}(\sigma_p - \sigma_{p-1})w_{p-1,t_{p-1}}\lambda$, and since

$$\begin{aligned}
&(1 - d_{j_r})w_{s-1,t_{s-1}}s_{w_\circ\overline{\beta_{j_r}^{\mathrm{OS}}}}\lambda - (1 - d_{j_r})w_{s-1,t_{s-1}}\lambda \\
&= -(1 - d_{j_r})w_{s-1,t_{s-1}}\langle\lambda, w_\circ\overline{\beta_{j_r}^{\mathrm{OS}}}\rangle w_\circ\overline{\beta_{j_r}^{\mathrm{OS}}}^\vee \\
&= -\frac{a_{j_r}}{\langle\lambda_-, \overline{\beta_{j_r}^{\mathrm{OS}}}\rangle}\langle\lambda_-, \overline{\beta_{j_r}^{\mathrm{OS}}}\rangle w_{s-1,t_{s-1}}w_\circ\overline{\beta_{j_r}^{\mathrm{OS}}}^\vee \quad \text{(by Remark 3.7)} \\
&= -a_{j_r}w_{s-1,t_{s-1}}w_\circ\overline{\beta_{j_r}^{\mathrm{OS}}}^\vee,
\end{aligned}$$

it follows that

$$\begin{aligned}
&\mathrm{wt}(\mathrm{pr}(\Xi(p_J^{\mathrm{OS}}))) \\
&= \sum_{p=1}^{s}(\sigma_p - \sigma_{p-1})w_{p-1,t_{p-1}}\lambda + (1 - d_{j_r})w_{s-1,t_{s-1}}s_{w_\circ\overline{\beta_{j_r}^{\mathrm{OS}}}}\lambda - (1 - d_{j_r})w_{s-1,t_{s-1}}\lambda \\
&= \mathrm{wt}(\mathrm{pr}(\Xi(p_K^{\mathrm{OS}}))) - a_{j_r}w_{s-1,t_{s-1}}w_\circ\overline{\beta_{j_r}^{\mathrm{OS}}}^\vee.
\end{aligned}$$

(2) We have

$$\begin{aligned}
\mathcal{R}(\Xi(p_J^{\mathrm{OS}})) &= \left(\prod_{p=0}^{s-1}\prod_{m=1}^{t_p}\mathrm{wt}_{\sigma_p\lambda}(w_{p,m-1} \xleftarrow{\beta_{p,m}} w_{p,m})\right) \\
&\qquad \times \deg_{d_{j_r}\lambda}\left(w_{s-1,t_{s-1}} \xleftarrow{w_\circ\overline{\beta_{j_r}^{\mathrm{OS}}}^\vee} w_{s-1,t_{s-1}}s_{w_\circ\overline{\beta_{j_r}^{\mathrm{OS}}}}\right) \\
&= \mathcal{R}(\Xi(p_K^{\mathrm{OS}}))\deg_{d_{j_r}\lambda}\left(w_{s-1,t_{s-1}} \xleftarrow{w_\circ\overline{\beta_{j_r}^{\mathrm{OS}}}^\vee} w_{s-1,t_{s-1}}s_{w_\circ\overline{\beta_{j_r}^{\mathrm{OS}}}}\right).
\end{aligned}$$

If $\mathrm{dir}(\mathrm{end}(p_K^{\mathrm{OS}})) \xrightarrow{-\overline{\beta_{j_r}^{\mathrm{OS}}}^\vee} \mathrm{dir}(\mathrm{end}(p_J^{\mathrm{OS}}))$ is a Bruhat (resp., quantum) edge, then $w_{s-1,t_{s-1}} \xleftarrow{w_\circ\overline{\beta_{j_r}^{\mathrm{OS}}}^\vee} w_{s-1,t_{s-1}}s_{w_\circ\overline{\beta_{j_r}^{\mathrm{OS}}}}$ is also a Bruhat (resp., quantum) edge by

Lemma 2.6 (2), since $\mathrm{dir}(\mathrm{end}(p_K^{\mathrm{OS}})) = w_{s-1,t_{s-1}} w_\circ$ and

$$\mathrm{dir}(\mathrm{end}(p_J^{\mathrm{OS}})) = \mathrm{dir}(\mathrm{end}(p_J^{\mathrm{OS}}) s_{\beta_{j_r}^{\mathrm{OS}}}) = w_{s-1,t_{s-1}} w_\circ s_{\overline{\beta_{j_r}^{\mathrm{OS}}}}.$$

Hence we see that

$$\mathcal{R}(\Xi(p_J^{\mathrm{OS}})) = \begin{cases} \mathcal{R}(\Xi(p_K^{\mathrm{OS}})) \frac{t^{-\frac{1}{2}}(1-t)}{1-q^{(1-d_{j_r})\langle \lambda, w_\circ \overline{\beta_{j_r}^{\mathrm{OS}}}\rangle} t^{\langle \rho, w_\circ \overline{\beta_{j_r}^{\mathrm{OS}}}\rangle}} \\ \qquad \text{if } \mathrm{dir}(\mathrm{end}(p_K^{\mathrm{OS}})) \xrightarrow{-\overline{\beta_{j_r}^{\mathrm{OS}}}^\vee} \mathrm{dir}(\mathrm{end}(p_J^{\mathrm{OS}})) \text{ is a Bruhat edge,} \\ \mathcal{R}(\Xi(p_K^{\mathrm{OS}})) \frac{q^{(1-d_{j_r})\langle \lambda, w_\circ \overline{\beta_{j_r}^{\mathrm{OS}}}\rangle} t^{\langle \rho, w_\circ \overline{\beta_{j_r}^{\mathrm{OS}}}\rangle - \frac{1}{2}}(1-t)}{1-q^{(1-d_{j_r})\langle \lambda, w_\circ \overline{\beta_{j_r}^{\mathrm{OS}}}\rangle} t^{\langle \rho, w_\circ \overline{\beta_{j_r}^{\mathrm{OS}}}\rangle}} \\ \qquad \text{if } \mathrm{dir}(\mathrm{end}(p_K^{\mathrm{OS}})) \xrightarrow{-\overline{\beta_{j_r}^{\mathrm{OS}}}^\vee} \mathrm{dir}(\mathrm{end}(p_J^{\mathrm{OS}})) \text{ is a quantum edge.} \end{cases}$$

Here, $\langle \rho, -\overline{\beta_{j_r}^{\mathrm{OS}}}\rangle = \langle \rho, w_\circ \overline{\beta_{j_r}^{\mathrm{OS}}}\rangle$ since $w_\circ \rho = -\rho$. Also, since $d_{j_r} = \frac{\langle \lambda_-, \overline{\beta_{j_r}^{\mathrm{OS}}}\rangle - a_{j_r}}{\langle \lambda_-, \overline{\beta_{j_r}^{\mathrm{OS}}}\rangle}$, we have $(1-d_{j_r})\langle \lambda, w_\circ \overline{\beta_{j_r}^{\mathrm{OS}}}\rangle = \left(1 - \frac{\langle \lambda, w_\circ \overline{\beta_{j_r}^{\mathrm{OS}}}\rangle - a_{j_r}}{\langle \lambda, w_\circ \overline{\beta_{j_r}^{\mathrm{OS}}}\rangle}\right) \langle \lambda, w_\circ \overline{\beta_{j_r}^{\mathrm{OS}}}\rangle = a_{j_r} = \deg(\beta_{j_r}^{\mathrm{OS}})$. Therefore, we deduce that

$$\mathcal{R}(\Xi(p_J^{\mathrm{OS}})) = \begin{cases} \mathcal{R}(\Xi(p_K^{\mathrm{OS}})) \frac{t^{-\frac{1}{2}}(1-t)}{1-q^{\deg(\beta_{j_r}^{\mathrm{OS}})} t^{\langle \rho, -\overline{\beta_{j_r}^{\mathrm{OS}}}\rangle}} \\ \qquad \text{if } \mathrm{dir}(\mathrm{end}(p_K^{\mathrm{OS}})) \xrightarrow{-\overline{\beta_{j_r}^{\mathrm{OS}}}^\vee} \mathrm{dir}(\mathrm{end}(p_J^{\mathrm{OS}})) \text{ is a Bruhat edge,} \\ \mathcal{R}(\Xi(p_K^{\mathrm{OS}})) \frac{q^{\deg(\beta_{j_r}^{\mathrm{OS}})} t^{\langle \rho, -\overline{\beta_{j_r}^{\mathrm{OS}}}\rangle - \frac{1}{2}}(1-t)}{1-q^{\deg(\beta_{j_r}^{\mathrm{OS}})} t^{\langle \rho, -\overline{\beta_{j_r}^{\mathrm{OS}}}\rangle}} \\ \qquad \text{if } \mathrm{dir}(\mathrm{end}(p_K^{\mathrm{OS}})) \xrightarrow{-\overline{\beta_{j_r}^{\mathrm{OS}}}^\vee} \mathrm{dir}(\mathrm{end}(p_J^{\mathrm{OS}})) \text{ is a quantum edge.} \end{cases}$$

This proves Claim 1 (2). ■

Proof of Claim 2. Let us prove part (1). Note that $\mathrm{end}(p_J^{\mathrm{OS}})) = \mathrm{end}(p_K^{\mathrm{OS}}) s_{\beta_{j_r}^{\mathrm{OS}}}$, and that

$$\mathrm{end}(p_K^{\mathrm{OS}}) = t(\mathrm{wt}(\mathrm{end}(p_K^{\mathrm{OS}})))\, \mathrm{dir}(\mathrm{end}(p_K^{\mathrm{OS}})) = t(\mathrm{wt}(\mathrm{end}(p_K^{\mathrm{OS}}))) w_{s-1,t_{s-1}} w_\circ;$$

the second equality follows from the comment at the beginning of the proof of Lemma 3.14. Also, we have $s_{\beta_{j_r}^{\mathrm{OS}}} = s_{a_{j_r}\tilde{\delta} + \overline{\beta_{j_r}^{\mathrm{OS}}}} = t\left(-a_{j_r} \overline{\beta_{j_r}^{\mathrm{OS}}}^\vee\right) s_{\overline{\beta_{j_r}^{\mathrm{OS}}}}$. Combining these, we see that

$$\begin{aligned} \mathrm{end}(p_J^{\mathrm{OS}}) &= \left(t(\mathrm{wt}(\mathrm{end}(p_K^{\mathrm{OS}}))) w_{s-1,t_{s-1}} w_\circ\right) \left(t\left(-a_{j_r} \overline{\beta_{j_r}^{\mathrm{OS}}}^\vee\right) s_{\overline{\beta_{j_r}^{\mathrm{OS}}}}\right) \\ &= t\left(\mathrm{wt}(\mathrm{end}(p_K^{\mathrm{OS}})) - a_{j_r} w_{s-1,t_{s-1}} w_\circ \overline{\beta_{j_r}^{\mathrm{OS}}}^\vee\right) w_{s-1,t_{s-1}} w_\circ s_{\overline{\beta_{j_r}^{\mathrm{OS}}}}, \end{aligned}$$

and hence that

$$\mathrm{wt}(\mathrm{end}(p_J^{\mathrm{OS}})) = \mathrm{wt}(\mathrm{end}(p_K^{\mathrm{OS}})) - a_{j_r} w_{s-1,t_{s-1}} w_\circ \overline{\beta_{j_r}^{\mathrm{OS}}}^\vee.$$

(2) If $\mathrm{dir}(\mathrm{end}(p_K^{\mathrm{OS}})) \xrightarrow{-\overline{\beta_{j_r}^{\mathrm{OS}}}^{\vee}} \mathrm{dir}(\mathrm{end}(p_J^{\mathrm{OS}}))$ is a Bruhat edge, then $J^- = K^-$ and $J^+ = K^+ \sqcup \{j_r\}$. Therefore, we see that

$$\begin{aligned}\mathcal{L}(p_J^{\mathrm{OS}}) &= t^{-\frac{1}{2}\#J}(1-t)^{\#J} \prod_{j\in J^+} \frac{1}{1-q^{\deg(\beta_j^{\mathrm{OS}})}t^{\langle\rho,-\overline{\beta_j^{\mathrm{OS}}}\rangle}} \prod_{j\in J^-} \frac{q^{\deg(\beta_j^{\mathrm{OS}})}t^{\langle\rho,-\overline{\beta_j^{\mathrm{OS}}}\rangle}}{1-q^{\deg(\beta_j^{\mathrm{OS}})}t^{\langle\rho,-\overline{\beta_j^{\mathrm{OS}}}\rangle}} \\ &= t^{-\frac{1}{2}\#K}(1-t)^{\#K} \prod_{j\in K^+} \frac{1}{1-q^{\deg(\beta_j^{\mathrm{OS}})}t^{\langle\rho,-\overline{\beta_j^{\mathrm{OS}}}\rangle}} \\ &\qquad \times \left(\prod_{j\in K^-} \frac{q^{\deg(\beta_j^{\mathrm{OS}})}t^{\langle\rho,-\overline{\beta_j^{\mathrm{OS}}}\rangle}}{1-q^{\deg(\beta_j^{\mathrm{OS}})}t^{\langle\rho,-\overline{\beta_j^{\mathrm{OS}}}\rangle}}\right) \frac{t^{-\frac{1}{2}}(1-t)}{1-q^{\deg(\beta_{j_r}^{\mathrm{OS}})}t^{\langle\rho,-\overline{\beta_{j_r}^{\mathrm{OS}}}\rangle}} \\ &= \mathcal{L}(\Xi(p_K^{\mathrm{OS}}))\frac{t^{-\frac{1}{2}}(1-t)}{1-q^{\deg(\beta_{j_r}^{\mathrm{OS}})}t^{\langle\rho,-\overline{\beta_{j_r}^{\mathrm{OS}}}\rangle}}.\end{aligned}$$

If $\mathrm{dir}(\mathrm{end}(p_K^{\mathrm{OS}})) \xrightarrow{-\overline{\beta_{j_r}^{\mathrm{OS}}}^{\vee}} \mathrm{dir}(\mathrm{end}(p_J^{\mathrm{OS}}))$ is a quantum edge, then $J^- = K^- \sqcup \{j_r\}$ and $J^+ = K^+$. Therefore, we see that

$$\begin{aligned}\mathcal{L}(p_J^{\mathrm{OS}}) &= t^{-\frac{1}{2}\#J}(1-t)^{\#J} \prod_{j\in J^+} \frac{1}{1-q^{\deg(\beta_j^{\mathrm{OS}})}t^{\langle\rho,-\overline{\beta_j^{\mathrm{OS}}}\rangle}} \prod_{j\in J^-} \frac{q^{\deg(\beta_j^{\mathrm{OS}})}t^{\langle\rho,-\overline{\beta_j^{\mathrm{OS}}}\rangle}}{1-q^{\deg(\beta_j^{\mathrm{OS}})}t^{\langle\rho,-\overline{\beta_j^{\mathrm{OS}}}\rangle}} \\ &= t^{-\frac{1}{2}\#K}(1-t)^{\#K} \prod_{j\in K^+} \frac{1}{1-q^{\deg(\beta_j^{\mathrm{OS}})}t^{\langle\rho,-\overline{\beta_j^{\mathrm{OS}}}\rangle}} \\ &\qquad \times \left(\prod_{j\in K^-} \frac{q^{\deg(\beta_j^{\mathrm{OS}})}t^{\langle\rho,-\overline{\beta_j^{\mathrm{OS}}}\rangle}}{1-q^{\deg(\beta_j^{\mathrm{OS}})}t^{\langle\rho,-\overline{\beta_j^{\mathrm{OS}}}\rangle}}\right) \frac{q^{\deg(\beta_{j_r}^{\mathrm{OS}})}t^{\langle\rho,-\overline{\beta_{j_r}^{\mathrm{OS}}}\rangle-\frac{1}{2}}(1-t)}{1-q^{\deg(\beta_{j_r}^{\mathrm{OS}})}t^{\langle\rho,-\overline{\beta_{j_r}^{\mathrm{OS}}}\rangle}} \\ &= \mathcal{L}(\Xi(p_K^{\mathrm{OS}}))\frac{q^{\deg(\beta_{j_r}^{\mathrm{OS}})}t^{\langle\rho,-\overline{\beta_{j_r}^{\mathrm{OS}}}\rangle-\frac{1}{2}}(1-t)}{1-q^{\deg(\beta_{j_r}^{\mathrm{OS}})}t^{\langle\rho,-\overline{\beta_{j_r}^{\mathrm{OS}}}\rangle}}.\end{aligned}$$

This proves Claim 2 (2). ■

This completes the proof of Lemma 3.14 by induction on $\#J$. □

PROOF OF THEOREM 2.10. (1) We know from Proposition 3.3 that

$$\begin{aligned}E_\mu(q,t) = \sum_{p_J^{\mathrm{OS}}\in\mathcal{B}(e;m_\mu)} & e^{\mathrm{wt}(\mathrm{end}(p_J^{\mathrm{OS}}))} t^{\frac{1}{2}(\ell(\mathrm{dir}(\mathrm{end}(p_J^{\mathrm{OS}})))-\ell(\mathrm{dir}(m_\mu))-\#J)}(1-t)^{\#J} \\ & \times \prod_{j\in J^+} \frac{1}{1-q^{\deg(\beta_j^{\mathrm{OS}})}t^{\langle\rho,-\overline{\beta_j^{\mathrm{OS}}}\rangle}} \prod_{j\in J^-} \frac{q^{\deg(\beta_j^{\mathrm{OS}})}t^{\langle\rho,-\overline{\beta_j^{\mathrm{OS}}}\rangle}}{1-q^{\deg(\beta_j^{\mathrm{OS}})}t^{\langle\rho,-\overline{\beta_j^{\mathrm{OS}}}\rangle}}.\end{aligned}$$

Therefore, it follows from Proposition 3.12 and Lemma 3.14 that

$$E_\mu(q,t) = \sum_{\overline{\eta}\in\mathrm{pQLS}^\mu(\lambda)} t^{\frac{1}{2}(\ell(v(\mu)w_\circ(S))-\ell(w_{s-1,t_{s-1}}))} e^{\mathrm{wt}(\mathrm{pr}(\overline{\eta}))}\mathcal{R}(\overline{\eta}).$$

Thus, Theorem 2.10 (1) is proved.

(2) We know from Proposition 3.4 that

$$P_\lambda(q,t) = \sum_{w\in W^S} \sum_{p_J^{\mathrm{OS}}\in\mathcal{B}(w;m_\lambda)} e^{\mathrm{wt}(\mathrm{end}(p_J^{\mathrm{OS}}))} t^{\frac{1}{2}(\ell(\mathrm{dir}(\mathrm{end}(p_J^{\mathrm{OS}})))-\ell(w\,\mathrm{dir}(m_\lambda))-\#J)}(1-t)^{\#J}$$

$$\times \prod_{j\in J^+} \frac{1}{1-q^{\deg(\beta_j^{\mathrm{OS}})}t^{\langle\rho,-\overline{\beta_j^{\mathrm{OS}}}\rangle}} \prod_{j\in J^-} \frac{q^{\deg(\beta_j^{\mathrm{OS}})}t^{\langle\rho,-\overline{\beta_j^{\mathrm{OS}}}\rangle}}{1-q^{\deg(\beta_j^{\mathrm{OS}})}t^{\langle\rho,-\overline{\beta_j^{\mathrm{OS}}}\rangle}}.$$

Therefore, it follows from Proposition 3.12 and Lemma 3.14 that

$$P_\lambda(q,t) = \sum_{w\in W^S} \sum_{\widetilde{\eta}\in w(\mathrm{pQLS}^\lambda(\lambda))} t^{\frac{1}{2}(\ell(ww_\circ(S))-\ell(w_{s-1,t_{s-1}}))} e^{\mathrm{wt}(\mathrm{pr}(\widetilde{\eta}))}\mathcal{R}(\widetilde{\eta})$$

$$= \sum_{\widetilde{\eta}\in\bigsqcup_{w\in W^S} w(\mathrm{pQLS}^\lambda(\lambda))} t^{\frac{1}{2}(\ell(ww_\circ(S))-\ell(w_{s-1,t_{s-1}}))} e^{\mathrm{wt}(\mathrm{pr}(\widetilde{\eta}))}\mathcal{R}(\widetilde{\eta}).$$

Thus, Theorem 2.10 (2) is proved. □

4. Pseudo-crystal structure on $\overline{\mathrm{pQLS}}(\lambda)$

4.1. Pseudo-crystals.

DEFINITION 4.1. A pseudo-crystal is a set $\mathbb{B}$, equipped with the maps wt : $\mathbb{B} \to P$, e_α, $f_\alpha : \mathbb{B}\sqcup\{\mathbf{0}\} \to \mathbb{B}\sqcup\{\mathbf{0}\}$ for $\alpha\in\Delta$, and ε_α, $\varphi_\alpha : \mathbb{B}\to\mathbb{Z}\sqcup\{-\infty\}$ for $\alpha\in\Delta$, satisfying the following conditions, where $\mathbf{0}$ is an extra element, which is not contained in $\mathbb{B}$, such that $e_\alpha\mathbf{0} = f_\alpha\mathbf{0} = \mathbf{0}$ for all $\alpha\in\Delta$, and $-\infty$ is an extra element, which is not contained in $\mathbb{Z}$, such that $a+(-\infty) = (-\infty)+a = -\infty$ for all $a\in\mathbb{Z}$:

(i) for $b\in\mathbb{B}$ and $\alpha\in\Delta$, $\varphi_\alpha(b) = \varepsilon_\alpha(b) + \langle\mathrm{wt}(b),\alpha^\vee\rangle$;
(ii) for $b\in\mathbb{B}$ and $\alpha\in\Delta$, if $e_\alpha b\neq\mathbf{0}$, then $\mathrm{wt}(e_\alpha b) = \mathrm{wt}(b)+\alpha$;
(iii) for $b\in\mathbb{B}$ and $\alpha\in\Delta$, if $f_\alpha b\neq\mathbf{0}$, then $\mathrm{wt}(f_\alpha b) = \mathrm{wt}(b)-\alpha$;
(iv) for $b\in\mathbb{B}$ and $\alpha\in\Delta$, if $e_\alpha b\neq\mathbf{0}$, then $\varepsilon_\alpha(e_\alpha b) = \varepsilon_\alpha(b)-1$, $\varphi_\alpha(e_\alpha b) = \varphi_\alpha(b)+1$;
(v) for $b\in\mathbb{B}$ and $\alpha\in\Delta$, if $f_\alpha b\neq\mathbf{0}$, then $\varepsilon_\alpha(f_\alpha b) = \varepsilon_\alpha(b)+1$, $\varphi_\alpha(f_\alpha b) = \varphi_\alpha(b)-1$;
(vi) for b, $b'\in\mathbb{B}$ and $\alpha\in\Delta$, $f_\alpha b = b'$ if and only if $b = e_\alpha b'$;
(vii) for $b\in\mathbb{B}$, if $\varphi_\alpha(b) = -\infty$, then $e_\alpha b = f_\alpha b = \mathbf{0}$.

DEFINITION 4.2. The pseudo-crystal graph of a pseudo-crystal $\mathbb{B}$ is the directed graph with vertex set $\mathbb{B}$ and directed edges labeled by arbitrary roots: for $b, b'\in\mathbb{B}$, and $\alpha\in\Delta$, $b\xrightarrow{\alpha} b'$ if $b' = f_\alpha b$. A pseudo-crystal is said to be connected if its pseudo-crystal graph is connected.

4.2. Root operators on $\overline{\mathrm{pQLS}}(\lambda)$. We identify an element

$$\eta = (w_1,\ldots,w_s;\sigma_0,\ldots,\sigma_s)\in\overline{\mathrm{pQLS}}(\lambda)$$

with the following piecewise linear, continuous map $\eta : [0,1]\to\mathfrak{h}_\mathbb{R}^*$:

$$\eta(t) = \sum_{k=1}^{p-1}(\sigma_k-\sigma_{k-1})w_k\lambda + (t-\sigma_{p-1})w_p\lambda \quad \text{for } \sigma_{p-1}\le t\le\sigma_p, 1\le p\le s.$$

LEMMA 4.3 (cf. [**Li95**, Lemma 4.5(a)]). *For* $\eta\in\overline{\mathrm{pQLS}}(\lambda)$, $\mathrm{wt}(\eta) := \eta(1)\in P$.

PROOF. Let $\eta = (w_1, \dots, w_s; \sigma_0, \dots, \sigma_s) \in \overline{\mathrm{pQLS}}(\lambda)$. We see that

$$\eta(1) = \sum_{u=1}^{s} (\sigma_u - \sigma_{u-1}) w_u \lambda = \underbrace{\sigma_s w_s \lambda}_{= w_s \lambda} + \sum_{u=1}^{s-1} \underbrace{\sigma_u (w_{u+1}\lambda - w_u \lambda)}_{\in P \text{ by Remark 2.3}} \in P.$$

This proves the lemma. □

We will define the root operator e_α for a root $\alpha \in \Delta$ (not necessarily simple nor positive) as in [**Li94**]. Fix $\alpha \in \Delta$. Let $\eta = (w_1, \dots, w_s; \sigma_0, \dots, \sigma_s) \in \overline{\mathrm{pQLS}}(\lambda)$. We set

$$H(t) = H_\alpha^\eta(t) := \langle \eta(t), \alpha^\vee \rangle \quad \text{for } t \in [0,1],$$
$$m = m_\alpha^\eta := \min\left(\{H_\alpha^\eta(t) \mid t \in [0,1]\} \cap \mathbb{Z}\right);$$

note that $m \in \mathbb{Z}_{\geq 0}$ since $H(0) = 0$. If $m = 0$, then we set $e_\alpha \eta := \mathbf{0}$. If $m \leq -1$, then we define $0 \leq t_0 < t_1 \leq 1$ by

$$t_1 := \min\{t \in [0,1] \mid H(t) = m\}, \tag{4.1}$$
$$t_0 := \max\{t \in [0,t_1] \mid H(t) = m+1\}; \tag{4.2}$$

observe that t_0 is a unique point in $[0, t_1]$ such that

$$H(t_0) = m+1 \quad \text{and} \quad m < H(t) < m+1 \text{ for } t_0 < t < t_1.$$

Now we define

$$(e_\alpha \eta)(t) := \begin{cases} \eta(t) & \text{for } 0 \leq t \leq t_0, \\ \eta(t_0) + s_\alpha(\eta(t) - \eta(t_0)) & \text{for } t_0 \leq t \leq t_1, \\ \eta(t) + \alpha & \text{for } t_1 \leq t \leq 1. \end{cases} \tag{4.3}$$

REMARK 4.4. Let $0 \leq p \leq q \leq s-1$ be such that $\sigma_{p-1} \leq t_0 < \sigma_p$ and $\sigma_{q-1} < t_1 \leq \sigma_q$; note that $p \leq q$. Then,

$$\begin{aligned} e_\alpha \eta = (&w_1, \dots, w_p, \lfloor s_\alpha w_p \rfloor, \lfloor s_\alpha w_{p+1} \rfloor, \dots, \lfloor s_\alpha w_q \rfloor, w_q, w_{q+1}, \dots, w_s; \\ &\sigma_0, \dots, \sigma_{p-1}, t_0, \sigma_p, \dots, \sigma_{q-1}, t_1, \sigma_q, \dots, \sigma_s). \end{aligned} \tag{4.4}$$

Here, if $t_0 = \sigma_{p-1}$, then we drop w_p and σ_{p-1}; if $t_1 = \sigma_q$ (and $\lfloor s_\alpha w_q \rfloor \neq w_{q+1}$), then we drop w_q and σ_q; if $t_1 = \sigma_q$ and $\lfloor s_\alpha w_q \rfloor = w_{q+1}$, then we drop w_{q+1} and t_1, σ_q.

PROPOSITION 4.5. *Let $\eta \in \overline{\mathrm{pQLS}}(\lambda)$, and $\alpha \in \Delta$. If $e_\alpha \eta \neq \mathbf{0}$, then $e_\alpha \eta \in \overline{\mathrm{pQLS}}(\lambda)$ and $\mathrm{wt}(e_\alpha \eta) = \mathrm{wt}(\eta) + \alpha$.*

PROOF. Keep the setting of Remark 4.4. We give a proof in the case that $t_1 \neq \sigma_q$ and $t_0 \neq \sigma_{p-1}$; the proofs in the other cases are similar. By (4.4) and Lemma 2.7, it suffices to show that

(i) there exists a path from $s_\alpha w_p$ to w_p in $\mathrm{DBG}_{t_0\lambda}$;
(ii) for each $p \leq u \leq q-1$, there exists a path from $s_\alpha w_{u+1}$ to $s_\alpha w_u$ in $\mathrm{DBG}_{\sigma_u \lambda}$;
(iii) there exists a path from w_q to $s_\alpha w_q$ in $\mathrm{DBG}_{t_1\lambda}$.

Let us show (i). We have

$$\eta(t_0) = \sum_{u=1}^{p-1} (\sigma_u - \sigma_{u-1}) w_u \lambda + (t_0 - \sigma_{p-1}) w_p \lambda = t_0 w_p \lambda + \sum_{u=1}^{p-1} \underbrace{\sigma_u (w_{u+1}\lambda - w_u \lambda)}_{\in P \text{ by Remark 2.3}}. \tag{4.5}$$

Since $\langle \eta(t_0), \alpha^\vee \rangle = H(t_0) = m + 1 \in \mathbb{Z}$, it follows that $\langle t_0 w_p \lambda, \alpha^\vee \rangle \in \mathbb{Z}$, and hence $t_0 \langle \lambda, w_p^{-1}\alpha^\vee \rangle \in \mathbb{Z}$; by the maximality of t_0, we see that $\langle w_p\lambda, \alpha^\vee \rangle = \langle \lambda, w_p^{-1}\alpha^\vee \rangle < 0$, which implies that $w_p^{-1}\alpha \in \Delta^- \setminus \Delta_S^-$. Hence we obtain the directed edge $w_p \xleftarrow{-w_p^{-1}\alpha} s_\alpha w_p$ in $\mathrm{DBG}_{t_0\lambda}$, which shows (i). Similarly, from the fact that $\langle \eta(t_1), \alpha^\vee \rangle = H(t_1) = m \in \mathbb{Z}$, we obtain the directed edge $s_\alpha w_q \xleftarrow{-w_q^{-1}\alpha} w_q$ in $\mathrm{DBG}_{t_1\lambda}$, which shows (iii). Part (ii) follows immediately from the definition of $\mathrm{DBG}_{\sigma_u\lambda}$. Indeed, if $w_u = w_{u,0} \xleftarrow{\beta_{u,1}} w_{u,1} \xleftarrow{\beta_{u,2}} \cdots \xleftarrow{\beta_{u,k_u}} w_{u,k_u} = w_{u+1}$ is a directed path from w_{u+1} to w_u in $\mathrm{DBG}_{\sigma_u\lambda}$, then

$$s_\alpha w_u = s_\alpha w_{u,0} \xleftarrow{\beta_{u,1}} s_\alpha w_{u,1} \xleftarrow{\beta_{u,2}} \cdots \xleftarrow{\beta_{u,k_u}} s_\alpha w_{u,k_u} = s_\alpha w_{u+1}$$

is a directed path from $s_\alpha w_{u+1}$ to $s_\alpha w_u$ in $\mathrm{DBG}_{\sigma_u\lambda}$. The equality $\mathrm{wt}(e_\alpha\eta) = \mathrm{wt}(\eta) + \alpha$ follows from (4.3). This proves the proposition. □

We can easily verify the following lemma by using the definition of root operators.

LEMMA 4.6. *Let $\eta \in \overline{\mathrm{pQLS}}(\lambda)$, and $\alpha \in \Delta$. Assume that $m_\alpha^\eta \le -1$, and define $0 \le t_0 < t_1 \le 1$ as in (4.1), (4.2) for the η and α. Then, $m_\alpha^{e_\alpha\eta} = m_\alpha^\eta + 1$, and $H_\alpha^{e_\alpha\eta}(t_0) = m_\alpha^\eta + 1 = m_\alpha^{e_\alpha\eta}$, which implies that $t_1' := \min\{t \in [0,1] \mid H_\alpha^{e_\alpha\eta}(t) = m_\alpha^{e_\alpha\eta}\}$ is less than or equal to t_0.*

The following is an immediate consequence of Lemma 4.6.

COROLLARY 4.7. *Let $\eta \in \overline{\mathrm{pQLS}}(\lambda)$, and $\alpha \in \Delta$.*

(1) *If we set $\varepsilon_\alpha(\eta) := \max\{n \ge 0 \mid e_\alpha^n \eta \ne \mathbf{0}\}$, then we have $\varepsilon_\alpha(\eta) = -m_\alpha^\eta$* (cf. [**Li94**, Proposition (a) in §1.5] and [**Li95**, Lemma 2.1 (c)]).
(2) *For each $1 \le k \le N := \varepsilon_\alpha(\eta) - 1$, define $0 \le t_0^{(k)} < t_1^{(k)} \le 1$ as in (4.1) and (4.2) for $e_\alpha^k\eta$ and α. Then, we have*

$$0 \le \underbrace{t_0^{(N)} < t_1^{(N)}}_{\text{for } e_\alpha^N\eta} \le \underbrace{t_0^{(N-1)} < t_1^{(N-1)}}_{\text{for } e_\alpha^{N-1}\eta} \le \cdots \le \underbrace{t_0^{(1)} < t_1^{(1)}}_{\text{for } e_\alpha\eta} \le \underbrace{t_0^{(0)} < t_1^{(0)}}_{\text{for } \eta} \le 1,$$

with

$$t_1^{(k)} = \min\{t \in [0,1] \mid H_\alpha^\eta(t) = m_\alpha^\eta + k\},$$
$$t_0^{(k)} = \max\{t \in [0, t_1^{(k)}] \mid H_\alpha^\eta(t) = m_\alpha^\eta + k + 1\};$$

note that $t_0^{(k)}$ is a unique point in $[0, t_1^{(k)}]$ such that

$$H_\alpha^\eta(t_0^{(k)}) = m_\alpha^\eta + k + 1, \text{ and } m_\alpha^\eta + k < H_\alpha^\eta(t) < m_\alpha^\eta + k + 1 \text{ for } t_0^{(k)} < t < t_1^{(k)}.$$

Similarly, we will define the root operator f_α for $\alpha \in \Delta$. Let $\eta \in \overline{\mathrm{pQLS}}(\lambda)$, and $\alpha \in \Delta$. If $H(1) = m$, then we set $f_\alpha\eta := \mathbf{0}$. If $H(1) - m \ge -1$, then we define $0 \le t_0 < t_1 \le 1$ by

$$t_0 := \max\{t \in [0,1] \mid H(t) = m\}, \tag{4.6}$$
$$t_1 := \min\{t \in [t_0, 1] \mid H(t) = m + 1\}. \tag{4.7}$$

Now we define

$$(4.8)\qquad (f_\alpha\eta)(t) := \begin{cases} \eta(t) & \text{for } 0 \le t \le t_0, \\ \eta(t_0) + s_\alpha(\eta(t) - \eta(t_0)) & \text{for } t_0 \le t \le t_1, \\ \eta(t) - \alpha & \text{for } t_1 \le t \le 1. \end{cases}$$

In the same way as for Proposition 4.5, we can show that if $f_\alpha\eta \ne \mathbf{0}$, then $f_\alpha\eta \in \overline{\mathrm{pQLS}}(\lambda)$ and $\mathrm{wt}(f_\alpha\eta) = \mathrm{wt}(\eta) - \alpha$. Also, we deduce that $\varphi_\alpha(\eta) := \max\{n \ge 0 \mid f_\alpha^n\eta \ne \mathbf{0}\}$ is equal to $H_\alpha^\eta(1) - m_\alpha^\eta$. The proof of the following proposition is the same as that of [**Li94**, Proposition (b) in §1.5].

PROPOSITION 4.8. *Let $\eta \in \overline{\mathrm{pQLS}}(\lambda)$, and $\alpha \in \Delta$. If $e_\alpha\eta \ne \mathbf{0}$, then $f_\alpha e_\alpha\eta = \eta$. If $f_\alpha\eta \ne \mathbf{0}$, then $e_\alpha f_\alpha\eta = \eta$.*

THEOREM 4.9. *The set $\overline{\mathrm{pQLS}}(\lambda)$, equipped with the maps $\mathrm{wt} : \overline{\mathrm{pQLS}}(\lambda) \to P$, $e_\alpha, f_\alpha : \overline{\mathrm{pQLS}}(\lambda) \sqcup \{\mathbf{0}\} \to \overline{\mathrm{pQLS}}(\lambda) \sqcup \{\mathbf{0}\}$ for $\alpha \in \Delta$, and $\varepsilon_\alpha, \varphi_\alpha : \overline{\mathrm{pQLS}}(\lambda) \to \mathbb{Z}_{\ge 0}$ for $\alpha \in \Delta$, becomes a pseudo-crystal.*

4.3. Connectedness of the pseudo-crystal $\overline{\mathrm{pQLS}}(\lambda)$. Fix a dominant weight λ. We set $\eta_w := (w; 0, 1)$ for $w \in W$.

THEOREM 4.10. *For all $\eta \in \overline{\mathrm{pQLS}}(\lambda)$, there exist $\zeta_1, \dots, \zeta_k \in \Delta$ such that $e_{\zeta_1} \cdots e_{\zeta_k}\eta = \eta_e$. In particular, the pseudo-crystal $\overline{\mathrm{pQLS}}(\lambda)$ is connected.*

In order to prove this theorem, we need a few lemmas given below. For $\eta \in \overline{\mathrm{pQLS}}(\lambda)$ and $\alpha \in \Delta$, we set $e_\alpha^{\max} := e_\alpha^{\varepsilon_\alpha(\eta)}\eta \in \overline{\mathrm{pQLS}}(\lambda)$; recall that $\varepsilon_\alpha(\eta) = -m_\alpha^\eta$.

LEMMA 4.11. *Let $\eta = (w_1, \dots, w_s; \sigma_0, \dots, \sigma_s) \in \overline{\mathrm{pQLS}}(\lambda)$, and $\alpha \in \Delta$. Assume that $H_\alpha^\eta(\sigma_1) = \sigma_1\langle w_1\lambda, \alpha^\vee\rangle \in \mathbb{Z}_{<0}$. Set $b := -H_\alpha^\eta(\sigma_1) \in \mathbb{Z}_{>0}$, $m := m_\alpha^\eta$, and $p := -m - b \in \mathbb{Z}_{\ge 0}$. For each $0 \le k \le \varepsilon_\alpha(\eta) - 1 = -m - 1 = p + b - 1$, define $0 \le t_0^{(k)} < t_1^{(k)} \le 1$ as in (4.1) and (4.2) for $e_\alpha^k\eta \in \overline{\mathrm{pQLS}}(\lambda)$ and α. Then, we have*

$$0 = \underbrace{t_0^{(p+b-1)} < t_1^{(p+b-1)}}_{\text{for } e_\alpha^{p+b-1}\eta} = \cdots = \underbrace{t_0^{(p+1)} < t_1^{(p+1)}}_{\text{for } e_\alpha^{p+1}\eta} = \underbrace{t_0^{(p)} < t_1^{(p)}}_{\text{for } e_\alpha^{p}\eta} = \sigma_1$$

$$\le \underbrace{t_0^{(p-1)} < t_1^{(p-1)}}_{\text{for } e_\alpha^{p-1}\eta} \le \cdots \le \underbrace{t_0^{(1)} < t_1^{(1)}}_{\text{for } e_\alpha\eta} \le \underbrace{t_0^{(0)} < t_1^{(0)}}_{\text{for } \eta} \le 1,$$

with

$$(4.9)\qquad t_1^{(p+k)} = \frac{(b-k)\sigma_1}{b}, \quad t_0^{(p+k)} = \frac{(b-k-1)\sigma_1}{b} \quad \text{for } 0 \le k \le b-1.$$

Hence, for $t \in [0, \sigma_1]$, we have

$$(4.10)\qquad (e_\alpha^{p+k}\eta)(t) = \begin{cases} t(w_1\lambda) & \text{for } t \in [0, \frac{(b-k)\sigma_1}{b}], \\ t(s_\alpha w_1\lambda) & \text{for } t \in [\frac{(b-k)\sigma_1}{b}, \sigma_1], \end{cases} \quad \text{for } 0 \le k \le b-1;$$

in particular,

$$(e_\alpha^{\max}\eta)(t) = t(s_\alpha w_1\lambda) \quad \text{for } t \in [0, \sigma_1].$$

PROOF. First we claim that

$$(4.11)\qquad (e_\alpha^p\eta)(t) = \eta(t) = t(w_1\lambda) \text{ for } t \in [0, \sigma_1],$$

which is equivalent to the inequalities: $\sigma_1 \le t_0^{(p-1)} < t_1^{(p-1)} \le \cdots \le t_0^{(1)} < t_1^{(1)} \le t_0^{(0)} < t_1^{(0)} \le 1$ (see Corollary 4.7). Suppose, for a contradiction, that $(e_\alpha^p\eta)(t) \ne \eta(t)$

for some $t \in [0, \sigma_1]$. Let $1 \leq q \leq p$ be the minimal element such that $(e_\alpha^q \eta)(t) \neq \eta(t)$ for some $t \in [0, \sigma_1]$. Since $(e_\alpha^{q-1}\eta)(t) = \eta(t)$ for all $t \in [0, \sigma_1]$ by the definition of q, we see from the definition of e_α that $t_0^{(q-1)} < \sigma_1$. Also, since

$$H_\alpha^\eta(t) = t\underbrace{\langle w_1\lambda, \alpha^\vee\rangle}_{<0} \text{ for } t \in [0, \sigma_1],$$

it follows that $H_\alpha^\eta(t_0^{(q-1)}) > H_\alpha^\eta(\sigma_1)$. However, by Corollary 4.7, we have

$$H_\alpha^\eta(t_0^{(q-1)}) = m_\alpha^\eta + (q-1) + 1 = m + q \leq m + p = H_\alpha^\eta(\sigma_1),$$

which is a contradiction. Thus we have shown (4.11). Now it follows from Corollary 4.7(1) that

$$-m_\alpha^{e_\alpha^p \eta} = \varepsilon_\alpha(e_\alpha^p \eta) = \varepsilon_\alpha(\eta) - p = -m - p = b.$$

By (4.11), together with the assumption that $\sigma_1\langle w_1\lambda, \alpha^\vee\rangle \in \mathbb{Z}_{<0}$, we see that the function $H_\alpha^{e_\alpha^p\eta}(t)$ is strictly decreasing on $[0, \sigma_1]$, and that $H_\alpha^{e_\alpha^p\eta}(\sigma_1) = -b = m_\alpha^{e_\alpha^p\eta}$. Hence we have $t_1^{(p)} = \sigma_1$ and $t_0^{(p)} = (b-1)\sigma_1/b$ by the definition of these points. Also, from the definition of the root operator e_α and (4.11), we deduce that for $t \in [0, \sigma_1]$,

$$(e_\alpha^{p+1}\eta)(t) = \begin{cases} t(w_1\lambda) & \text{for } t \in [0, \frac{(b-1)\sigma_1}{b}], \\ t(s_\alpha w_1\lambda) & \text{for } t \in [\frac{(b-1)\sigma_1}{b}, \sigma_1]. \end{cases} \tag{4.12}$$

The same argument as above shows that the function $H_\alpha^{e_\alpha^{p+1}\eta}(t)$ is strictly decreasing on $[0, t_0^{(p)}]$, and that $H_\alpha^{e_\alpha^{p+1}\eta}(t_0^{(p)}) = -b + 1 = m_\alpha^{e_\alpha^{p+1}\eta}$. Hence we have $t_1^{(p+1)} = t_0^{(p)}$ and $t_0^{(p+1)} = (b-2)\sigma_1/b$ by the definition of these points. From the definition of the root operator e_α and (4.12), we deduce that

$$(e_\alpha^{p+2}\eta)(t) = \begin{cases} t(w_1\lambda) & \text{for } t \in [0, \frac{(b-2)\sigma_1}{b}], \\ t(s_\alpha w_1\lambda) & \text{for } t \in [\frac{(b-2)\sigma_1}{b}, \sigma_1]. \end{cases}$$

By repeating this argument, we obtain (4.9) and (4.10). This proves the lemma. □

LEMMA 4.12. *Let $\eta = (w_1, \ldots, w_s; \sigma_1 \ldots, \sigma_s) \in \overline{\mathrm{pQLS}}(\lambda)$, with $s \geq 2$. Let $v_2 \in w_2W_S$ be such that there exists a directed path $w_1 = x_0 \xleftarrow{\gamma_1} x_1 \xleftarrow{\gamma_2} \cdots \xleftarrow{\gamma_p} x_p = v_2$ from v_2 to w_1 in $\mathrm{DBG}_{\sigma_1\lambda}$ whose edge labels lie in $\Delta^+ \setminus \Delta_S^+$; see Lemma 2.7(1). Then, for all $1 \leq q \leq p$, we have $H_{-w_1\gamma_q}^\eta(\sigma_1) \in \mathbb{Z}_{<0}$.*

PROOF. Since λ is a dominant weight, and $\gamma_q \in \Delta^+ \setminus \Delta_S^+$, we see from the definition of $\mathrm{DBG}_{\sigma_1\lambda}$ that $\sigma_1\langle\lambda, \gamma_q^\vee\rangle \in \mathbb{Z}_{>0}$. Hence it follows that

$$H_{-w_1\gamma_q}^\eta(\sigma_1) = \sigma_1\langle w_1\lambda, -w_1\gamma_q^\vee\rangle = \sigma_1\langle\lambda, -\gamma_q^\vee\rangle \in \mathbb{Z}_{<0}.$$

This proves the lemma. □

LEMMA 4.13. *Let $\eta = (w_1, \ldots, w_s; \sigma_1 \ldots, \sigma_s) \in \overline{\mathrm{pQLS}}(\lambda)$, with $s \geq 2$. Then, there exists a monomial X in the root operators e_α, $\alpha \in \Delta$, such that $X\eta$ is of the form: $(v_1, \ldots; \sigma_0, \sigma_1', \ldots)$, with $\sigma_1 < \sigma_1' \leq 1$.*

PROOF. As in Lemma 4.12, let $v_2 \in w_2 W_S$ be such that there exists a directed path $w_1 = x_0 \xleftarrow{\gamma_1} x_1 \xleftarrow{\gamma_2} \cdots \xleftarrow{\gamma_p} x_p = v_2$ from v_2 to w_1 in $\mathrm{DBG}_{\sigma_1\lambda}$ whose edge labels lie in $\Delta^+ \setminus \Delta_S^+$. We will prove the assertion of the lemma by induction on $p \geq 1$. Assume first that $p = 1$. Set $\alpha = -w_1\gamma_1 \in \Delta$; notice that $w_2 = \lfloor w_1 s_{\gamma_1} \rfloor = \lfloor s_\alpha w_1 \rfloor$. Then, by Lemma 4.12, we see that $H^\eta_\alpha(\sigma_1) = \sigma_1\langle w_1\lambda, \alpha^\vee\rangle \in \mathbb{Z}_{<0}$. As in Lemma 4.11, we set $b := -H^\eta_\alpha(\sigma_1) \in \mathbb{Z}_{>0}$, $m := m^\eta_\alpha$, and $p := -m - b \in \mathbb{Z}_{\geq 0}$. Also, for each $0 \leq k \leq \varepsilon_\alpha(\eta) - 1 = -m - 1 = p + b - 1$, we define $0 \leq t_0^{(k)} < t_1^{(k)} \leq 1$ as in (4.1), (4.2) for $e_\alpha^k \eta \in \overline{\mathrm{pQLS}}(\lambda)$ and α. Then, by Lemma 4.11, we see that

$$0 = \underbrace{t_0^{(p+b-1)} < t_1^{(p+b-1)}}_{\text{for } e_\alpha^{p+b-1}\eta} = \cdots = \underbrace{t_0^{(p+1)} < t_1^{(p+1)}}_{\text{for } e_\alpha^{p+1}\eta} = \underbrace{t_0^{(p)} < t_1^{(p)}}_{\text{for } e_\alpha^{p}\eta} = \sigma_1$$
$$\leq \underbrace{t_0^{(p-1)} < t_1^{(p-1)}}_{\text{for } e_\alpha^{p-1}\eta} \leq \cdots \leq \underbrace{t_0^{(1)} < t_1^{(1)}}_{\text{for } e_\alpha\eta} \leq \underbrace{t_0^{(0)} < t_1^{(0)}}_{\text{for } \eta} \leq 1,$$

and that $(e_\alpha^{\max}\eta)(t) = t(s_\alpha w_1\lambda)$ for $t \in [0, \sigma_1]$. Since $\langle w_2\lambda, \alpha^\vee\rangle = \langle s_\alpha w_1\lambda, \alpha^\vee\rangle > 0$, it follows that $H^\eta_\alpha(\sigma_1 + \epsilon) > H^\eta_\alpha(\sigma_1) = m + p$ for sufficiently small $\epsilon > 0$. Also, since $m + p - 1 < H^\eta_\alpha(t) < m + p$ for $t_0^{(p-1)} < t < t_1^{(p-1)}$ (see Corollary 4.7), we obtain $t_0^{(p-1)} \neq \sigma_1$, and hence $\sigma_1 < t_0^{(p-1)}$. Therefore, we deduce that

$$(e_\alpha^{\max}\eta)(\sigma_1 + \epsilon) = w_2\lambda = s_\alpha w_1\lambda$$

for sufficiently small $\epsilon > 0$, which implies that $e_\alpha^{\max}\eta$ is of the form:

$$(\lfloor s_\alpha w_1 \rfloor, \ldots; \sigma_0, \sigma_1', \ldots), \quad \text{with } \sigma_1' > \sigma_1.$$

Assume next that $p > 1$. As above, if we set $\alpha := -w_1\gamma_1$, then we see by Lemma 4.12 that $H^\eta_\alpha(\sigma_1) = \sigma_1\langle w_1\lambda, \alpha^\vee\rangle \in \mathbb{Z}_{<0}$, and then by Lemma 4.11 that $(e_\alpha^{\max}\eta)(t) = t(s_\alpha w_1\lambda)$ for $t \in [0, \sigma_1]$. Hence it follows that $e_\alpha^{\max}\eta$ is of the form: $e_\alpha^{\max}\eta = (\lfloor s_\alpha w_1 \rfloor, \ldots; \sigma_0, \sigma_1', \ldots)$, with $\sigma_1' \geq \sigma_1$. If $\sigma_1' > \sigma_1$, then the assertion is obvious. Assume now that $\sigma_1' = \sigma_1$. Then, $e_\alpha^{\max}\eta$ is of the form:

$$e_\alpha^{\max}\eta = (\lfloor s_\alpha w_1 \rfloor, w_2' \ldots; \sigma_0, \sigma_1, \ldots);$$

notice that w_2' is either w_2 or $\lfloor s_\alpha w_2 \rfloor$ by the definition of the root operator e_α. Assume that $w_2' = w_2$. Since

$$s_\alpha w_1 = w_1 s_{\gamma_1} = x_1 \xleftarrow{\gamma_2} \cdots \xleftarrow{\gamma_p} x_p = v_2$$

is a directed path (of length $p - 1$) from $v_2 \in w_2 W_S = w_2' W_S$ to $s_\alpha w_1$ in $\mathrm{DBG}_{\sigma_1\lambda}$ whose labels lie in $\Delta^+ \setminus \Delta_S^+$, the assertion follows from our induction hypothesis (applied to $e_\alpha^{\max}\eta$). Assume that $w_2' = \lfloor s_\alpha w_2 \rfloor$. If we set $p := H^\eta_\alpha(\sigma_1) - m^\eta_\alpha$, then we deduce from Lemma 4.11 that $e_\alpha^p\eta$ is of the form:

$$e_\alpha^p\eta = (w_1, w_2', \ldots; \sigma_0, \sigma_1, \ldots).$$

Since $s_\alpha v_2 = s_{-w_1\gamma_1} w_1 s_{\gamma_1} s_{\gamma_2} \cdots s_{\gamma_p} = w_1 s_{\gamma_2} \cdots s_{\gamma_p}$, it follows that

$$w_1 \xleftarrow{\gamma_2} w_1 s_{\gamma_2} \xleftarrow{\gamma_3} \cdots \xleftarrow{\gamma_p} w_1 s_{\gamma_2} \cdots s_{\gamma_p} = s_\alpha v_2$$

is a directed path (of length $p - 1$) from $s_\alpha v_2 \in w_2' W_S$ to w_1 in $\mathrm{DBG}_{\sigma_1\lambda}$ whose labels lie in $\Delta^+ \setminus \Delta_S^+$. Hence the assertion follows from our induction hypothesis (applied to $e_\alpha^p\eta$). This completes the proof of the lemma. $\square$

PROOF OF THEOREM 4.10. We set

$$T_\lambda := \{b \in [0,1] \cap \mathbb{Q} \mid b\langle \lambda, \alpha^\vee \rangle \in \mathbb{Z} \text{ for some } \alpha \in \Delta\};$$

notice that T_λ is a finite set. Let $\eta = (w_1, \ldots, w_s; \sigma_0, \ldots, \sigma_s) \in \overline{\mathrm{pQLS}}(\lambda)$; remark that $\sigma_u \in T_\lambda$ for all $0 \le u \le s$. We prove the assertion of the theorem by (descending) induction on $\sigma_1 \in T_\lambda$. Assume that $\sigma_1 = 1$; note that η is of the form $\eta = \eta_w = (w; 0, 1)$. We show the assertion in this case by induction on the length $\ell(w)$ of $w \in W^S$. If $\ell(w) = 0$, i.e., if $w = e$, then there is nothing to show. Assume that $\ell(w) > 0$. Let us take $i \in I$ such that $\langle w\lambda, \alpha_i^\vee \rangle < 0$; note that $s_i w \in W^S$ in this case. Then, by Lemma 4.11, we deduce that $e_{\alpha_i}^{\max}\eta = e_{\alpha_i}^{\max}\eta_w = (s_i w; 0, 1)$. Since $\ell(s_i w) < \ell(w)$, the assertion follows by the induction hypothesis. Assume now that $\sigma_1 < 1$; notice that $s \ge 2$. By Lemma 4.13, there exists a monomial X in the root operators e_α, $\alpha \in \Delta$, such that $X\eta$ is of the form: $(v_1, \ldots; \sigma_0, \sigma_1', \ldots)$, with $\sigma_1 < \sigma_1' \le 1$. The assertion of the theorem follows from our induction hypothesis (applied to $X\eta$). This completes the proof of Theorem 4.10. □

References

[CO15] I. Cherednik and D. Orr, *Nonsymmetric difference Whittaker functions*, Math. Z. **279** (2015), no. 3-4, 879–938, DOI 10.1007/s00209-014-1397-0. MR3318254

[CHM+22] S. Corteel, J. Haglund, O. Mandelshtam, S. Mason, and L. Williams, *Compact formulas for Macdonald polynomials and quasisymmetric Macdonald polynomials*, Selecta Math. (N.S.) **28** (2022), no. 2, Paper No. 32, 33, DOI 10.1007/s00029-021-00721-7. MR4359567

[CMW22] S. Corteel, O. Mandelshtam, and L. Williams, *From multiline queues to Macdonald polynomials via the exclusion process*, Amer. J. Math. **144** (2022), no. 2, 395–436, DOI 10.1353/ajm.2022.0007. MR4401508

[GR21] W. Guo and A. Ram, *Comparing formulas for type GL_n Macdonald polynomials*, Algebr. Comb. **5** (2022), no. 5, 849–883. MR4511154

[GW20] A. Garbali and M. Wheeler, *Modified Macdonald polynomials and integrability*, Comm. Math. Phys. **374** (2020), no. 3, 1809–1876, DOI 10.1007/s00220-020-03680-w. MR4076089

[HHL05] J. Haglund, M. Haiman, and N. Loehr, *A combinatorial formula for Macdonald polynomials*, J. Amer. Math. Soc. **18** (2005), no. 3, 735–761, DOI 10.1090/S0894-0347-05-00485-6. MR2138143

[HHL08] J. Haglund, M. Haiman, and N. Loehr, *A combinatorial formula for nonsymmetric Macdonald polynomials*, Amer. J. Math. **130** (2008), no. 2, 359–383, DOI 10.1353/ajm.2008.0015. MR2405160

[K91] M. Kashiwara, *On crystal bases of the Q-analogue of universal enveloping algebras*, Duke Math. J. **63** (1991), no. 2, 465–516, DOI 10.1215/S0012-7094-91-06321-0. MR1115118

[Le09] C. Lenart, *On combinatorial formulas for Macdonald polynomials*, Adv. Math. **220** (2009), no. 1, 324–340, DOI 10.1016/j.aim.2008.09.007. MR2462843

[Le12] C. Lenart, *From Macdonald polynomials to a charge statistic beyond type A*, J. Combin. Theory Ser. A **119** (2012), no. 3, 683–712, DOI 10.1016/j.jcta.2011.11.013. MR2871757

[Li94] P. Littelmann, *A Littlewood-Richardson rule for symmetrizable Kac-Moody algebras*, Invent. Math. **116** (1994), no. 1-3, 329–346, DOI 10.1007/BF01231564. MR1253196

[Li95] P. Littelmann, *Paths and root operators in representation theory*, Ann. of Math. (2) **142** (1995), no. 3, 499–525, DOI 10.2307/2118553. MR1356780

[LNS+15] C. Lenart, S. Naito, D. Sagaki, A. Schilling, and M. Shimozono, *A uniform model for Kirillov-Reshetikhin crystals I: Lifting the parabolic quantum Bruhat graph*, Int. Math. Res. Not. IMRN **7** (2015), 1848–1901, DOI 10.1093/imrn/rnt263. MR3335235

[LNS$^+$16] C. Lenart, S. Naito, D. Sagaki, A. Schilling, and M. Shimozono, *Quantum Lakshmibai-Seshadri paths and root operators*, Schubert calculus—Osaka 2012, Adv. Stud. Pure Math., vol. 71, Math. Soc. Japan, [Tokyo], 2016, pp. 267–294, DOI 10.2969/aspm/07110267. MR3644827

[LNS$^+$17a] C. Lenart, S. Naito, D. Sagaki, A. Schilling, and M. Shimozono, *A uniform model for Kirillov-Reshetikhin crystals II. Alcove model, path model, and $P = X$*, Int. Math. Res. Not. IMRN **14** (2017), 4259–4319, DOI 10.1093/imrn/rnw129. MR3674171

[LNS$^+$17b] C. Lenart, S. Naito, D. Sagaki, A. Schilling, and M. Shimozono, *A uniform model for Kirillov-Reshetikhin crystals III: nonsymmetric Macdonald polynomials at $t = 0$ and Demazure characters*, Transform. Groups **22** (2017), no. 4, 1041–1079, DOI 10.1007/s00031-017-9421-1. MR3717224

[LP] C. Lenart and A. Postnikov, *A combinatorial model for crystals of Kac-Moody algebras*, Trans. Amer. Math. Soc. **360** (2008), no. 8, 4349–4381, DOI 10.1090/S0002-9947-08-04419-X. MR2395176

[LS91] V. Lakshmibai and C. S. Seshadri, *Standard monomial theory*, Proceedings of the Hyderabad Conference on Algebraic Groups (Hyderabad, 1989), Manoj Prakashan, Madras, 1991, pp. 279–322. MR1131317

[M95] I. G. Macdonald, *Symmetric functions and Hall polynomials*, 2nd ed., Oxford Mathematical Monographs, The Clarendon Press, Oxford University Press, New York, 1995. With contributions by A. Zelevinsky; Oxford Science Publications. MR1354144

[M03] I. G. Macdonald, *Affine Hecke algebras and orthogonal polynomials*, Cambridge Tracts in Mathematics, vol. 157, Cambridge University Press, Cambridge, 2003, DOI 10.1017/CBO9780511542824. MR1976581

[NNS18] S. Naito, F. Nomoto, and D. Sagaki, *Specialization of nonsymmetric Macdonald polynomials at $t = \infty$ and Demazure submodules of level-zero extremal weight modules*, Trans. Amer. Math. Soc. **370** (2018), no. 4, 2739–2783, DOI 10.1090/tran/7114. MR3748584

[NNS20] S. Naito, F. Nomoto, and D. Sagaki, *Tensor product decomposition theorem for quantum Lakshmibai-Seshadri paths and standard monomial theory for semi-infinite Lakshmibai-Seshadri paths*, J. Combin. Theory Ser. A **169** (2020), 105122, 36, DOI 10.1016/j.jcta.2019.105122. MR3983099

[NS05] S. Naito and D. Sagaki, *Crystal of Lakshmibai-Seshadri paths associated to an integral weight of level zero for an affine Lie algebra*, Int. Math. Res. Not. **14** (2005), 815–840, DOI 10.1155/IMRN.2005.815. MR2146858

[NS16] S. Naito and D. Sagaki, *Demazure submodules of level-zero extremal weight modules and specializations of Macdonald polynomials*, Math. Z. **283** (2016), no. 3-4, 937–978, DOI 10.1007/s00209-016-1628-7. MR3519989

[NS21] S. Naito and D. Sagaki, *Level-zero van der Kallen modules and specialization of nonsymmetric Macdonald polynomials at $t = \infty$*, Transform. Groups **26** (2021), no. 3, 1077–1111, DOI 10.1007/s00031-020-09586-0. MR4309559

[N16] F. Nomoto, *Quantum Lakshmibai-Seshadri paths and the specialization of Macdonald polynomials at $t = 0$ in type $A_{2n}^{(2)}$*, arXiv:1606.01067, 2016.

[N17] F. Nomoto, Generalized Weyl modules and Demazure submodules of level-zero extremal weight modules, arXiv:1701.08377, 2017.

[OS18] D. Orr and M. Shimozono, *Specializations of nonsymmetric Macdonald-Koornwinder polynomials*, J. Algebraic Combin. **47** (2018), no. 1, 91–127, DOI 10.1007/s10801-017-0770-6. MR3757150

[P05] A. Postnikov, *Quantum Bruhat graph and Schubert polynomials*, Proc. Amer. Math. Soc. **133** (2005), no. 3, 699–709, DOI 10.1090/S0002-9939-04-07614-2. MR2113918

[RY11] A. Ram and M. Yip, *A combinatorial formula for Macdonald polynomials*, Adv. Math. **226** (2011), no. 1, 309–331, DOI 10.1016/j.aim.2010.06.022. MR2735761

[Sa22] J. Saied, *A combinatorial formula for Sahi, Stokman, and Venkateswaran's generalization of Macdonald polynomials.* Part B, Adv. Math. **404** (2022), Paper No. 108440, 51, DOI 10.1016/j.aim.2022.108440. MR4420440

[Sc06] C. Schwer, *Galleries, Hall-Littlewood polynomials, and structure constants of the spherical Hecke algebra*, Int. Math. Res. Not., posted on 2006, Art. ID 75395, 31, DOI 10.1155/IMRN/2006/75395. MR2264725

[SSV21] S. Sahi, J. V. Stokman, and V. Venkateswaran, *Metaplectic representations of Hecke algebras, Weyl group actions, and associated polynomials*, Selecta Math. (N.S.) **27** (2021), no. 3, Paper No. 47, 42, DOI 10.1007/s00029-021-00654-1. MR4273644

[SSV22] S. Sahi, J. Stokman, and V. Venkateswaran, Quasi-polynomial representations of double affine Hecke algebras, arXiv:2205.12509, 2022.

[Y12] M. Yip, *A Littlewood-Richardson rule for Macdonald polynomials*, Math. Z. **272** (2012), no. 3-4, 1259–1290, DOI 10.1007/s00209-012-0986-z. MR2995168

Department of Mathematics and Statistics, State University of New York at Albany, Albany, New York 12222

Email address: clenart@albany.edu

Department of Mathematics, Institute of Science Tokyo, 2-12-1 Oh-Okayama, Meguro-ku, Tokyo 152-8551, Japan

Email address: naito@math.titech.ac.jp

Institute of Science Tokyo High School, 3-3-6 Shibaura, Minato-ku, Tokyo 108-0023, Japan

Email address: fnomoto@hst.titech.ac.jp

Department of Mathematics, Institute of Pure and Applied Sciences, University of Tsukuba, 1-1-1 Tennodai, Tsukuba, Ibaraki 305-8571, Japan

Email address: sagaki@math.tsukuba.ac.jp

Contemporary Mathematics
Volume **815**, 2025
https://doi.org/10.1090/conm/815/16319

Shuffle approach to wreath Pieri operators

Joshua Jeishing Wen

ABSTRACT. We describe a way to study and compute Pieri rules for wreath Macdonald polynomials using the quantum toroidal algebra. The Macdonald pairing can be naturally generalized to the wreath setting, but the wreath Macdonald polynomials are no longer collinear with their duals. We establish the relationship between these dual polynomials and the quantum toroidal algebra, and we outline a way to compute norm formulas. None of the aforementioned formulas are successfully computed in this paper.

1. Introduction

In symmetric function theory, one finds tremendous utility in having precise control over creation and annihilation. Specifically, by creation we mean some form of *Pieri rules.* The point of this paper is to establish a way to study Pieri rules for *wreath Macdonald polynomials.* To motivate this, let us first consider the simplest case. Let Λ be the ring of symmetric functions in infinitely many variables, e_n be the nth elemetary symmetric function, h_n be the nth complete symmetric function, and s_λ be the Schur function indexed by the partition λ. The Pieri rules take the form (cf. [**12**]):

$$h_n s_\mu = \sum_\lambda s_\lambda \tag{1.1}$$

$$e_n s_\mu = \sum_\lambda s_\lambda \tag{1.2}$$

In both formulas, the summation is over λ containing μ such that $|\lambda\backslash\mu| = n$. For (1.1), we require $\lambda\backslash\mu$ to contain no *vertically* adjacent boxes, whereas for (1.2), we require $\lambda\backslash\mu$ to contain no *horizontally* adjacent boxes.

For annihilation, we consider the *Hall pairing* $\langle -,-\rangle$ on Λ under which $\{s_\lambda\}$ form an orthonormal basis and take the adjoint to the Pieri rules. Although it is not usually presented in this fashion, one can use (1.1) and the fact that the complete symmetric function h_λ is dual to the *monomial symmetric function* m_λ under $\langle -,-\rangle$ to obtain the *tableaux sum formula* for s_λ. Namely, after specializing to n variables, we have

$$s_\lambda(x_1,\ldots,x_n) = \sum_T x^T$$

2020 *Mathematics Subject Classification.* Primary 05E05, 17B37; Secondary 81R10.

This work received support from NSF-RTG grant "Algebraic Geometry and Representation Theory at Northeastern University" (DMS-1645877).

Here, the summation is over column-strict Young tableaux of shape λ. Such a tableau T can be viewed as a nested sequence of subpartitions:

$$\varnothing = \lambda^{(0)} \subset \lambda^{(1)} \subset \cdots \subset \lambda^{(n)} = \lambda$$

We set x^T as the monomial where the exponent of x_i is $|\lambda^{(i)} \backslash \lambda^{(i-1)}|$. This is but one facet of the rich theory of skew Schur functions.

1.1. The Macdonald case. The *Macdonald polynomials* $\{P_\lambda\}$ are a basis of the twice-deformed ring of symmetric functions $\Lambda_{q,t} := \Lambda \otimes \mathbb{C}(q,t)$. They can be obtained by deforming $\langle -, - \rangle$ into the *Macdonald pairing* $\langle -, - \rangle_{q,t}$ and performing the Gram-Schmidt process on $\{s_\lambda\}$ with respect to the new pairing and the dominance order on partitions. P_λ is normalized so that its leading term s_λ has coefficient 1, and thus its norm is not necessarily 1—its dual is denoted Q_λ. Here, the Pieri rules involve e_n and the *transformed* complete symmetric functions g_n:

(1.3)

$$g_n Q_\mu = \sum_\lambda \prod_{i<j\leq \ell(\lambda)} \frac{\left(q^{\mu_i-\lambda_j+1}t^{j-i-1};q\right)_{\lambda_i-\mu_i}}{\left(q^{\mu_i-\lambda_j}t^{j-i};q\right)_{\lambda_i-\mu_i}} \prod_{i\leq j\leq \ell(\mu)} \frac{\left(q^{\mu_i-\mu_j}t^{j-i+1};q\right)_{\lambda_i-\mu_i}}{\left(q^{\mu_i-\mu_j+1}t^{j-i};q\right)_{\lambda_i-\mu_i}} Q_\lambda$$

(1.4)

$$e_n P_\mu = \sum_\lambda \prod_{\substack{i<j \\ \lambda_i=\mu_i \\ \lambda_j=\mu_j+1}} \frac{\left(1-q^{\mu_i-\mu_j}t^{j-i-1}\right)\left(1-q^{\lambda_i-\lambda_j}t^{j-i+1}\right)}{\left(1-q^{\mu_i-\mu_j}t^{j-i}\right)\left(1-q^{\lambda_i-\lambda_j}t^{j-i}\right)} P_\lambda$$

Here, the partitions appearing in (1.3) are as in (1.1) and those appearing in (1.4) are as in (1.2). The factor $(x;q)_k$ appearing in (1.3) denotes the q-Pochhammer symbol:

$$(x;q)_k := \prod_{i=0}^{k-1} \left(1-q^i x\right)$$

The basis $\{g_\lambda\}$ built out of $\{g_n\}$ is dual under $\langle -, - \rangle_{q,t}$ to $\{m_\lambda\}$, and thus (1.3) yields the tableaux sum formula for P_λ specialized to n variables:

$$P_\lambda(x_1, \ldots, x_n) = \sum_T \prod_{k=1}^n \psi_{\lambda^{(k)}/\lambda^{(k-1)}} x^T$$

where $\psi_{\lambda/\mu}$ is the coefficient from (1.3). On the other hand, Macdonald [**12**] gives a clever proof of the e_n Pieri rule (1.4) that simultaneously establishes three other fundamental formulas for Macdonald polynomials:

- *Principal evaluation*: For λ with $\ell(\lambda) \leq n$,

$$P_\lambda(t^{n-1}, t^{n-2}, \cdots, 1) = t^{n(\lambda)} \prod_{\square\in\lambda} \frac{1-q^{a'(\square)}t^{n-\ell'(\square)}}{1-q^{a(\square)}t^{\ell(\square)+1}}$$

  Here, $n(\lambda) = \sum_i (i-1)\lambda_i$, and for $\square = (a,b) \in \lambda$ (cf. 2.2.1 for notation), $a'(\square) = a$, $\ell'(\square) = b$, $a(\square) = \lambda_b - a$, and $\ell(\square) = {}^t\lambda_a - b$, where ${}^t\lambda$ is the transposed partition.
- *Norm formula*: We have:

$$\langle P_\lambda, P_\lambda \rangle_{q,t} = \prod_{\square\in\lambda} \frac{1-q^{a(\square)+1}t^{\ell(\square)}}{1-q^{a(\square)}t^{\ell(\square)+1}} \tag{1.5}$$

- *Evaluation duality*: For λ and μ with $\ell(\lambda), \ell(\mu) \leq n$,

$$\frac{P_\lambda(q^{\mu_1}t^{n-1}, q^{\mu_2}t^{n-2}, \ldots, q^{\mu_n})}{P_\lambda(t^{n-1}, t^{n-2}, \ldots 1)} = \frac{P_\mu(q^{\lambda_1}t^{n-1}, q^{\lambda_2}t^{n-2}, \ldots, q^{\lambda_n})}{P_\mu(t^{n-1}, t^{n-2}, \ldots 1)}$$

Thus, in this case, the Pieri rules play a crucial role in establishing these widely-used aspects of Macdonald polynomials.

1.2. Quantum toroidal algebras. With a view towards generalization, we recall the action of the rank 1 *quantum toroidal algebra* $U_{q,t}(\ddot{\mathfrak{gl}}_1)$ on $\Lambda_{q,t}$. This action is was constructed by Feigin *et. al.* [**3**] via *vertex operators* that were already present in prior work of Jing [**10**] and Garsia-Haiman-Tesler [**6**]. Haiman's proof of Macdonald positivity (cf. [**7**]) showed that it was natural to assign Macdonald polynomials to fixed point classes in torus-equivariant K-theory of the Hilbert scheme of points on $\mathbb{C}^2$, and this yielded a different perspective on the action of $U_{q,t}(\ddot{\mathfrak{gl}}_1)$ on Macdonald polynomials. Feigin-Tsymbaliuk [**5**] and Schiffmann-Vasserot [**22**] reconstructed this action using correspondences, and from this perspective, the generators of $U_{q,t}(\ddot{\mathfrak{gl}}_1)$ act by very fine creation and annihilation operators.

At first glance, these creation and annihilation operators seem to have little to do with symmetric functions—they are just operators on a vector space indexed by partitions. However, Feigin-Tsymbaliuk were able to recover the Pieri rules through a very roundabout manner. Through work of Neguț [**16**], $U_{q,t}(\ddot{\mathfrak{gl}}_1)$ is isomorphic to a *shuffle algebra*, a space of functions endowed with an exotic product structure—by abuse of notation, we will denote the function variables using $\{x_i\}$. Neguț's isomorphism provides a completely different way to think about $U_{q,t}(\ddot{\mathfrak{gl}}_1)$, and elements that are easy to express on one side of the isomorphism can be very complicated to write down in the other side. From [**3**], one can expect that a scalar multiple of the shuffle element

$$K_n := \prod_{1\leq i<j\leq n} \frac{(x_i - qx_j)(x_j - qx_i)}{(x_i - x_j)^2} \tag{1.6}$$

acts as multlpication by e_n, and in [**5**], the authors show that the matrix elements of its action on Macdonald polynomials agree with the Pieri rule (1.4). Following the ideas outlined in Section 4 below, one can also re-derive the norm formula (1.5). Using a toroidal approach to the Cherednik-Macdonald-Mehta identity (cf. [**2**]), we expect a toroidal derivation of principal evaluation and evaluation duality, but that is outside the scope of this paper. Overall, we wish to convey that one can derive the basic elements of Macdonald theory using $U_{q,t}(\ddot{\mathfrak{gl}}_1)$, although this path lacks the elegance of Macdonald's simultaneous proof.

1.3. Wreath Macdonald polynomials. Defined by Haiman, the wreath Macdonald polynomials are a generalization of the Macdonald polynomials from the symmetric groups Σ_n to their wreath products with a fixed cyclic group $\mathbb{Z}/\ell\mathbb{Z}$. Using a wreath analogue of the Frobenius characteristic, we can view them as elements of $\Lambda_{q,t}^{\otimes\ell}$, whose bases are naturally indexed by ℓ-tuples of partitions. A basic component of their definition is the decomposition of an ordinary partition into its ℓ-core and ℓ-quotient, which we review in 2.2 below. Roughly speaking, the ℓ-quotient is an ℓ-tuple of partitions that records the ribbons of length ℓ we can peel of an ordinary partition and the ℓ-core is the partition that remains when all ribbons have been peeled off. The wreath Macdonald polynomials are again indexed by ordinary partitions, and we will recycle notation and denote them by $\{P_\lambda\}$. Letting λ range over partitions with the same ℓ-core, we obtain a basis of $\Lambda_{q,t}^{\otimes\ell}$. We have elementary symmetric functions $e_n(p)$ for each tensorand in $\Lambda_{q,t}^{\otimes\ell}$, and thus the e_n Pieri rules come in ℓ *colors.*

Wreath analogues of the standard elements of Macdonald theory are still largely unexplored, and we suspect part of the reason is that things are just more complicated in this setting. A key component of Macdonald's simultaeneous proof that we have yet to mention are the *Macdonald operators*, a family of difference operators diagonalized by the Macdonald polynomials. Wreath analogues of Macdonald operators were discovered in our work with Daniel Orr and Mark Shimozono [**20**], and we invite the reader to compare the wreath Macdonald operators with their predecessors to get a sense of the difference in complexity. Even if these operators were simple, the inductive argument used by Macdonald fails on the spot. The main problem here is that one has little control over the partitions λ that appear in the expression

$$e_n(p)P_\mu = \sum_\lambda c_{\lambda,\mu}(q,t)P_\lambda,$$

especially with respect to dominance order.

Something that does generalize is the quantum toroidal algebra. In our previous work [**24**], we laid the groundwork for studying wreath Macdonald polynomials using an action of the higher rank quantum toroidal algebra $U_{\mathfrak{q},\mathfrak{d}}(\ddot{\mathfrak{sl}}_\ell)$. These higher rank algebras are also isomorphic to certain shuffle algebras [**18**], and we found shuffle elements $E_{p,n}$ generalizing K_n (1.6) whose action corresponds to multiplication by $e_n(p)$. Finding wreath Pieri rules then amounts to computing matrix elements for $E_{p,n}$ with respect to the wreath Macdonald basis. This paper lays the necessary preparation for such a calculation. We also discuss a natural wreath generalization of the Macdonald pairing and give a pathway towards computing norm formulas. An interesting feature here is that P_λ is no longer collinear to its dual and the two are related by the square of the *Miki automorphism.* Despite all this preparation, we were unable to carry out these calculations—the formulas are too complicated and the combinatorics too chaotic. Thus, this paper is the site of an abandoned expedition.

1.4. Outline. Section 1 goes over the necessary elements for defining and working with wreath Macdonald polynomials: wreath products, the core-quotient decomposition, etc. We also introduce the wreath Macdonald pairing and dual wreath Macdonald polynomials. Section 2 covers the quantum toroidal algebra and its realization as a shuffle algebra. The key points are formulas for shuffle elements corresponding to $e_n(p)$ and the appropriate generalization of g_n as well as a way to compute matrix elements of shuffle elements in the *Fock representation.* Finally, Section 3 isolates a key computation that yields the norm formula and is a prerequisite to computing wreath Pieri rules using $U_{\mathfrak{q},\mathfrak{d}}(\ddot{\mathfrak{sl}}_\ell)$.

2. Wreath Macdonald polynomials

Here, we introduce the basics of wreath Macdonald polynomials. We recommend the excellent article of Orr and Shimozono [**19**] for a deeper and broader introduction.

2.1. Wreath products. Let

$$\Gamma_n := (\mathbb{Z}/\ell\mathbb{Z})^n \rtimes \Sigma_n$$

denote the wreath product of $\mathbb{Z}/\ell\mathbb{Z}$ and the symmetric group Σ_n. We denote by $(\mathbb{Z}/\ell\mathbb{Z})_*$ and $(\mathbb{Z}/\ell\mathbb{Z})^*$ the set of conjugacy classes and characters, respectively, of $\mathbb{Z}/\ell\mathbb{Z}$. To distinguish between the two, we will write elements of $(\mathbb{Z}/\ell\mathbb{Z})_*$ as powers of a generator ϵ, whereas we will index $(\mathbb{Z}/\ell\mathbb{Z})^*$ additively by simply using integers modulo ℓ. Specifically, for $i \in \mathbb{Z}/\ell\mathbb{Z}$, let $\gamma_i \in (\mathbb{Z}/\ell\mathbb{Z})^*$ be the character

$$\gamma_i(\epsilon^j) = \zeta^{ij}$$

where $\zeta = e^{\frac{2\pi i}{\ell}}$. Finally, let

$$\Lambda \cong \mathbb{C}[p_n]_{n=1}^{\infty}$$

denote the ring of symmetric functions. We will apply the general framework of Chapter I, Appendix B of [**12**] to study representations of Γ_n. All results reviewed in this subsection appear in *loc. cit.*.

2.1.1. *Symmetric functions.* Consider the ring

$$\Lambda^{\otimes \ell} \cong \mathbb{C}[p_n(c)]_{c \in (\mathbb{Z}/\ell\mathbb{Z})_*}$$

For a partition λ, we set

$$p_\lambda(\epsilon^i) := \prod_k p_{\lambda_k}(\epsilon^i)$$

We will also need the set of ring generators indexed instead by $(\mathbb{Z}/\ell\mathbb{Z})^*$ and given by

$$p_n(i) := \sum_{j=0}^{\ell-1} \zeta^{ij} p_n(\epsilon^j)$$

For each $i \in \mathbb{Z}/\ell\mathbb{Z}$, we have an isomorphism $\Lambda \cong \mathbb{C}[p_n(i)]$ given by $p_n \mapsto p_n(i)$. We define $e_n(i)$ and $h_n(i)$ to be the images of e_n and h_n, respectively, under this isomorphism. Similarly, we define $p_\lambda(i)$, $e_\lambda(i)$, $h_\lambda(i)$, and $s_\lambda(i)$ as the images of p_λ, e_λ, h_λ, and s_λ, respectively. We will follow the conventions of [**24**], so we have

$$e_\lambda = \prod_{k=1}^{\ell(\lambda)} e_{\iota\lambda_k} \tag{2.1}$$

where $\ell(\lambda)$ is the length. For an ℓ-multipartition $\vec{\lambda} = (\lambda^0, \ldots, \lambda^{\ell-1})$, we define

$$p_{\vec{\lambda}} := \prod_{i \in \mathbb{Z}/\ell\mathbb{Z}} p_{\lambda^i}(i)$$

We similarly define $e_{\vec{\lambda}}$, $h_{\vec{\lambda}}$, and $s_{\vec{\lambda}}$. Finally, we set

$$p^{\mathbf{c}}_{\vec{\lambda}} := \prod_{i \in \mathbb{Z}/\ell\mathbb{Z}} p_{\lambda^i}(\epsilon^i)$$

2.1.2. *Wreath Hall pairing.* The *Hall pairing* on $\Lambda^{\otimes \ell}$ is given by

$$\langle p_{\vec{\lambda}}, p_{\vec{\mu}} \rangle = \delta_{\vec{\lambda},\vec{\mu}} z_{\vec{\lambda}}$$

where if we write $\lambda^i = (1^{m_1^i} 2^{m_2^i} \ldots)$, then

$$z_{\vec{\lambda}} = \prod_{i \in \mathbb{Z}/\ell\mathbb{Z}} \prod_k k^{m_k^i} m_k^i!$$

Under this pairing, $\{s_{\vec{\lambda}}\}$ forms an orthonormal basis.

REMARK 2.1. In Chapter I, Appendix B of [**12**], the Hall pairing is written in terms of the conjugacy class basis $\{p^{\mathrm{c}}_{\vec{\lambda}}\}$. However, the same functions $\{s_{\vec{\lambda}}\}$ still form an orthogonal basis under the pairing in *loc. cit.*, from which it follows that we can define the Hall pairing as we do here.

2.1.3. *Wreath Frobenius characteristic.* Let

$$R_n := \mathrm{Rep}(\Gamma_n)$$

In analogy with the study of the representation theory of symmetric groups, it is useful to consider the representations of all Γ_n:

$$R := \bigoplus_n R_n$$

We can endow R with the structure of a $\mathbb{C}$-algebra under the *induction product*: if $[V] \in \Gamma_n$ and $[W] \in \Gamma_m$,

$$[V] * [W] := \left[\mathrm{Ind}_{\Gamma_n \times \Gamma_m}^{\Gamma_{m+n}} (V \boxtimes W)\right] \in R_{n+m}$$

Observe as well that each R_n has an inner product given by the *Hom pairing*:

$$\langle [V], [W] \rangle = \dim \mathrm{Hom}_{\Gamma_n}(V, W)$$

for actual (i.e. nonvirtual) representations V, W of Γ_n.

The irreducible representations of Γ_n are indexed by ℓ-multipartitions $\vec{\lambda}$ with $|\vec{\lambda}| = n$. For such $\vec{\lambda}$, we denote the corresponding irrep by $V_{\vec{\lambda}}$. The classes $[V_{\vec{\lambda}}]$ form an orthonormal basis under the Hom pairing. We have the following result:

PROPOSITION 2.2. *The linear map $R \mapsto \Lambda^{\otimes \ell}$ induced by*

$$[V_{\vec{\lambda}}] \mapsto s_{\vec{\lambda}}$$

is both an isometry and a ring isomorphism.

We call this map the *(wreath) Frobenius characteristic.*

2.1.4. *Matrix plethysm*[1]. It will be useful to introduce plethystic notation for $\Lambda^{\otimes \ell}$. However, we will have ℓ alphabets $\{X^{(i)}\}$ indexed by $i \in \mathbb{Z}/\ell\mathbb{Z}$:

$$p_n(i) = p_n\left[X^{(i)}\right]$$

The index i will be called the *color* of the alphabet. We will use $X^\bullet$ to denote the alphabets over all colors and reserve $X^{(i)}$ for the specific color i. For example,

$$s_{\vec{\lambda}} = s_{\vec{\lambda}}[X^\bullet]$$

but $s_{\vec{\lambda}}\left[X^{(i)}\right]$ would only make sense if the only nonempty component of $\vec{\lambda}$ was its ith coordinate.

Aside from the usual plethystic automorphisms of $\Lambda^{\otimes \ell}$ given by scaling each alphabet, we will also have automorphisms that mix colors in a linear fashion. We will only consider such transformations built out of two basic operators. First, we will use the *upward color shift* σ:

$$\sigma X^{(i)} = X^{(i+1)}$$

[1] We are indebted to D. Orr and M. Shimozono for this notion.

for all $i \in \mathbb{Z}/\ell\mathbb{Z}$. For an example of how we will use this operator, consider for an indeterminate s the transformations

$$p_n\left[X^{(i)}\right] \mapsto p_n\left[(1-s\sigma^{\pm 1})X^{(i)}\right] = p_n\left[X^{(i)}\right] - s^n p_n\left[X^{(i\pm 1)}\right]$$

One can directly compute that the inverse of the transformations above are given by

$$p_n\left[X^{(i)}\right] \mapsto \frac{\sum_{j=0}^{\ell-1} s^{nj} p_n\left[X^{(i\pm j)}\right]}{1-s^{n\ell}} =: p_n\left[(1-s\sigma^{\pm 1})^{-1}X^{(i)}\right]$$

The second basic operator is the *color negation* ι:

$$\iota X^{(i)} = X^{(-i)}$$

This corresponds to taking the dual representation in R. We close by noting that σ and ι satisfy the following commutation relation:

$$\iota\sigma = \sigma^{-1}\iota$$

2.2. Cores and quotients. We will now review the core-quotient decomposition of a partition.

2.2.1. *Young-Maya correspondence.* In this paper, we will utilize two visual representations of a partition: its *Young diagram* and its *Maya diagram.* We will follow the French convention for Young diagrams. For example, below is the Young diagram for $(5,4,1)$:

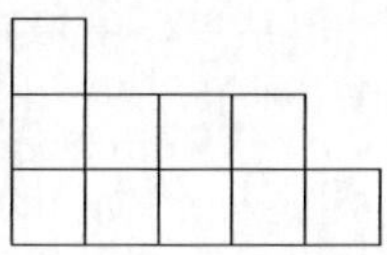

We call the boxes in the diagram *nodes* and index them according to their upper right coordinate, wherein each square has dimensions 1×1 and the bottom left corner of the partition has coordinate $(0,0)$. Thus, the topmost node in $(5,4,1)$ has coordinate $(1,3)$. For a node $\square = (i,j)$, we define its *content* by $c(\square) := j - i$. Thus, the content is constant along integer translates of the line $y = x$; we call these translates *content lines.* For $i \in \mathbb{Z}/\ell\mathbb{Z}$, a node $\square$ is called an i-node if $c(\square) \equiv i \bmod \ell$. Finally, for a positive integer k, we define a *k-strip* in λ to be a contiguous set of k nodes on the outer rim of λ whose removal leaves behind another partition. We call its node with the greatest content (northwesternmost) its *initial node* and its node with the least content (southeasternmost) its *final node.*

A *Maya diagram* is a function $m : \mathbb{Z} \to \{\pm 1\}$ such that $m(n) = -1$ for all $n \gg 0$ and $m(n) = 1$ for all $n \ll 0$. We can visualize this as a string of beads indexed by the integers, wherein the bead assigned to n is black if $m(n) = 1$ and white if $m(n) = -1$. In our conventions, we will always draw our string with index increasing towards the *left.* Moreover, we will draw a notch between the indices 0 and -1; we call this the *central line.* The Maya diagram where all beads right of the central line are black and all beads left of the central line are white is called the *vacuum diagram.* We give an example below:

○ ● ○ ● | ● ● ○ ● ○ ●
⋯ 3 2 1 0 -1 -2 -3 -4 -5 -6 ⋯

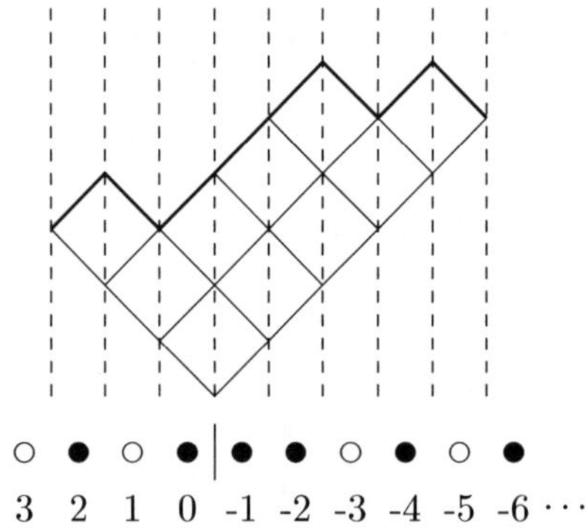

FIGURE 1. The Young-Maya correspondence for $(5, 4, 1)$.

For a Maya diagram m, we would expect beads to the left of the central line to be white and beads to the right of the central line to be black. The *charge* $c(m)$ measures the discrepancy between the number of white beads on the right and the number of black beads on the left:

$$c(m) = |\{n < 0 : m(n) = -1\}| - |\{n \geq 0 : m(n) = 1\}|$$

The *Young-Maya correspondence* assigns to a partition λ a Maya diagram $m(\lambda)$. It is given as follows. First, tilt the Young diagram of λ counter-clockwise by 45 degrees so that it is in the Russian convention. Draw the content lines and index the gap left of the content n line by n. Within each of these gaps, there is a segment of the outer border of the Young diagram. We set $m(\lambda)(n) = 1$ if the segment in the gap for n has slope 1 and $m(\lambda)(n) = -1$ if it has slope -1. For the indices n less than the least content of a node in λ minus 1, we set $m(\lambda)(n) = 1$, and likewise for indices greater than the greatest content of a node in λ, we set $m(\lambda)(n) = -1$. In Figure 1, we give the example of $\lambda = (5, 4, 1)$.

PROPOSITION 2.3. *The Young-Maya correspondence gives a bijection between partitions and Maya diagrams with charge zero.*

REMARK 2.4. Observe that adding a node to λ corresponds to the following switch on adjacent beads in $m(\lambda)$:

○ ● → ● ○

On the other hand, removing a node corresponds to the opposite switch:

● ○ → ○ ●

Notice now that doing the same swaps above in $m(\lambda)$ to beads indexed at n and $n + k$ instead adds/removes k-strips to/from λ.

2.2.2. *Core-quotient decomposition.* For a partition λ and its Maya diagram $m(\lambda)$, consider its *quotient Maya diagrams*

$$m_i(\lambda)(n) = m(\lambda)(i + n\ell)$$

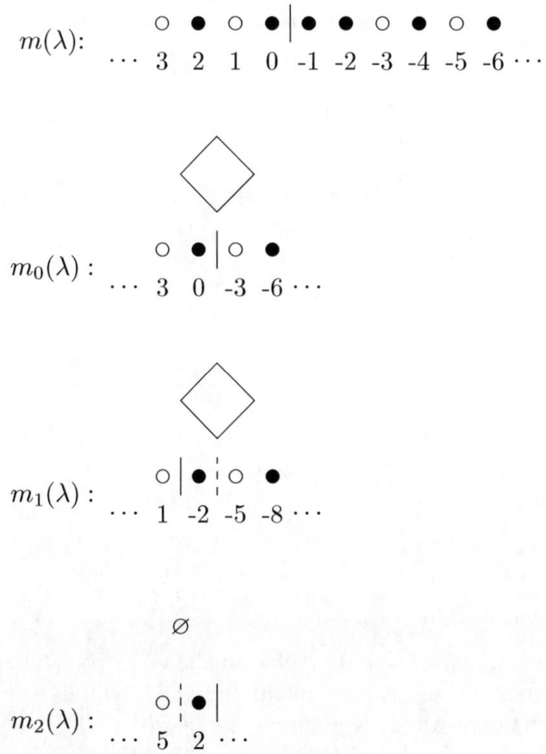

FIGURE 2. Computing the 3-quotient of $(5,4,1)$. We index the beads in $m_i(\lambda)$ with their original indices in $m(\lambda)$. The solid line is the central line, whereas the dashed line is the notch right of $-c_i$.

for $0 \leq i \leq \ell - 1$. In general, $m_i(\lambda)$ will have a charge c_i that may not be zero. If we draw a notch immediately right of the bead indexed by $-c_i$ in $m_i(\lambda)$, then we can perform the Young-Maya correspondence where the content zero line of the Young diagram is lined up with this notch instead. This yields a partition λ^i. We call

$$\mathrm{quot}(\lambda) := (\lambda^0, \ldots, \lambda^{\ell-1})$$

the *ℓ-quotient.* The example of $\ell = 3$ and $\lambda = (5,4,1)$ is show in Figure 2. Applying Remark 2.4 to $m_i(\lambda)$ and then to $m(\lambda)$, we can see that each node of $m_i(\lambda)$ corresponds to an ℓ-strip in $m(\lambda)$ whose initial node is an i-node. Thus, $\mathrm{quot}(\lambda)$ records the ℓ-strips of λ.

Once all the ℓ-strips are removed from λ, we are left with its *ℓ-core* $\mathrm{core}(\lambda)$. An *ℓ-core partition* is one that has no ℓ-strips. To obtain the ℓ-core from $m(\lambda)$, we change each $m_i(\lambda)$ into the diagram where all beads $< -c_i$ are black and all beads $\geq -c_i$ are white. This is like the vacuum diagram except that the central line is replaced by our notch right of $-c_i$. Performing corresponding changes in the total Maya diagram $m(\lambda)$, we obtain $\mathrm{core}(\lambda)$. This is illustrated for $\ell = 3$ and

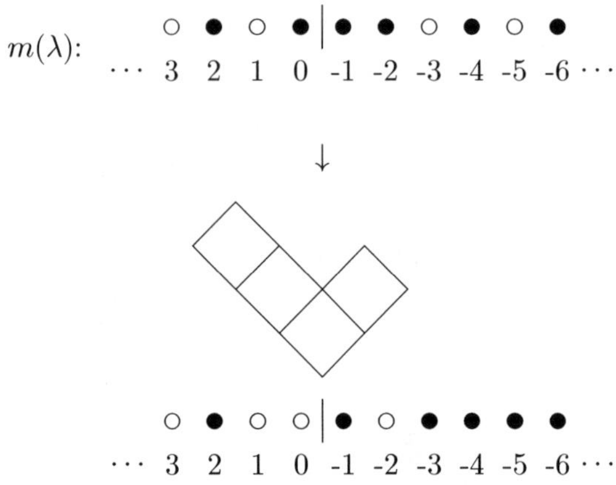

FIGURE 3. The 3-core of $\lambda = (5, 4, 1)$.

$\lambda = (5, 4, 1)$ in Figure 3. From this, it follows that $\text{core}(\lambda)$ is determined by the charges c_i. Since $m(\lambda)$ has charge zero, it follows that

$$c_0 + \cdots + c_{\ell-1} = 0$$

Thus, we can view $(c_0, \ldots, c_{\ell-1})$ as a vector in the $A_{\ell-1}$ root lattice Q, and in fact all root lattice vectors can be realized in this way. We will abuse notation and also denote this vector as $\text{core}(\lambda)$. This assignment $\lambda \mapsto (\text{core}(\lambda), \text{quot}(\lambda))$ is called the *core-quotient decomposition* of λ. We have the following:

PROPOSITION 2.5 ([**8**]). *The core-quotient decomposition yields a bijection*

$$\{\textit{partitions}\} \leftrightarrow \{\ell\textit{-core partitions}\} \times \{\ell\textit{-multipartitions}\}$$

2.2.3. *Orders on columns and rows.* To conclude this subsection, we will review an ordering on the columns and rows of $\text{quot}(\lambda)$ introduced in [**24**]. First observe that every Maya diagram has a maximal infinite sequence of consecutive black beads and a maximal infinite sequence of consecutive white beads; we call them the *black sea* and *white sea*, respectively. By Remark 2.4, adding a column of length k to μ^i that is no longer than any of the existing columns of μ^i entails the following swap in $m_i(\mu)$.

$$\underset{j+k}{\circ} \ \circ \cdots \circ \ \underset{j}{\bullet} \quad \rightarrow \quad \underset{j+k}{\bullet} \ \circ \cdots \circ \ \underset{j}{\circ}$$

Here, all the beads in the middle are white and the black bead is the leftmost bead in the black sea of $m_i(\mu)$. Putting this in the total Maya diagram, one can see that we have added a $k\ell$-strip to μ wherein the consecutive i and $(i+1)$ nodes are vertically adjacent. Similarly, adding a row of length k to μ^i that is no longer than any of the previous nodes entails swapping the rightmost bead in the white sea of $m_i(\mu)$ with a black bead k places to the right, and this is equivalent to adding a $k\ell$-strip to μ wherein consecutive i and $(i+1)$-nodes are horizontally adjacent.

For a partition λ, we define the *left-to-right order* on the columns of $\text{quot}(\lambda)$ to be the unique order such that if we perform the corresponding strip additions to $\text{core}(\lambda)$, the initial nodes for earlier columns are northwest of the initial nodes for later columns. The existence of this order comes from our observation above: to each column of λ^i, we can assign a black bead that came from the black sea, and we are merely stipulating that, viewing these beads in $m(\lambda)$, those of earlier columns are left of those of later columns. In the example of of the 3-quotient of $\lambda = (5, 4, 1)$ depicted in Figure 2, we have that the box in λ^0 comes before the box in λ^1. Similarly, the *right-to-left order* on the rows of quot(λ) is the unique one in which performing the strip additions to core(λ) in that order results in the final nodes of earlier rows lying southeast of those of later rows. We note that while we call these orderings on the columns and rows of quot(λ), they are defined in terms of the original partition λ.

2.3. Wreath Macdonald polynomials. With the combinatorics of ℓ-cores and ℓ-quotients reviewed, we are ready to define the transformed and ordinary wreath Macdonald polynomials.

2.3.1. *Plethysm.* Now, for two indeterminates q and t, let

$$\Lambda_{q,t} := \mathbb{C}(q,t) \otimes \Lambda$$
$$R_{q,t} := \mathbb{C}(q,t) \otimes R$$

The group Γ_n has a natural reflection representation $\mathfrak{h}_n$. For an indeterminate s, we define the operators $\bigwedge_s^+$ and $\bigwedge_s^-$ on $\mathbb{C}(s) \otimes R$ such that

$$\left.\bigwedge\nolimits_s^+\right|_{R_n} = (-) \otimes \sum_{i=0}^{n} (-s)^i \left[\bigwedge\nolimits^i \mathfrak{h}_n\right]$$
$$\left.\bigwedge\nolimits_s^-\right|_{R_n} = (-) \otimes \sum_{i=0}^{n} (-s)^i \left[\bigwedge\nolimits^i \mathfrak{h}_n^*\right]$$

The following was proved in [**24**]:

PROPOSITION 2.6. *$\bigwedge_s^\pm$ is the ring automorphisms on $\mathbb{C}(s) \otimes \Lambda^{\otimes \ell}$ given by*

$$\bigwedge\nolimits_s^\pm f\left[X^\bullet\right] = f\left[(1 - s\sigma^{\pm 1})X^\bullet\right] \tag{2.2}$$

2.3.2. *Transformed polynomials.* For $\lambda, \mu \vdash n$, we let $\lambda \geq_\ell \mu$ denote that $\lambda \geq \mu$ and core(λ) = core(μ).

DEFINITION 2.7. For λ with quot(λ) $\vdash n$, the *transformed wreath Macdonald polynomial* $H_\lambda(q,t) = H_\lambda\left[X^\bullet; q, t\right]$ is the element of $\Lambda_{q,t}^{\otimes \ell}$ characterized by

(1) $H_\lambda\left[(1 - q\sigma^{-1})X^\bullet; q, t\right]$ lies in the span of $\{s_{\mathrm{quot}(\mu)} : \mu \geq_\ell \lambda\}$;
(2) $H_\lambda\left[(1 - t^{-1}\sigma^{-1})X^\bullet; q, t\right]$ lies in the span of $\{s_{\mathrm{quot}(\mu)} : \mu \leq_\ell \lambda\}$;
(3) the coefficient of the trivial representation of Γ_n is 1.

When appropriate, we will abbreviate it by H_λ.

The transformed wreath Macdonald polynomials are the ones defined in [**7**] and studied in our previous work [**24**]. Their existence was proved by Bezrukavnikov and Finkelberg [**1**]. We have rewritten the definition in terms of matrix plethysms using (2.2).

2.3.3. *Sectors.* Recall from 2.2.2 that Q denotes the $A_{\ell-1}$ root lattice, and we can use it to index ℓ-cores. For $\alpha \in Q$, we denote the corresponding basis element in $\mathbb{C}[Q]$ by e^α. We will work in the ring

$$\Lambda^{\otimes \ell} \otimes \mathbb{C}[Q]$$

PROPOSITION 2.8. *$\{H_\lambda \otimes e^{\mathrm{core}(\lambda)}\}$ is a basis of $\Lambda^{\otimes \ell} \otimes \mathbb{C}[Q]$.*

Finally, if $\alpha = (a_0, \ldots, a_{\ell-1})$, then we define

$$^t\alpha = (-a_{\ell-1}, -a_{\ell-2}, \ldots, -a_1, -a_0)$$

This operation is defined so that if $\mathrm{core}(\lambda) = \alpha$, then $\mathrm{core}(^t\lambda) = {}^t\alpha$.

2.3.4. *Transformed pairing.* Next, we consider a pairing under which $\{H_\lambda\}$ is some kind of orthogonal basis. Recall that the Hall pairing $\langle -, - \rangle$ corresponds to the Hom pairing in R. Alternatively, we can twist with ι to obtain the pairing corresponding to tensoring and taking invariants:

$$\langle f, g \rangle^* := \langle f\left[\iota X^\bullet\right], g \rangle$$

Note that under $\langle -, - \rangle^*$, the shift plethysm σ is self-adjoint. On the other hand, under $\langle -, - \rangle$, σ is adjoint to σ^{-1}. The *transformed wreath Macdonald pairing* is defined as follows:

$$\begin{aligned}\langle f, g \rangle'_{q,t} &:= \left\langle f, g\left[\sigma(1 - t^{-1}\sigma^{-1})(1 - q\sigma^{-1})X^\bullet\right]\right\rangle^* \\ &= \left\langle f\left[\iota X^\bullet\right], g\left[\sigma(1 - t^{-1}\sigma^{-1})(1 - q\sigma^{-1})X^\bullet\right]\right\rangle\end{aligned}$$

A priori, this is only defined on $\Lambda_{q,t}^{\otimes \ell}$. We extend it to $\Lambda_{q,t}^{\otimes \ell} \otimes \mathbb{C}[Q]$ by setting

$$\langle f \otimes e^\alpha, g \otimes e^\beta \rangle'_{q,t} = \delta_{\alpha, {}^t\beta} \langle f, g \rangle'_{q,t}$$

where ${}^t\beta$ was defined in 2.3.3. To find a dual basis to $\{H_\lambda \otimes e^{\mathrm{core}(\lambda)}\}$, we introduce the following *orthogonal variant*:

$$H_\lambda^*\left[X^\bullet; q, t\right] := H_\lambda\left[-X^\bullet; t^{-1}, q^{-1}\right]$$

As with H_λ, we abbreviate it by H_λ^*.

PROPOSITION 2.9. *The pairing $\langle H_{^t\lambda}^*, H_\mu \rangle'_{q,t}$ is nonzero if and only if $\lambda = \mu$.*

PROOF. If $\lambda \not\geq \mu$, then by adjunction, we have

$$\begin{aligned}\langle H_{^t\lambda}^*, H_\mu \rangle'_{q,t} &= \left\langle H_{^t\lambda}^*, H_\mu\left[\sigma(1 - t^{-1}\sigma^{-1})(1 - q\sigma^{-1})X^\bullet\right]\right\rangle^* \\ &= \left\langle H_{^t\lambda}^*\left[\sigma(1 - t^{-1}\sigma^{-1})X^\bullet\right], H_\mu\left[(1 - q\sigma^{-1})X^\bullet\right]\right\rangle^* \\ &= \left\langle H_{^t\lambda}\left[-\iota\sigma(1 - t^{-1}\sigma^{-1})X^\bullet; t^{-1}, q^{-1}\right], H_\mu\left[(1 - q\sigma^{-1})X^\bullet; q, t\right]\right\rangle \qquad (2.3)\end{aligned}$$

By the first triangularity condition in Definition 2.7, the second argument of (2.3) is a linear combination of $s_{\mathrm{quot}(\nu)}$ for $\nu \geq_\ell \mu$. Similarly, the first argument of (2.3) is a linear combination of $s_{\mathrm{quot}(^t\nu)}\left[-\iota\sigma X^\bullet\right]$ for ${}^t\nu \geq_\ell {}^t\lambda$. The key observation is that

$$s_{\mathrm{quot}(^t\nu)}\left[-\iota\sigma X^\bullet\right] = (-1)^{|\mathrm{quot}(\nu)|} s_{\mathrm{quot}(\nu)}$$

To see this, note that $\iota\sigma$ will send color i to color $-i-1$, while transposing an ℓ-strip starting at an i-node yields an ℓ-strip starting at an $-i-1$ node. From this, it follows that the pairing (2.3) is zero when $\lambda \not\geq \mu$. The case $\lambda \not\leq \mu$ is similar, and thus the pairing is always zero when $\lambda \neq \mu$. Finally, because $\left\{H_\lambda \otimes e^{\mathrm{core}(\lambda)}\right\}$ and $\left\{H_\lambda^* \otimes e^{\mathrm{core}(\lambda)}\right\}$ are bases and $\langle -, - \rangle'_{q,t}$ is nondegenerate, the pairing for $\lambda = \mu$ must be nonzero. □

REMARK 2.10. A cleaner approach would be to define $\langle -,- \rangle'_{q,t}$ without the extra twist by $\iota\sigma$ and move it into the definition of H^*_λ. Namely, one can check that $H_{{}^t\lambda}\left[-\iota\sigma X^\bullet; t^{-1}, q^{-1}\right]$ satisfies the triangularity conditions of Definition 2.7 but with σ^{-1} replaced by σ:

(1) $H_{{}^t\lambda}\left[(1-q\sigma)(-\iota\sigma X^\bullet); t^{-1}, q^{-1}\right]$ lies in the span of $\{s_{\mathrm{quot}(\mu)} : \mu \geq_\ell \lambda\}$;
(2) $H_{{}^t\lambda}\left[(1-t^{-1}\sigma)(-\iota\sigma X^\bullet); t^{-1}, q^{-1}\right]$ lies in the span of $\{s_{\mathrm{quot}(\mu)} : \mu \leq_\ell \lambda\}$.

Thus, we can instead use these altered triangularity conditions (along with a natural normalization condition) to define the positive variants. While this approach is more natural, it leads to "one twist/swap too many" for us (the author) to handle when we include the quantum toroidal algebra later on.

2.3.5. *Ordinary polynomials.* Recall that for $\ell = 1$, the *ordinary* Macdonald polynomial P_λ is obtained from $H_\lambda(q,t^{-1})$ by renormalizing $H_\lambda\left[(1-t)X; q, t^{-1}\right]$ so that the leading term s_λ has coefficient 1. The second triangularity condition (2) of Definition 2.7 ensures that the latter is upper triangular with respect to the Schur function basis ordered by dominance. We similarly define the ordinary polynomials for general ℓ. Note that the proof of Proposition 2.9 implies that the coefficient of $s_{\mathrm{quot}(\lambda)}$ in $H_\lambda\left[(1-t^{-1}\sigma^{-1})X^\bullet\right]$ is nonzero.

DEFINITION 2.11. The *ordinary wreath Macdonald polynomial* $P_\lambda(q,t)$ is the renormalization of $H_\lambda\left[(1-t\sigma^{-1})X^\bullet; q, t^{-1}\right]$ such that the coefficient of $s_{\mathrm{quot}(\lambda)}$ is 1. We also set $P^*_\lambda(q,t) := P_\lambda\left[-X^\bullet; t, q\right]$.

Inspired by the $\ell = 1$ case, we define the (ordinary) *wreath Macdonald pairing* by

$$\langle f, g\rangle_{q,t} := \left\langle f, g\left[\frac{\sigma(1-q\sigma^{-1})}{(1-t\sigma^{-1})}X^\bullet\right]\right\rangle^*$$

We extend it to $\Lambda^{\otimes \ell} \otimes \mathbb{C}[Q]$ like we did with $\langle -,- \rangle'_{q,t}$, namely by making e^α dual to $e^{{}^t\alpha}$. The *dual (ordinary) wreath Macdonald polynomial* $Q_\lambda(q,t)$ is the renormalization of $H_\lambda\left[(1-t\sigma^{-1})X^\bullet; q, t^{-1}\right]$ such that the coefficient of $s_{\mathrm{quot}(\lambda)}$ in

$$Q_\lambda\left[\frac{(1-q\sigma^{-1})}{(1-t\sigma^{-1})}X^\bullet\right]$$

is 1. Finally, we set $Q^*_\lambda := Q_\lambda\left[-X^\bullet; t, q\right]$. As before, we will drop (q,t) for P_λ, P^*_λ, Q_λ, and Q^*_λ if it causes no confusion, although care is certainly necessary as these parameters are often swapped and inverted. The following is proved much like Proposition 2.9 (note the transpose):

PROPOSITION 2.12. *We have*

$$\langle P^*_{{}^t\lambda}, Q_\mu\rangle_{q,t} = \langle Q^*_{{}^t\lambda}, P_\mu\rangle_{q,t} = \delta_{\lambda,\mu}$$

In this paper, we will not work directly with P_λ or Q_λ, but rather with

$$\tilde{P}_\lambda[X^\bullet; q, t] := P_\lambda\left[(1-t\sigma^{-1})^{-1}X^\bullet; q, t\right]$$
$$\tilde{Q}_\lambda[X^\bullet; q, t] := Q_\lambda\left[(1-t\sigma^{-1})^{-1}X^\bullet; q, t\right]$$

Note that $\tilde{P}_\lambda(q,t^{-1})$ and $\tilde{Q}_\lambda(q,t^{-1})$ are scalar multiples of H_λ.

We end by stating a natural wreath generalization of the Macdonald norm formula (1.5) that has been confirmed by computations of Orr and Shimozono:

CONJECTURE 2.13 (Conjecture 3.36 of [**19**]). *For a node* $\square = (a, b) \in \lambda$, *let:*

- $a(\square) = \lambda_b - a$ *denote its arm length;*
- $\ell(\square) = {}^t\lambda_a - b$ *denote its leg length;*
- $h(\square) = a(\square) + \ell(\square) + 1$ *denote its hook length.*

We then have the norm formula:

$$\langle P^*_{{}^t\lambda}, P_\lambda \rangle_{q,t} = \prod_{\substack{\square \in \lambda \\ h(\square) \equiv 0 \bmod \ell}} \frac{1 - q^{a(\square)+1} t^{\ell(\square)}}{1 - q^{a(\square)} t^{\ell(\square)+1}}$$

3. Quantum toroidal and shuffle algebras

In this section, $\ell \geq 3$. Let $\mathfrak{q}, \mathfrak{d}$ be indeterminates and set $\mathbb{F} := \mathbb{C}(\mathfrak{q}, \mathfrak{d})$—we will specify their relationship to q and t later on. We will make use of the delta function $\delta(z)$:

$$\delta(z) = \sum_{n \in \mathbb{Z}} z^n$$

3.1. Quantum toroidal algebra.

For $i, j \in \mathbb{Z}/\ell\mathbb{Z}$, we let

$$a_{i,i} = 2,\ a_{i,i\pm 1} = -1,\ m_{i,i\pm 1} = \mp 1,\ \text{and } a_{i,j} = m_{i,j} = 0 \text{ otherwise}$$

Also, define

$$g_{i,j}(z) := \frac{\mathfrak{q}^{a_{i,j}} z - 1}{z - \mathfrak{q}^{a_{i,j}}}$$

3.1.1. *Definition.* The *quantum toroidal algebra* $U_{\mathfrak{q},\mathfrak{d}}(\ddot{\mathfrak{sl}}_\ell)$ is a $\mathbb{F}$-algebra with generators

$$\{e_{i,n}, f_{i,n}, \psi_{i,n}, \psi_{i,0}^{-1}, \gamma^{\pm\frac{1}{2}}, \mathfrak{q}^{d_1}, \mathfrak{q}^{d_2}\}_{i \in \mathbb{Z}/\ell\mathbb{Z}}^{n \in \mathbb{Z}}$$

To describe its relations, we define the following generating series of generators, called *currents*:

$$e_i(z) = \sum_{n \in \mathbb{Z}} e_{i,n} z^{-n}$$
$$f_i(z) = \sum_{n \in \mathbb{Z}} f_{i,n} z^{-n}$$
$$\psi_i^{\pm}(z) = \psi_{i,0}^{\pm 1} + \sum_{n > 0} \psi_{i,\pm n} z^{\mp n}$$

The relations are then:

$$[\psi_i^\pm(z), \psi_j^\pm(w)] = 0,\ \gamma^{\pm\frac{1}{2}} \text{ are central},$$
$$\psi_{i,0}^{\pm 1}\psi_{i,0}^{\mp 1} = \gamma^{\pm\frac{1}{2}}\gamma^{\mp\frac{1}{2}} = \mathfrak{q}^{\pm d_1}\mathfrak{q}^{\mp d_1} = \mathfrak{q}^{\pm d_2}\mathfrak{q}^{\mp d_2} = 1,$$
$$\mathfrak{q}^{d_1}e_i(z)\mathfrak{q}^{-d_1} = e_i(\mathfrak{q}z),\ \mathfrak{q}^{d_1}f_i(z)\mathfrak{q}^{-d_1} = f_i(\mathfrak{q}z),\ \mathfrak{q}^{d_1}\psi_i^\pm(z)\mathfrak{q}^{-d_1} = \psi_i^\pm(\mathfrak{q}z),$$
$$\mathfrak{q}^{d_2}e_i(z)\mathfrak{q}^{-d_2} = \mathfrak{q}e_i(z),\ \mathfrak{q}^{d_2}f_i(z)\mathfrak{q}^{-d_2} = \mathfrak{q}^{-1}f_i(z),\ \mathfrak{q}^{d_2}\psi_i^\pm(z)\mathfrak{q}^{-d_2} = \psi_i^\pm(z),$$
$$g_{i,j}(\gamma^{-1}\mathfrak{d}^{m_{i,j}}z/w)\psi_i^+(z)\psi_j^-(w) = g_{i,j}(\gamma\mathfrak{d}^{m_{i,j}}z/w)\psi_j^-(w)\psi_i^+(z),$$
$$e_i(z)e_j(w) = g_{i,j}(\mathfrak{d}^{m_{i,j}}z/w)e_j(w)e_i(z),$$
$$f_i(z)f_j(w) = g_{i,j}(\mathfrak{d}^{m_{i,j}}z/w)^{-1}f_j(w)f_i(z),$$
$$(\mathfrak{q}-\mathfrak{q}^{-1})[e_i(z), f_j(w)] = \delta_{i,j}\left(\delta(\gamma w/z)\psi_i^+(\gamma^{\frac{1}{2}}w) - \delta(\gamma z/w)\psi_i^-(\gamma^{\frac{1}{2}}z)\right),$$
$$\psi_i^\pm(z)e_j(w) = g_{i,j}(\gamma^{\pm\frac{1}{2}}\mathfrak{d}^{m_{i,j}}z/w)e_j(w)\psi_i^\pm(z),$$
$$\psi_i^\pm(z)f_j(w) = g_{i,j}(\gamma^{\mp\frac{1}{2}}\mathfrak{d}^{m_{i,j}}z/w)^{-1}f_j(w)\psi_i^\pm(z),$$
$$\mathrm{Sym}_{z_1,z_2}[e_i(z_1), [e_i(z_2), e_{i\pm 1}(w)]_{\mathfrak{q}}]_{\mathfrak{q}^{-1}} = 0,\ [e_i(z), e_j(w)] = 0 \text{ for } j \neq i, i\pm 1,$$
$$\mathrm{Sym}_{z_1,z_2}[f_i(z_1), [f_i(z_2), f_{i\pm 1}(w)]_{\mathfrak{q}}]_{\mathfrak{q}^{-1}} = 0,\ [f_i(z), f_j(w)] = 0 \text{ for } j \neq i, i\pm 1,$$

In the above, $[a,b]_{\mathfrak{q}} = ab - \mathfrak{q}ba$ is the $\mathfrak{q}$-commutator. We will also need the elements $\{b_{i,n}\}_{i\in\mathbb{Z}/\ell\mathbb{Z}}^{n\neq 0}$ defined by

$$\psi_i^\pm(z) = \psi_{i,0}^{\pm 1}\exp\left(\pm(\mathfrak{q}-\mathfrak{q}^{-1})\sum_{n>1} b_{i,\pm n}z^{\mp n}\right)$$

Denote by $\mathcal{H}$ the subalgebra generated by $\{b_{i,n}\} \cup \{\gamma^{\pm\frac{1}{2}}\}$.

3.1.2. *Miki automorphism.* One feature making the quantum toroidal algebra deserving of its name is that it contains two copies of the quantum affine algebra $U_{\mathfrak{q}}(\dot{\mathfrak{sl}}_\ell)$. The first, often called the *horizontal subalgebra*, is generated by the zero modes:

$$\{e_{i,0}, f_{i,0}, \psi_{i,0}^{\pm 1}\}_{i\in\mathbb{Z}/\ell\mathbb{Z}}$$

The second copy, the *vertical subalgebra*, is roughly-speaking generated by currents where $i \neq 0$:

$$\{e_i(z), f_i(z)\}_{i\neq 0}^{i\in\mathbb{Z}/\ell\mathbb{Z}} \cup \{\psi_i^\pm(z)\}_{i\in\mathbb{Z}/\ell\mathbb{Z}}$$

We will not go into detail on these subalgebras as they will not be used in the sequel. However, they motivate the following automorphism of Miki, which swaps the "two loops" of the torus.

Let $'\ddot{U}$ denote the subalgebra of $U_{\mathfrak{q},\mathfrak{d}}(\ddot{\mathfrak{sl}}_\ell)$ obtained by omitting $\mathfrak{q}^{d_1}$, $\ddot{U}'$ denote the subalgebra obtained by omitting $\mathfrak{q}^{d_2}$, and $'\ddot{U}'$ denote the subalgebra obtained by omitting both elements. Next, let η denote the $\mathbb{C}(\mathfrak{q})$-linear antiautomorphism of $'\ddot{U}'$ defined by

$$\eta(\mathfrak{d}) = \mathfrak{d}^{-1}$$
$$\eta(e_{i,k}) = e_{i,-k},\ \eta(f_{i,k}) = f_{i,-k},\ \eta(h_{i,k}) = -\gamma^k h_{i,-k}, \tag{3.1}$$
$$\eta(\psi_{i,0}) = \psi_{i,0}^{-1},\ \eta(\gamma^{\frac{1}{2}}) = \gamma^{\frac{1}{2}}$$

We have the following beautiful result of Miki:

THEOREM 3.1 ([**13**], [**14**]). *There is an isomorphism* $\varpi : {}'\ddot{U} \to \ddot{U}'$ *sending the horizontal subalgebra to the vertical subalgebra. Moreover, when restricted to* ${}'\ddot{U}'$, *we have:*

$$\varpi^{-1} = \eta\varpi\eta$$

3.2. Representations. We will go over two representations of $U_{\mathfrak{q},\mathfrak{d}}(\ddot{\mathfrak{sl}}_\ell)$: the *vertex* and *Fock representations.*

3.2.1. *Heisenberg subalgebra.* Let $[n]_\mathfrak{q}$ denote the quantum integer:

$$[n]_\mathfrak{q} := \frac{\mathfrak{q}^n - \mathfrak{q}^{-n}}{\mathfrak{q} - \mathfrak{q}^{-1}}$$

The subalgebra $\mathcal{H}$ satisfies the relation

$$[b_{i,n}, b_{j,m}] = \delta_{m+n,0}\frac{\mathfrak{d}^{-nm_{i,j}}[na_{i,j}]_\mathfrak{q}(\gamma^n - \gamma^{-n})}{n(\mathfrak{q} - \mathfrak{q}^{-1})} \tag{3.2}$$

Let

- $\mathcal{H}^+$ be the subalgebra generated by $\{b_{i,n}\}_{i\in\mathbb{Z}/\ell\mathbb{Z}}^{n>0}$;
- $\mathcal{H}^-$ be the subalgebra generated by $\{b_{i,-n}\}_{i\in\mathbb{Z}/\ell\mathbb{Z}}^{n>0}$;
- and $\mathcal{H}^\geq$ be the subalgebra generated by $\{b_{i,n}\}_{i\in\mathbb{Z}/\ell\mathbb{Z}}^{n>0} \cup \left\{\gamma^{\pm\frac{1}{2}}\right\}$;

$\mathcal{H}^\geq$ has a one-dimensional representation $\mathbb{F}\mathbb{1}$ where

$$\gamma^{\pm\frac{1}{2}}\mathbb{1} = \mathfrak{q}^{\pm\frac{1}{2}}\mathbb{1}$$
$$b_{i,n}\mathbb{1} = 0$$

We then have

$$\mathrm{Ind}_{\mathcal{H}^\geq}^{\mathcal{H}}\mathbb{F}\mathbb{1} \cong \mathbb{C}[b_{i,-n}]_{i\in\mathbb{Z}/\ell\mathbb{Z}}^{n>0}$$

as $\mathcal{H}^-$-modules. If we identify $\mathbb{F}\otimes\Lambda^{\otimes\ell}$ with $\mathcal{H}^-$ by specifying

$$p_k(i) \mapsto \frac{k}{[k]_\mathfrak{q}}b_{i,-k} \tag{3.3}$$

then we also obtain an identification

$$\mathbb{F}\otimes\Lambda^{\otimes\ell} \cong \mathrm{Ind}_{\mathcal{H}^\geq}^{\mathcal{H}}\mathbb{F}\mathbb{1}$$

3.2.2. *Vertex representation.* We will now review what one can consider a homogeneous bosonic Fock space for $U_{\mathfrak{q},\mathfrak{d}}(\ddot{\mathfrak{sl}}_\ell)$. For the $A_{\ell-1}$ root system, let $\{\alpha_j\}_{j=1}^{\ell-1}$ be the simple roots, let $\{h_j\}_{j=1}^{\ell-1}$ be the corresponding simple coroots, and let $\langle -,-\rangle$ denote the natural pairing between them. We set

$$\alpha_0 := -\sum_{j=1}^{\ell-1}\alpha_j,$$
$$h_0 := -\sum_{j=1}^{\ell-1}h_j,$$

The *twisted group algebra* $\mathbb{F}\{Q\}$ is the $\mathbb{F}$-algebra generated by $\{e^{\pm\alpha_j}\}_{j=1}^{\ell-1}$ satisfying the relations

$$e^{\alpha_i}e^{\alpha_j} = (-1)^{\langle h_i,\alpha_j\rangle}e^{\alpha_j}e^{\alpha_i}$$

For $\alpha \in Q$ with $\alpha = \sum_{j=1}^{\ell-1} m_j\alpha_j$, we set

$$e^{\alpha} = e^{m_1\alpha_1} \cdots e^{m_{\ell-1}\alpha_{\ell-1}}$$

We will identify $\mathbb{F}[Q]$ and $\mathbb{F}\{Q\}$ by the 'identity map' $e^{\alpha} \mapsto e^{\alpha}$. Let

$$W := \operatorname{Ind}_{\mathcal{H}^{\geq}}^{\mathcal{H}} \mathbb{F}\mathbb{1} \otimes \mathbb{F}\{Q\}$$

Altogether, we have an identification

$$\Lambda^{\otimes \ell} \otimes \mathbb{F}[Q] \cong W$$

Finally, we will need to define some additional operators. Let $v \otimes e^{\alpha} \in W$ where

$$v = b_{i_1,-k_1} \cdots b_{i_N,-k_N} v_0$$

$$\alpha = \sum_{j=1}^{\ell-1} m_j\alpha_j$$

The operators ∂_{α_i}, $z^{H_{i,0}}$, and d are defined by

$$\partial_{\alpha_i}(v \otimes e^{\alpha}) := \langle h_i, \alpha\rangle v \otimes e^{\alpha},$$

$$z^{H_{i,0}}(v \otimes e^{\alpha}) := z^{\langle h_i,\alpha\rangle} \mathfrak{d}^{\frac{1}{2}\sum_{j=1}^{\ell-1}\langle h_i, m_j\alpha_j\rangle m_{i,j}} v \otimes e^{\alpha},$$

$$d(v \otimes e^{\alpha}) := -\left(\frac{(\alpha,\alpha)}{2} + (\alpha, \Lambda_p) + \sum_{i=1}^{N} k_i\right) v \otimes e^{\alpha}$$

The following theorem of Yoshihisa Saito defines the *vertex representation*:

THEOREM 3.2 ([**21**]). *For any $\vec{\mathfrak{c}} = (\mathfrak{c}_0, \ldots, \mathfrak{c}_{\ell-1}) \in (\mathbb{C}^{\times})^{\ell}$, the following formulas endow W with an action of $\ddot{U}'$*

$$\rho_{\vec{\mathfrak{c}}}(e_i(z)) = \mathfrak{c}_i \exp\left(\sum_{k>0} \frac{\mathfrak{q}^{-\frac{k}{2}}}{[k]_{\mathfrak{q}}} b_{i,-k} z^k\right)$$

$$\times \exp\left(-\sum_{k>0} \frac{\mathfrak{q}^{-\frac{k}{2}}}{[k]_{\mathfrak{q}}} b_{i,k} z^{-k}\right) e^{\alpha_i} z^{1+H_{i,0}},$$

$$\rho_{\vec{\mathfrak{c}}}(f_i(z)) = \frac{(-1)^{\ell\delta_{i,0}}}{\mathfrak{c}_i} \exp\left(-\sum_{k>0} \frac{\mathfrak{q}^{\frac{k}{2}}}{[k]_{\mathfrak{q}}} b_{i,-k} z^k\right)$$

$$\times \exp\left(\sum_{k>0} \frac{\mathfrak{q}^{\frac{k}{2}}}{[k]_{\mathfrak{q}}} b_{i,k} z^{-k}\right) e^{-\alpha_i} z^{1-H_{i,0}},$$

$$\rho_{\vec{\mathfrak{c}}}(\psi_i^{\pm}(z)) = \exp\left(\pm(\mathfrak{q} - \mathfrak{q}^{-1}) \sum_{k>0} b_{i,\pm k} z^{\mp k}\right) \mathfrak{q}^{\pm \partial_{\alpha_i}},$$

$$\rho_{\vec{\mathfrak{c}}}(\gamma^{\frac{1}{2}}) = \mathfrak{q}^{\frac{1}{2}},\ \rho_{\vec{\mathfrak{c}}}(\mathfrak{q}^{d_1}) = \mathfrak{q}^{d}$$

The representation W is irreducible.

3.2.3. *Fock representations.* The fermionic Fock space for $U_{\mathfrak{q},\mathfrak{d}}(\ddot{\mathfrak{sl}}_{\ell})$ is often simply called the *Fock representation* without ambiguity. We denote its underlying vector space by $\mathcal{F}$. It has a basis $\{|\lambda\rangle\}$ indexed by partitions. We denote by $\langle\lambda|$ the dual element to $|\lambda\rangle$.

Let $\square = (a,b)$ be a node which may or may not be in a partition λ. We will denote by:

(1) $\chi_\square = q^{a-1}t^{b-1}$ the character of the node;
(2) $c_\square$ its content and $\bar{c}_\square$ its content mod ℓ;
(3) $d_i(\lambda)$ the number of i-nodes in λ;
(4) and $A_i(\lambda)$ and $R_i(\lambda)$ the addable and removable i-nodes of λ, respectively.

Finally, we will abbreviate $a \equiv b \bmod \ell$ by simply $a \equiv b$ and use the Kronecker delta function

$$\delta_{a\equiv b} = \begin{cases} 1 & \text{if } a \equiv b \\ 0 & \text{otherwise} \end{cases}$$

In the proposition below, we consider a partition λ and its enlargement $\lambda + \square$ into another partition by a single node $\square$.

PROPOSITION 3.3 ([**15**],[**4**],[**24**]). *Let $v \in \mathbb{F}^\times$. If we identify*

$$q = \mathfrak{q}\mathfrak{d},\ t = \mathfrak{q}\mathfrak{d}^{-1}$$

we can define a $'\ddot{U}$-action τ_v^- on $\mathcal{F}$ where the only nonzero matrix elements of the generators are

$$\langle\lambda|\tau_v^-(e_i(z))|\lambda+\square\rangle = \delta_{c_\square\equiv i}(-\mathfrak{d})^{d_{i+1}(\lambda)}\delta\left(\frac{z}{\chi_\square v}\right)\frac{\prod\limits_{\blacksquare\in R_i(\lambda)}(\chi_\square - \mathfrak{q}^2\chi_\blacksquare)}{\prod\limits_{\substack{\blacksquare\in A_i(\lambda)\\ \blacksquare\neq\square}}(\chi_\square - \chi_\blacksquare)}$$

$$\langle\lambda+\square|\tau_v^-(f_i(z))|\lambda\rangle = \delta_{c_\square\equiv i}(-\mathfrak{d})^{-d_{i+1}(\lambda)}\delta\left(\frac{z}{\chi_\square v}\right)\frac{\prod\limits_{\substack{\blacksquare\in A_i(\lambda)\\ \blacksquare\neq\square}}(\mathfrak{q}\chi_\square - \mathfrak{q}^{-1}\chi_\blacksquare)}{\prod\limits_{\blacksquare\in R_i(\lambda)}\mathfrak{q}(\chi_\square - \chi_\blacksquare)}$$

$$\langle\lambda|\tau_v^-(\psi_i^\pm(z))|\lambda\rangle = \prod_{\blacksquare\in A_i(\lambda)}\frac{(\mathfrak{q}z - \mathfrak{q}^{-1}\chi_\blacksquare v)}{(z-\chi_\blacksquare v)}\prod_{\blacksquare\in R_i(\lambda)}\frac{(\mathfrak{q}^{-1}z - \mathfrak{q}\chi_\blacksquare v)}{(z-\chi_\blacksquare v)},$$

$$\langle\lambda|\tau_v^-(\gamma^{\frac{1}{2}})|\lambda\rangle = 1,\ \langle\lambda|\tau_v^-(\mathfrak{q}^{d_2})|\lambda\rangle = \mathfrak{q}^{-|\lambda|}$$

On the other hand, if we identify

$$q = \mathfrak{q}^{-1}\mathfrak{d},\ t = \mathfrak{q}^{-1}\mathfrak{d}^{-1}$$

then we can define another $'\ddot{U}$*-action* τ_v^+ *on* $\mathcal{F}$ *with matrix elements*

$$\langle\lambda+\square|\tau_v^+(e_i(z))|\lambda\rangle = \delta_{c_\square\equiv i}(-\mathfrak{d})^{-d_{i+1}(\lambda)}\delta\left(\frac{z}{\chi_\square v}\right)\frac{\prod\limits_{\substack{\blacksquare\in A_i(\lambda)\\ \blacksquare\neq\square}}(\chi_\square-\mathfrak{q}^2\chi_\blacksquare)}{\prod\limits_{\blacksquare\in R_i(\lambda)}(\chi_\square-\chi_\blacksquare)}$$

$$\langle\lambda|\tau_v^+(f_i(z))|\lambda+\square\rangle = \delta_{c_\square\equiv i}(-\mathfrak{d})^{d_{i+1}(\lambda)}\delta\left(\frac{z}{\chi_\square v}\right)\frac{\prod\limits_{\blacksquare\in R_i(\lambda)}(\mathfrak{q}\chi_\square-\mathfrak{q}^{-1}\chi_\blacksquare)}{\prod\limits_{\substack{\blacksquare\in A_i(\lambda)\\ \blacksquare\neq\square}}\mathfrak{q}\,(\chi_\square-\chi_\blacksquare)}$$

$$\langle\lambda|\tau_v^+(\psi_i^\pm(z))|\lambda\rangle = \prod_{\blacksquare\in A_i(\lambda)}\frac{(\mathfrak{q}^{-1}z-\mathfrak{q}\chi_\blacksquare v)}{(z-\chi_\blacksquare v)}\prod_{\blacksquare\in R_i(\lambda)}\frac{(\mathfrak{q}z-\mathfrak{q}^{-1}\chi_\blacksquare v)}{(z-\chi_\blacksquare v)},$$

$$\langle\lambda|\tau_v^+(\gamma^{\frac{1}{2}})|\lambda\rangle = 1,\ \langle\lambda|\tau_v^+(\mathfrak{q}^{d_2})|\lambda\rangle = \mathfrak{q}^{|\lambda|}$$

Both representations are irreducible.

We call τ_v^- the *highest weight* Fock representation and τ_v^+ the *lowest weight* Fock representation.

REMARK 3.4. We have presented $\tau_v^\pm$ in a way that hides its fermionic nature. Following ideas from [**15**], one can show that the both representations are isomorphic to representations on the $\mathfrak{q}$-deformed Fock space of [**11**]. Our basis here corresponds to a 'spinon' basis rather than to $\mathfrak{q}$-deformed semiinfinite wedges (cf. [**9**]).

3.2.4. *Tsymbaliuk isomorphisms.* The relationship between the vertex and Fock representations was established by Tsymbaliuk [**23**] (cf. 3.5.2 of [**24**]):

THEOREM 3.5. *The vacuum-to-vacuum map*

$$\mathbb{1}\otimes 1\mapsto|\varnothing\rangle$$

induces an isomorphism $\mathrm{T}_-: W\to\mathcal{F}$ *between the* $'\ddot{U}$*-modules* $\rho_{\vec{\mathfrak{c}}}\circ\varpi$ *and* τ_v^-*, where the parameters* $\vec{c}$ *and* v *are related by*

$$v=(-1)^{\frac{(\ell-1)(\ell-2)}{2}}\frac{\mathfrak{q}\mathfrak{d}^{-\frac{\ell}{2}}}{\mathfrak{c}_0\cdots\mathfrak{c}_{\ell-1}}$$

It also induces an isomorphism $\mathrm{T}_+: W\to\mathcal{F}$ *between the* $'\ddot{U}'$*-modules* $\rho_{\vec{\mathfrak{c}}}\circ\varpi^{-1}$ *and* τ_v^+*, where the parameters are instead related by*

$$v=(-1)^{\frac{(\ell-1)(\ell-2)}{2}}\mathfrak{q}^{-1}\mathfrak{d}^{-\frac{\ell}{2}}\mathfrak{c}_0\cdots\mathfrak{c}_{\ell-1}$$

We note that while both isomorphisms are induced from the vacuum-to-vacuum map, they are propagated to the rest of W using different actions of the quantum toroidal algebra. Our previous result [**24**] adds more detail to T_-. Recall that we have identfied W and $\Lambda_{q,t}^{\otimes\ell}\otimes\mathbb{F}[Q]$ using (3.3), and thus we can view the wreath Macdonald polynomials $H_\lambda\otimes e^{\mathrm{core}(\lambda)}$ as elements of W.

THEOREM 3.6. *The Tsymabliuk isomorphisms* $\mathrm{T}_\pm$ *satisfy the following:*

(1) *Under the identification* (3.3) *and the matching of parameters*

$$q = \mathfrak{q}\mathfrak{d},\ t = \mathfrak{q}\mathfrak{d}^{-1} \tag{3.4}$$

the Tsymbaliuk isomorphism T_- *sends* $\mathbb{F}(H_\lambda \otimes e^{\mathrm{core}(\lambda)})$ *to* $\mathbb{F}|\lambda\rangle$.

(2) *On the other hand, under the identification* (3.3) *and the matching of parameters*

$$q = \mathfrak{q}^{-1}\mathfrak{d},\ t = \mathfrak{q}^{-1}\mathfrak{d}^{-1}$$

the Tsymbaliuk isomorphism T_+ *sends* $\mathbb{F}(H_\lambda[-X^\bullet] \otimes e^{\mathrm{core}(\lambda)})$ *to* $\mathbb{F}|\lambda\rangle$.

Part (1) was proved in [**24**]. The proof of part (2) is similar once we have established Propositions 3.12 and 3.13 below. We have emphasized that the matching of toroidal parameters $(\mathfrak{q}, \mathfrak{d})$ and symmetric function parameters (q, t) differ between the two isomorphisms $\mathrm{T}_\pm$. We will take the approach of adhering to the different parameter matchings when making computations and then reverting to (3.4) when stating final results about wreath Macdonald polynomials. This amounts to performing the swap $q \leftrightarrow t^{-1}$. Doing this makes Theorem 3.6(2) really a statement about H_λ^*. We highlight here that in W,

- the basis $\left\{H_\lambda \otimes e^{\mathrm{core}(\lambda)}\right\}$ diagonalizes $\varpi(\mathcal{H})$;
- the basis $\left\{H_\lambda^* \otimes e^{\mathrm{core}(\lambda)}\right\}$ diagonalizes $\varpi^{-1}(\mathcal{H})$.

3.3. Shuffle algebra. Utilizing the Tsymbaliuk isomorphisms, we can transfer the study of W to that of $\mathcal{F}$ after applying $\varpi^{\pm 1}$. However, studying the images of elements of $U_{\mathfrak{q},\mathfrak{d}}(\ddot{\mathfrak{sl}}_\ell)$ under Miki's automorphism in terms of generators and relations is extremely difficult. To access these images and their actions, we will utilize the *shuffle algebra*.

3.3.1. *Definition.* Let $\vec{k} = (k_0, \ldots, k_{\ell-1}) \in (\mathbb{Z}_{\geq 0})^{\mathbb{Z}/\ell\mathbb{Z}}$. The space of rational functions

$$\mathbb{F}(x_{i,r})_{i\in\mathbb{Z}/\ell\mathbb{Z}, 1\leq r\leq k_i}$$

carries an action of the group

$$\Sigma_{\vec{k}} := \prod_i \Sigma_{k_i}$$

wherein the factor Σ_{k_i} only permutes the variables $\{x_{i,r}\}_{1\leq r\leq k_i}$. For the variable $x_{i,r}$, we call i its *color*, so $\Sigma_{\vec{k}}$ acts via *color-preserving* permuations. We will be concerned with the space of *color-symmetric* rational functions:

$$\mathbb{S}_{\vec{k}} := \mathbb{F}(x_{i,r})^{\Sigma_{\vec{k}}}_{i\in\mathbb{Z}/\ell\mathbb{Z}, 1\leq r\leq k_i}, \qquad \mathbb{S} := \bigoplus_{\vec{k}\in(\mathbb{Z}_{\geq 0})^{\mathbb{Z}/\ell\mathbb{Z}}} \mathbb{S}_{\vec{k}}$$

Finally, we will need the notation

$$|\vec{k}| := \sum_{i\in\mathbb{Z}/\ell\mathbb{Z}} k_i$$

$$\vec{k}! := \prod_{i\in\mathbb{Z}/\ell\mathbb{Z}} k_i!$$

On $\mathbb{S}$, we can define the *shuffle product* $\star$. First, we need the *mixing terms*:

$$\omega_{i,j}(z,w) := \begin{cases} \left(z - \mathfrak{q}^2 w\right)^{-1}(z-w)^{-1} & \text{if } i = j \\ \left(\mathfrak{q}w - \mathfrak{d}^{-1}z\right) & \text{if } i+1=j \\ \left(z - \mathfrak{q}\mathfrak{d}^{-1}w\right) & \text{if } i-1=j \\ 1 & \text{otherwise} \end{cases}$$

For $F \in \mathbb{S}_{\vec{n}}$ and $G \in \mathbb{S}_{\vec{m}}$, $F \star G \in \mathbb{S}_{\vec{n}+\vec{m}}$ is given by

$$\frac{1}{\vec{n}!\vec{m}!}\operatorname{Sym}\left(F(\{x_{i,r}\}_{r\le n_i})G(\{x_{j,s}\}_{n_j<s}) \prod_{i,j\in\mathbb{Z}/\ell\mathbb{Z}} \prod_{\substack{r\le n_i\\ n_j<s}} \omega_{i,j}(x_{i,r}/x_{j,s}) \right)$$

where Sym denotes the *color symmetrization*: for $f \in \mathbb{F}(\{x_{i,1},\dots,x_{i,k_i}\}_{i\in\mathbb{Z}/\ell\mathbb{Z}})$,

$$\operatorname{Sym}(f) := \sum_{(\sigma_0,\dots,\sigma_{\ell-1})\in\prod\Sigma_{\vec{k}}} f(\{x_{i,\sigma_i(r)}\})$$

Let $\mathcal{S}_{\vec{k}} \subset \mathbb{S}_{\vec{k}}$ be the subspace of functions F satisfying:

(1) *Pole conditions:* F is of the form

$$F = \frac{f(\{x_{i,r}\})}{\prod_{i\in\mathbb{Z}/\ell\mathbb{Z}}\prod_{r\neq r'}(x_{i,r} - \mathfrak{q}^2 x_{i,r'})}$$

for a color-symmetric Laurent polynomial f.

(2) *Wheel conditions:* F has a well-defined finite limit when

$$\frac{x_{i,r_1}}{x_{i+\epsilon,s}} \to \mathfrak{q}\mathfrak{d}^{\epsilon} \text{ and } \frac{x_{i+\epsilon,s}}{x_{i,r_2}} \to \mathfrak{q}\mathfrak{d}^{-\epsilon}$$

for any choice of i, r_1, r_2, s, and ϵ, where $\epsilon \in \{\pm 1\}$. This is equivalent to specifying that the Laurent polynomial f in the pole conditions evaluates to zero.

We set

$$\mathcal{S} := \bigoplus_{\vec{k}\in(\mathbb{Z}_{\ge 0})^{\mathbb{Z}/\ell\mathbb{Z}}} \mathcal{S}_{\vec{k}}$$

PROPOSITION 3.7. *The product $\star$ is associative and $\mathcal{S}$ is closed under it.*

$\mathcal{S}$ is called the *shuffle algebra of type $\hat{A}_{\ell-1}$*. Let $\mathcal{S}^+ := \mathcal{S}$ and $\mathcal{S}^- := \mathcal{S}^{\mathrm{opp}}$.

3.3.2. *Relation to $U_{\mathfrak{q},\mathfrak{d}}(\ddot{\mathfrak{sl}}_\ell)$.* Let $\ddot{U}^+, \ddot{U}^- \subset U_{\mathfrak{q},\mathfrak{d}}(\ddot{\mathfrak{sl}}_\ell)$ be the subalgebras generated by

$$\{e_{i,n}\}_{i\in\mathbb{Z}/\ell\mathbb{Z}}^{n\in\mathbb{Z}} \text{ and } \{f_{i,n}\}_{i\in\mathbb{Z}/\ell\mathbb{Z}}^{n\in\mathbb{Z}}$$

respectively. The following fundamental result was proved by Neguț:

THEOREM 3.8 ([**18**]). *There exist algebra isomorphisms $\Psi_+ : \mathcal{S}^+ \to \ddot{U}^+$ and $\Psi_- : \mathcal{S}^- \to \ddot{U}^-$ such that*

$$\Psi_+(x_{i,1}^k) = e_{i,k}$$
$$\Psi_-(x_{i,1}^k) = f_{i,k}$$

Therefore, we can use $\mathcal{S}$ to give alternative presentations of elements of $\ddot{U}^{\pm}$. The utility of this is bolstered by the following:

PROPOSITION 3.9 ([**17**], cf. Proposition 4.8 of [**24**]). *For $F \in \mathcal{S}^+_{\vec{k}}$ and $G \in \mathcal{S}^-_{\vec{k}}$, the actions of $\Psi_+(F)$ and $\Psi_-(G)$ on $\mathcal{F}$ via $\tau_\upsilon^{\pm}$ are such that the only nonzero matrix elements involve pairs of partitions μ and λ where μ adds k_i i-nodes to λ for all $i \in \mathbb{Z}/\ell\mathbb{Z}$. In this case, assign to each node $\square \in \mu\backslash\lambda$ its own variable $x_\square$ of color $\bar{c}_\square$. We then have*

$$\langle\mu|(\tau_\upsilon^+ \circ \Psi_+)(F)|\lambda\rangle = \left(F(\{x_{i,r}\})\Big|_{x_\square \to \chi_\square \upsilon}\right) \times \prod_{\square\in\mu\backslash\lambda} \frac{\left(\chi_\square - \mathfrak{q}^2\right)^{\delta_{c_\square \equiv 0}}}{\chi_\square(1-\mathfrak{q}^2)} \prod_{\blacksquare\in\lambda} \omega_{\bar{c}_\square,\bar{c}_\blacksquare}(\chi_\square,\chi_\blacksquare) \tag{3.5}$$

$$\langle\mu|(\tau_\upsilon^- \circ \Psi_-)(G)|\lambda\rangle = \left(G(\{x_{i,r}\})\Big|_{x_\square \to \chi_\square \upsilon}\right) \times \prod_{\square\in\mu\backslash\lambda} \frac{\left(\mathfrak{q}\chi_\square - \mathfrak{q}^{-1}\right)^{\delta_{c_\square \equiv 0}}}{\chi_\square(\mathfrak{q}-\mathfrak{q}^{-1})} \prod_{\blacksquare\in\lambda} \omega_{\bar{c}_\blacksquare,\bar{c}_\square}(\chi_\blacksquare,\chi_\square) \tag{3.6}$$

Formula (3.6) was proven in detail in [**24**]. The proof of (3.5) follows the same scheme, although we emphasize that the matching of parameters $(\mathfrak{q},\mathfrak{d})$ with (q,t) differ in the two Fock representations.

REMARK 3.10. Note that the formulas (3.5) and (3.6) are only for box *additions*. Analogous formulas for box removals exist—the case of τ_υ^- was worked out in [**24**].

3.3.3. *Pieri operators.* Wreath analogues of Pieri and dual Pieri rules for Macdonald polynomials would entail writing the products

$$e_n\left[X^{(i)}\right]P_\lambda \text{ and } h_n\left[\frac{(1-t\sigma^{-1})}{(1-q\sigma^{-1})}X^{(i)}\right]P_\lambda$$

in terms of the ordinary (wreath) Macdonald basis $\{P_\mu\}$. Our chosen transformation for h_n is the wreath analogue of the transformation sending h_n to the g_n of [**12**]. Inverting t and performing a plethysm to obtain scalar multiples of $\{H_\mu\}$, we can equivalently study

$$e_n\left[(1-t^{-1}\sigma^{-1})^{-1}X^{(i)}\right]\tilde{P}_\lambda(q,t^{-1})$$
$$\text{and } h_n\left[(1-q\sigma^{-1})^{-1}X^{(i)}\right]\tilde{P}_\lambda(q,t^{-1})$$

If we match parameters using

$$q = \mathfrak{q}\mathfrak{d},\ t = \mathfrak{q}\mathfrak{d}^{-1}$$

then by Theorem 3.6, we can use T_- to carry this study over to $\mathcal{F}$. Using the identification (3.3), we would be interested in studying the action on $\mathcal{F}$ of

$$(\tau_\upsilon^- \circ \varpi^{-1})\left(e_n\left[(1-t^{-1}\sigma^{-1})^{-1}X^{(i)}\right]\right)$$
$$\text{and } (\tau_\upsilon^- \circ \varpi^{-1})\left(h_n\left[(1-q\sigma^{-1})^{-1}X^{(i)}\right]\right)$$

We can then use (3.6) provided that these elements (without τ_v^-) are in the image of Ψ_- and we have formulas for their preimages.

To that end, let

$$E_{p,n} := \mathrm{Sym}\left(\prod_{1\le r<s\le n} \left\{ \frac{x_{p+1,r} - \mathfrak{q}^{-1}\mathfrak{d}^{-1}x_{p,s}}{x_{p+1,r} - \mathfrak{q}\mathfrak{d}^{-1}x_{p,s}} \prod_{i,j\in\mathbb{Z}/\ell\mathbb{Z}} \omega_{i,j}\left(x_{i,r}, x_{j,s}\right) \right\} \times \prod_{r=1}^{n} \left\{ \left(\mathfrak{q}^{-1}\mathfrak{d}^{-1}\frac{x_{0,r}}{x_{p+1,r}} - \frac{x_{0,r}}{x_{p,r}} \right) \prod_{i\in\mathbb{Z}/\ell\mathbb{Z}} x_{i,r} \right\} \right) \tag{3.7}$$

$$H_{p,n} := \mathrm{Sym}\left(\prod_{1\le r<s\le n} \left\{ \frac{\mathfrak{q}^{-1}\mathfrak{d}x_{p+1,s} - x_{p,r}}{\mathfrak{q}\mathfrak{d}x_{p+1,s} - x_{p,r}} \prod_{i,j\in\mathbb{Z}/\ell\mathbb{Z}} \omega_{i,j}\left(x_{i,r}, x_{j,s}\right) \right\} \times \prod_{r=1}^{n} \left\{ \left(\mathfrak{q}\mathfrak{d}^{-1}\frac{x_{0,r}}{x_{p+1,r}} - \frac{x_{0,r}}{x_{p,r}} \right) \prod_{i\in\mathbb{Z}/\ell\mathbb{Z}} x_{i,r} \right\} \right) \tag{3.8}$$

The following is Theorem 4.31 of [**24**]:

PROPOSITION 3.11. *The functions $E_{p,n}$ and $H_{p,n}$ lie in $\mathcal{S}$. Under the matching of parameters*

$$q = \mathfrak{q}\mathfrak{d},\ t = \mathfrak{q}\mathfrak{d}^{-1}$$

we have

$$\varpi^{-1}e_n\left[(1-t^{-1}\sigma^{-1})^{-1}X^{(p)}\right] = \frac{(1-qt)^{n\ell}}{\prod_{r=1}^{n}(1-(qt)^{-r})}\Psi_-(E_{p,n})$$
$$\varpi^{-1}h_n\left[(1-q\sigma^{-1})^{-1}X^{(p)}\right] = \frac{(qt)^{-n}(1-qt)^{n\ell}}{\prod_{r=1}^{n}(1-(qt)^{-r})}\Psi_-(H_{p,n})$$

Alternatively, we can use the matching

$$q = \mathfrak{q}^{-1}\mathfrak{d},\ t = \mathfrak{q}^{-1}\mathfrak{d}^{-1}$$

and use Theorem 3.6(2). To carry this out, we would use plethysms to transfer the Pieri rules over to

$$e_n\left[(1-t^{-1}\sigma^{-1})^{-1}(-X^{(i)})\right]\tilde{P}_\lambda\left[-X^\bullet; q, t^{-1}\right]$$
$$\text{and } h_n\left[(1-q\sigma^{-1})^{-1}(-X^{(i)})\right]\tilde{P}_\lambda\left[-X^\bullet; q, t^{-1}\right]$$

We would then be interested in the action on $\mathcal{F}$ of

$$(\tau_v^+\circ\varpi)\left(e_n\left[(1-t^{-1}\sigma^{-1})^{-1}(-X^{(i)})\right]\right)$$
$$\text{and } (\tau_v^+\circ\varpi)\left(h_n\left[(1-q\sigma^{-1})^{-1}(-X^{(i)})\right]\right)$$

PROPOSITION 3.12. *Under the matching of parameters*

$$q = \mathfrak{q}^{-1}\mathfrak{d},\, t = \mathfrak{q}^{-1}\mathfrak{d}^{-1}$$

we have

$$\varpi e_n\left[(1-t^{-1}\sigma^{-1})^{-1}(-X^{(p)})\right] = \frac{(-1)^n(qt)^n(\mathfrak{q}-\mathfrak{q}^{-1})^{n\ell}}{\prod_{r=1}^n(1-(qt)^r)}\Psi_+(H_{p,n}) \tag{3.9}$$

$$\varpi h_n\left[(1-q\sigma^{-1})^{-1}(-X^{(p)})\right] = \frac{(-1)^n(\mathfrak{q}-\mathfrak{q}^{-1})^{n\ell}}{\prod_{r=1}^n(1-(qt)^r)}\Psi_+(E_{p,n}) \tag{3.10}$$

PROOF. First consider (3.9). Our goal is to utilize Proposition B.1 of [**24**]. Although not explicitly mentioned in the proposition, we are using (3.3):

$$\sum_{n=0}^\infty \varpi e_n\left[(1-t^{-1}\sigma^{-1})^{-1}(-X^{(p)})\right](-z)^n = \varpi\exp\left(\sum_{k=1}^\infty \frac{\sum_{i=0}^{\ell-1} t^{-ki} b_{p+i,-k}}{[k]_{\mathfrak{q}}(1-t^{-k\ell})} z^k\right) \tag{3.11}$$

By Theorem 3.1, we have $\varpi = \eta\varpi^{-1}\eta$. Applying this to (3.11) yields

$$= \eta\exp\left(-\sum_{k=1}^\infty \frac{\sum_{i=0}^{\ell-1} q^{-ki}\varpi^{-1}(b_{p+i,k})}{[k]_{\mathfrak{q}}(1-q^{-k\ell})} z^k\right)$$

Through computations similar to those of Lemma 3.15 of [**24**], we can see that

$$\frac{-\sum_{i=0}^{\ell-1} q^{-ki}\varpi^{-1}(b_{p+i,k})}{[k]_{\mathfrak{q}}(1-q^{-k\ell})} = \frac{\mathfrak{q}^{-k}\varpi^{-1}(b_{p,k}^\perp) - \mathfrak{d}^k\varpi^{-1}(b_{p+1,k}^\perp)}{\mathfrak{q}-\mathfrak{q}^{-1}}$$

Here, $b_{i,k}^\perp$ is defined in 3.1.3 of *loc. cit.* in terms of a pairing. Note that Proposition B.1 of *loc. cit.* is written in terms of the other matching of parameters

$$q = \mathfrak{q}\mathfrak{d},\, t = \mathfrak{q}\mathfrak{d}^{-1}$$

Applying it correctly here gives us

$$\begin{aligned} &\sum_{n=0}^\infty \varpi e_n\left[(1-t^{-1}\sigma^{-1})^{-1}(-X^{(p)})\right] z^n \\ &= \eta\left(\sum_{n=0}^\infty \frac{(-1)^{n\ell}q^{-n\ell}\left(1-\mathfrak{q}^{-2}\right)^{n\ell}}{\mathfrak{q}^{2n}\prod_{r=1}^n(1-(qt)^r)}\Psi_+(E_{p,n}^\dagger)\right) \end{aligned} \tag{3.12}$$

where

$$E_{p,n}^\dagger = \mathrm{Sym}\left(\prod_{1\le r<s\le n}\left\{\frac{x_{p+1,r}-\mathfrak{q}^{-1}\mathfrak{d}^{-1}x_{p,s}}{x_{p+1,r}-\mathfrak{q}\mathfrak{d}^{-1}x_{p,s}}\prod_{i,j\in\mathbb{Z}/\ell\mathbb{Z}}\omega_{i,j}\left(x_{i,r},x_{j,s}\right)\right\}\right.$$
$$\left.\times\prod_{r=1}^n\left\{\left(\frac{x_{p,r}}{x_{0,r}}-\mathfrak{q}\mathfrak{d}\frac{x_{p+1,r}}{x_{0,r}}\right)\prod_{i\in\mathbb{Z}/\ell\mathbb{Z}}x_{i,r}\right\}\right)$$

Finally, we observe that η induces the following map on $\mathcal{S}^+$:

$$\Psi_+^{-1}\eta\Psi_+(F) = F(x_{i,r}^{-1})\prod_{i\in\mathbb{Z}\ell\mathbb{Z}}\prod_{r=1}^{k_i}(-\mathfrak{d})^{k_{i+1}k_i}x_{i,r}^{k_{i+1}+k_{i-1}-2(k_i-1)}\Bigg|_{\mathfrak{d}\mapsto\mathfrak{d}^{-1}}$$

Using this to deal with η, the right hand side of (3.12) becomes

$$= \sum_{n=0}^{\infty} \frac{t^{-n\ell}(1-\mathfrak{q}^{-2})^{n\ell}}{\mathfrak{q}^{2n}\prod_{r=1}^{n}(1-(qt)^r)}(-1)^n \mathfrak{d}^{-n\ell}\Psi_+(H_{p,n})$$

from which (3.9) follows. The proof of (3.10) is similar. □

We end by applying Proposition 3.9 to $E_{p,n}$ and $H_{p,n}$. The zeroes of $E_{p,n}$ and $H_{p,n}$ will force certain matrix elements to vanish. Specifically, we obtain:

PROPOSITION 3.13. *The matrix elements*

$$\langle\mu|(\tau_v^- \circ \Psi_-)(E_{p,n})|\lambda\rangle \text{ and } \langle\mu|(\tau_v^+ \circ \Psi_+)(H_{p,n})|\lambda\rangle$$

are nonzero only if $\mu\backslash\lambda$ *has exactly* n *nodes of each color and no horizontally adjacent* p*- and* $(p+1)$*-nodes. Similarly,*

$$\langle\mu|(\tau_v^- \circ \Psi_-)(H_{p,n})|\lambda\rangle \text{ and } \langle\mu|(\tau_v^+ \circ \Psi_+)(E_{p,n})|\lambda\rangle$$

are nonzero only if $\mu\backslash\lambda$ *has exactly* n *nodes of each color and no vertically adjacent* p*- and* $(p+1)$*-nodes.*

4. Normalization

Here, we highlight a normalization quantity that directly yields the inner products $\langle P^*_{\iota\lambda}, P_\lambda\rangle_{q,t}$. Computing this normalization is also key for applying the shuffle algebra to studying Pieri rules.

4.1. Setup. When

$$q = \mathfrak{q}^{-1}\mathfrak{d},\ t = \mathfrak{q}^{-1}\mathfrak{d}^{-1}$$

we define the scalars $N^+_\lambda(q,t)$ and $M^+_\lambda(q,t)$ by

$$(\tau_v^+ \circ \varpi)\left(\tilde{P}_\lambda[-X^\bullet; q, t^{-1}]\right)|\mathrm{core}(\lambda)\rangle = N^+_\lambda(q,t^{-1})|\lambda\rangle$$

$$(\tau_v^+ \circ \varpi)\left(\tilde{Q}_\lambda[-X^\bullet; q, t^{-1}]\right)|\mathrm{core}(\lambda)\rangle = M^+_\lambda(q,t^{-1})|\lambda\rangle$$

Similarly, with

$$q = \mathfrak{q}\mathfrak{d},\ t = \mathfrak{q}\mathfrak{d}^{-1}$$

we define $N^-_\lambda(q,t)$ and $M^-_\lambda(q,t)$ as

$$(\tau_v^- \circ \varpi^{-1})\left(\tilde{P}_\lambda(q, t^{-1})\right)|\mathrm{core}(\lambda)\rangle = N^-_\lambda(q,t^{-1})|\lambda\rangle$$

$$(\tau_v^- \circ \varpi^{-1})\left(\tilde{Q}_\lambda(q, t^{-1})\right)|\mathrm{core}(\lambda)\rangle = M^-_\lambda(q,t^{-1})|\lambda\rangle$$

Our interest in these scalars comes from the following trivial observation:

PROPOSITION 4.1. *The norm formulas can be computed as follows:*

$$\langle P^*_{\iota\lambda}, P_\lambda\rangle_{q,t} = \frac{N^+_{\iota\lambda}(t,q)}{M^+_{\iota\lambda}(t,q)} = \frac{N^-_\lambda(q,t)}{M^-_\lambda(q,t)}$$

$$\langle Q^*_{\iota\lambda}, Q_\lambda\rangle_{q,t} = \frac{M^+_{\iota\lambda}(t,q)}{N^+_{\iota\lambda}(t,q)} = \frac{M^-_\lambda(q,t)}{N^-_\lambda(q,t)}$$

PROOF. The key observation is:

$$P_\lambda = \frac{N_\lambda^-(q,t)}{M_\lambda^-(q,t)} Q_\lambda$$

This yields:

$$\langle P_{\iota\lambda}^*, P_\lambda\rangle_{q,t} = \frac{N_\lambda^-(q,t)}{M_\lambda^-(q,t)}\langle P_{\iota\lambda}^*, Q_\lambda\rangle_{q,t} = \frac{N_\lambda^-(q,t)}{M_\lambda^-(q,t)} \cdot 1$$

The proof of the other equalities is similar. □

4.2. Pieri triangularity. Recall that in the case $\ell = 1$, the $\{s_\lambda\}$, $\{e_\lambda\}$ and $\{h_\lambda\}$ satisfy the following triangularity conditions with respect to dominance order (cf. [**12**] I.6):

$$\begin{aligned} s_\lambda &= e_\lambda + \sum_{\mu<\lambda} \alpha_{\lambda\mu} e_\mu \\ &= h_\lambda + \sum_{\mu>\lambda} \alpha'_{\lambda\mu} h_\mu \end{aligned}$$

for some $\alpha_{\lambda\mu}, \alpha'_{\lambda\mu} \in \mathbb{Z}$ (recall our convention (2.1) for e_λ). It is not difficult to see that for general ℓ, this implies (cf. [**24**] Remark 2.7(3)):

$$\begin{aligned} s_{\mathrm{quot}(\lambda)} &= e_{\mathrm{quot}(\lambda)} + \sum_{\mu<_\ell\lambda} \beta_{\lambda\mu} e_{\mathrm{quot}(\mu)} \\ &= h_{\mathrm{quot}(\lambda)} + \sum_{\mu>_\ell\lambda} \beta'_{\lambda\mu} h_{\mathrm{quot}(\mu)} \end{aligned}$$

for some $\beta_{\lambda\mu}, \beta'_{\lambda\mu} \in \mathbb{Z}$. Consequently,

$$\begin{aligned} \tilde{P}_\lambda(q,t^{-1}) &= e_{\mathrm{quot}(\lambda)}\left[(1-t^{-1}\sigma^{-1})X^\bullet\right] + \sum_{\mu<_\ell\lambda} \gamma_{\lambda\mu}(q,t) e_{\mathrm{quot}(\mu)}\left[(1-t^{-1}\sigma^{-1})X^\bullet\right] \\ \tilde{Q}_\lambda(q,t^{-1}) &= h_{\mathrm{quot}(\lambda)}\left[(1-q\sigma^{-1})X^\bullet\right] + \sum_{\mu>_\ell\lambda} \gamma'_{\lambda\mu}(q,t) h_{\mathrm{quot}(\mu)}\left[(1-q\sigma^{-1})X^\bullet\right] \end{aligned}$$

for some $\gamma_{\lambda\mu}(q,t), \gamma'_{\lambda\mu}(q,t) \in \mathbb{C}(q,t)$. Since the inverse of a unitriangular matrix is also unitriangular, we obtain

$$\begin{aligned} e_{\mathrm{quot}(\lambda)}\left[(1-t^{-1}\sigma^{-1})X^\bullet\right] &= \tilde{P}_\lambda(q,t^{-1}) + \sum_{\mu<_\ell\lambda} \delta_{\lambda\mu}(q,t)\tilde{P}_\mu(q,t^{-1}) \\ h_{\mathrm{quot}(\lambda)}\left[(1-q\sigma^{-1})X^\bullet\right] &= \tilde{Q}_\lambda(q,t^{-1}) + \sum_{\mu>_\ell\lambda} \delta'_{\lambda\mu}(q,t)\tilde{Q}_\mu(q,t^{-1}) \end{aligned} \tag{4.1}$$

for some $\delta_{\lambda\mu}(q,t), \delta'_{\lambda\mu}(q,t) \in \mathbb{C}(q,t)$. Thus,

$$
\begin{aligned}
N^+_\lambda(q,t^{-1}) &= \left\langle \lambda \left| (\tau^+_v \circ \varpi) \left(\tilde{P}_\lambda \left[-X^\bullet; q, t^{-1} \right] \right) \right| \mathrm{core}(\lambda) \right\rangle \\
&= \left\langle \lambda \left| (\tau^+_v \circ \varpi) \left(e_{\mathrm{quot}(\lambda)} \left[(1 - t^{-1}\sigma^{-1})^{-1} (-X^\bullet) \right] \right) \right| \mathrm{core}(\lambda) \right\rangle \\
M^+_\lambda(q,t^{-1}) &= \left\langle \lambda \left| (\tau^+_v \circ \varpi) \left(\tilde{Q}_\lambda \left[-X^\bullet; q, t^{-1} \right] \right) \right| \mathrm{core}(\lambda) \right\rangle \\
&= \left\langle \lambda \left| (\tau^+_v \circ \varpi) \left(h_{\mathrm{quot}(\lambda)} \left[(1 - q\sigma^{-1})^{-1} (-X^\bullet) \right] \right) \right| \mathrm{core}(\lambda) \right\rangle \\
N^-_\lambda(q,t^{-1}) &= \left\langle \lambda \left| (\tau^-_v \circ \varpi^{-1}) \left(\tilde{P}_\lambda(q,t^{-1}) \right) \right| \mathrm{core}(\lambda) \right\rangle \\
&= \left\langle \lambda \left| (\tau^-_v \circ \varpi^{-1}) \left(e_{\mathrm{quot}(\lambda)} \left[(1 - t^{-1}\sigma^{-1})^{-1} X^\bullet \right] \right) \right| \mathrm{core}(\lambda) \right\rangle \\
M^-_\lambda(q,t^{-1}) &= \left\langle \lambda \left| (\tau^-_v \circ \varpi^{-1}) \left(\tilde{Q}_\lambda(q,t^{-1}) \right) \right| \mathrm{core}(\lambda) \right\rangle \\
&= \left\langle \lambda \left| (\tau^-_v \circ \varpi^{-1}) \left(h_{\mathrm{quot}(\lambda)} \left[(1 - q\sigma^{-1})^{-1} X^\bullet \right] \right) \right| \mathrm{core}(\lambda) \right\rangle
\end{aligned}
\tag{4.2}
$$

The equations (4.2) give a pathway to using our shuffle elements $E_{p,n}$ and $H_{p,n}$ to compute $N^\pm_\lambda(q,t)$ and $M^\pm_\lambda(q,t)$. We would like to compute these matrix elements by an inductive procedure along the rows or columns of $\mathrm{quot}(\lambda)$. The combinatorics of doing so is subtle, and this is where the orders reviewed in 2.2.3 come in.

LEMMA 4.2. *For a partition λ, suppose the final column in the left-to-right order on columns of* $\mathrm{quot}(\lambda)$ *has length n and lies in component p. Let $\lambda' \subset \lambda$ be the partition obtained by removing this column. Then*

$$
\begin{aligned}
N^+_\lambda(q,t^{-1}) &= N^+_{\lambda'}(q,t^{-1}) \left\langle \lambda \left| (\tau^+_v \circ \varpi) \left(e_n \left[(1 - t^{-1}\sigma^{-1})^{-1} (-X^{(p)}) \right] \right) \right| \lambda' \right\rangle \\
N^-_\lambda(q,t^{-1}) &= N^-_{\lambda'}(q,t^{-1}) \left\langle \lambda \left| (\tau^-_v \circ \varpi^{-1}) \left(e_n \left[(1 - t^{-1}\sigma^{-1})^{-1} X^{(p)} \right] \right) \right| \lambda' \right\rangle
\end{aligned}
$$

On the other hand, suppose the final row in the right-to-left order on rows of $\mathrm{quot}(\lambda)$ *has length n^* and lies in component p^*. Letting $\lambda'' \subset \lambda$ be the partition obtained by removing this row, we have*

$$
\begin{aligned}
M^+_\lambda(q,t^{-1}) &= M^+_{\lambda''}(q,t^{-1}) \left\langle \lambda \left| (\tau^+_v \circ \varpi) \left(h_{n^*} \left[(1 - q\sigma^{-1})^{-1} (-X^{(p^*)}) \right] \right) \right| \lambda'' \right\rangle \\
M^-_\lambda(q,t^{-1}) &= M^-_{\lambda''}(q,t^{-1}) \left\langle \lambda \left| (\tau^-_v \circ \varpi^{-1}) \left(h_{n^*} \left[(1 - q\sigma^{-1})^{-1} X^{(p^*)} \right] \right) \right| \lambda'' \right\rangle
\end{aligned}
$$

PROOF. We will prove in detail the case of $N^-_\lambda(q,t^{-1})$. First note that we have $e_{\mathrm{quot}(\lambda)} = e_n[X^{(p)}] e_{\mathrm{quot}(\lambda')}$. Using (4.1) for λ', we have

$$
\begin{aligned}
&(\tau^-_v \circ \varpi^{-1}) \left(e_{\mathrm{quot}(\lambda)} \left[(1 - t^{-1}\sigma^{-1})^{-1} X^\bullet \right] \right) |\mathrm{core}(\lambda)\rangle \\
&= (\tau^-_v \circ \varpi^{-1}) \left(e_n \left[(1 - t^{-1}\sigma^{-1})^{-1} X^{(p)} \right] \right) \\
&\quad \times \left\{ N^-_{\lambda'}(q,t) |\lambda'\rangle + \sum_{\mu' <_\ell \lambda'} \gamma_{\lambda'\mu'}(q,t) N^-_{\mu'}(q,t) |\mu'\rangle \right\}
\end{aligned}
\tag{4.3}
$$

By Proposition 3.11, we can study (4.3) using the shuffle element $E_{p,n}$. Proposition 3.13 states that to a Fock basis element, the action of $E_{p,n}$ adds n boxes of each color such that no p- and $(p+1)$-colored boxes are horizontally adjacent. The following technical result is part of Lemma 5.2 of [**24**]:

LEMMA. *Suppose that λ' is obtained from λ by removing the last column of* quot(λ) *in accordance with the left-to-right order of columns. Let that column be in the p-th component with length n and let $\mu' \leq_\ell \lambda'$. If we obtain μ by adding n i-nodes for each $i \in \mathbb{Z}/\ell\mathbb{Z}$ to μ' in a manner where no added p- and $(p+1)$-nodes are horizontally adjacent, then $\mu = \lambda$ only if $\mu' = \lambda'$.*

Thus, the only contribution to $|\lambda\rangle$ in (4.3) comes from the summand $|\lambda'\rangle$. The desired equation follows. For $M_\lambda^\pm(q,t^{-1})$, we use Lemma 5.3 of [**24**]. □

4.3. Conclusion. Combining Lemma 4.2 with Propositions 3.11 and 3.12, the computation of $N_\lambda^\pm(q,t^{-1})$ reduces to computing

$$\frac{(-1)^n(qt)^n(\mathfrak{q}-\mathfrak{q}^{-1})^{n\ell}}{\prod_{r=1}^n(1-(qt)^r)}\langle\lambda|(\tau_v^+\circ\Psi_+)(H_{p,n})|\lambda'\rangle$$
$$\text{and}\quad\frac{(1-qt)^{n\ell}}{\prod_{r=1}^n(1-(qt)^{-r})}\langle\lambda|(\tau_v^-\circ\Psi_-)(E_{p,n})|\lambda'\rangle$$

Here, $\lambda' \subset \lambda$ is obtained by removing the final column in the left-to-right order of columns of quot(λ) and said column is length n in coordinate p. Similarly, the computation of $M_\lambda^\pm(q,t^{-1})$ reduces to computing

$$\frac{(-1)^n(\mathfrak{q}-\mathfrak{q}^{-1})^{n\ell}}{\prod_{r=1}^n(1-(qt)^r)}\langle\lambda|(\tau_v^+\circ\Psi_+)(E_{p^*,n^*})|\lambda''\rangle$$
$$\text{and}\quad\frac{(qt)^{-n}(1-qt)^{n\ell}}{\prod_{r=1}^n(1-(qt)^{-r})}\langle\lambda|(\tau_v^-\circ\Psi_-)(H_{p^*,n^*})|\lambda''\rangle$$

where $\lambda'' \subset \lambda$ is obtained by removing the final row in the right-to-left order of rows of quot(λ) and said row is length n^* in coordinate p^*. These formulas are nice in that they are inductive, but computing them using Proposition 3.9 is still a formidable task due to the symmetrization in the formulas for $H_{p,n}$ and $E_{p,n}$. Moreover, one needs to compute these normalization factors before using Proposition 3.9 to compute Pieri rules.

We end by computing an extended example. Let $\ell = 3$ and consider $\lambda = (4,3,1)$, as depicted in Figure 4. We then have core$(\lambda) = (2)$ and quot$(\lambda) = (\varnothing,\varnothing,(2))$. For $N_\lambda^-(q,t^{-1})$, we will need two applications of $(\tau_v^-\circ\Psi_-)(E_{2,1})$, whereas $M_\lambda^-(q,t^{-1})$ will only need one application of $(\tau_v^-\circ\Psi_-)(H_{2,2})$. In the first application of $E_{2,1}$, we will add the boxes with characters t, t^2, qt to yield

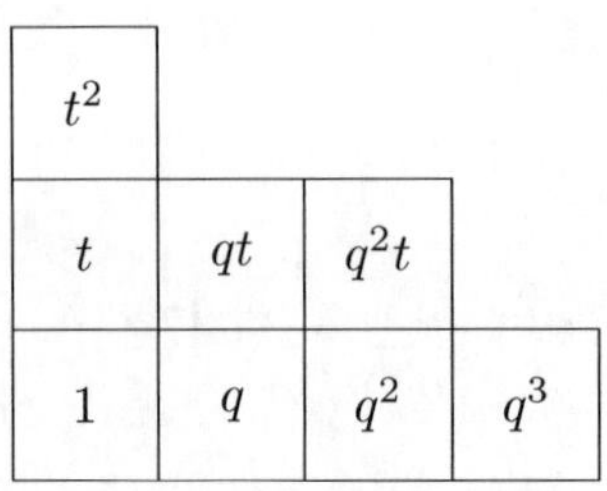

FIGURE 4. The partition $\lambda = (4,3,1)$ with the characters of its boxes.

$\lambda' = (2,2,1)$. Altogether, we have

$$N^-_{\lambda'}(q,t^{-1}) = \frac{(1-qt)^3}{(1-(qt)^{-1})}\langle \lambda' | (\tau_v^- \circ \Psi_-)(E_{2,1}) | \operatorname{core}(\lambda)\rangle = -(q^{-1} - qt^{-1})qt^3$$

It is worth noting that

$$M^-_{\lambda'}(q,t^{-1}) = \frac{(1-qt)^3}{qt(1-(qt)^{-1})}\langle \lambda' | (\tau_v^- \circ \Psi_-)(H_{2,1}) | \operatorname{core}(\lambda)\rangle = -(q^{-1} - t^{-2})qt^3$$

and thus

$$\langle P^*_{t\lambda'}, P_{\lambda'}\rangle_{q,t} = \frac{N^-_{\lambda'}(q,t)}{M^-_{\lambda'}(q,t)} = \frac{1-q^2t}{1-qt^2}$$

which confirms Conjecture 2.13 for $\lambda' = (2,2,1)$.

Moving onwards, we perform the second application of $E_{2,1}$, which appends the boxes with characters q^2, q^2t, q^3:

$$\begin{aligned} N^-_{\lambda}(q,t^{-1}) &= N^-_{\lambda'}(q,t^{-1})\frac{(1-qt)^3}{(1-(qt)^{-1})}\langle \lambda | (\tau_v^- \circ \Psi_-)(E_{2,1}) | \lambda'\rangle \\ &= -\mathfrak{d}^{-5}t^4(1-q^4t^{-2})(1-q^2t^{-1}) \end{aligned}$$

Finally, we compute $M^-_{\lambda}(q,t^{-1})$. Due to the mixing terms in the formula (3.8) for $H_{2,2}$, all but two summands of the symmetrization will vanish upon evaluation.[2] We obtain:

$$\begin{aligned} &M^-_{\lambda}(q,t^{-1}) \\ &= \frac{(1-qt)^6}{q^2t^2\,(1-(qt)^{-1})\,(1-(qt)^{-2})}\langle \lambda | (\tau_v \circ \Psi_-)(H_{2,2}) | \operatorname{core}(\lambda)\rangle \\ &= \mathfrak{q}^6(-\mathfrak{d})^{-7}t^2\frac{(1-q^3)(1-qt^{-2})^2(1-q^3t^{-3})(1-qt)}{(1-q^2t^2)(1-q^2t^{-1})}\left(\frac{1}{1-q^3} + \frac{1}{qt(1-qt^{-2})}\right) \\ &= \mathfrak{q}^6(-\mathfrak{d})^{-7}t^2\frac{(1-q^3)(1-qt^{-2})^2(1-q^3t^{-3})(1-qt)}{(1-q^2t^2)(1-q^2t^{-1})}\cdot\frac{1-q^3+qt-q^2t^{-1}}{qt(1-q^3)(1-qt^{-2})} \\ &= \mathfrak{q}^4(-\mathfrak{d})^{-7}t^2\frac{(1-qt^{-2})(1-q^3t^{-3})(1-qt)}{(1-q^2t^2)(1-q^2t^{-1})}\cdot(1+qt)(1-q^2t^{-1}) \\ &= -\mathfrak{d}^{-5}t^4(1-qt^{-2})(1-q^3t^{-3}) \end{aligned}$$

Therefore, we have

$$\langle P^*_{t\lambda}, P_{\lambda}\rangle_{q,t} = \frac{N^-_{\lambda}(q,t)}{M^-_{\lambda}(q,t)} = \frac{(1-q^4t^2)(1-q^2t)}{(1-q^3t^3)(1-qt^2)}$$

which again confirms Conjecture 2.13.

[2]More summands can appear for general partitions, even for degree two Pieri operators. For degree n, there will be at least $n!$ summands.

Acknowledgments

I would like to thank Daniel Orr and Mark Shimozono for their collaboration—this paper would not see the light of day if not for their interest.

References

[1] Roman Bezrukavnikov and Michael Finkelberg, *Wreath Macdonald polynomials and the categorical McKay correspondence*, Camb. J. Math. **2** (2014), no. 2, 163–190, DOI 10.4310/CJM.2014.v2.n2.a1. With an appendix by Vadim Vologodsky. MR3295916

[2] Erik Carlsson, Nikita Nekrasov, and Andrei Okounkov, *Five dimensional gauge theories and vertex operators* (English, with English and Russian summaries), Mosc. Math. J. **14** (2014), no. 1, 39–61, 170, DOI 10.17323/1609-4514-2014-14-1-39-61. MR3221946

[3] B. Feigin, K. Hashizume, A. Hoshino, J. Shiraishi, and S. Yanagida, *A commutative algebra on degenerate* $\mathbb{CP}^1$ *and Macdonald polynomials*, J. Math. Phys. **50** (2009), no. 9, 095215, 42, DOI 10.1063/1.3192773. MR2566895

[4] B. Feigin, M. Jimbo, T. Miwa, and E. Mukhin, *Representations of quantum toroidal* $\mathfrak{gl}_n$, J. Algebra **380** (2013), 78–108, DOI 10.1016/j.jalgebra.2012.12.029. MR3023228

[5] B. L. Feigin and A. I. Tsymbaliuk, *Equivariant K-theory of Hilbert schemes via shuffle algebra*, Kyoto J. Math. **51** (2011), no. 4, 831–854, DOI 10.1215/21562261-1424875. MR2854154

[6] A. M. Garsia, M. Haiman, and G. Tesler, *Explicit plethystic formulas for Macdonald* q,t*-Kostka coefficients*, Sém. Lothar. Combin. **42** (1999), Art. B42m, 45. The Andrews Festschrift (Maratea, 1998). MR1701592

[7] Mark Haiman, *Combinatorics, symmetric functions, and Hilbert schemes*, Current developments in mathematics, 2002, Int. Press, Somerville, MA, 2003, pp. 39–111. MR2051783

[8] G. D. James, *Some combinatorial results involving Young diagrams*, Math. Proc. Cambridge Philos. Soc. **83** (1978), no. 1, 1–10, DOI 10.1017/S0305004100054220. MR463280

[9] Michio Jimbo, Rinat Kedem, Hitoshi Konno, Tetsuji Miwa, and Jens-Ulrik H. Petersen, *Level-0 structure of level-1* $U_q(\widehat{\mathrm{sl}}_2)$*-modules and Macdonald polynomials*, J. Phys. A **28** (1995), no. 19, 5589–5606. MR1364373

[10] Nai Huan Jing, *q-hypergeometric series and Macdonald functions*, J. Algebraic Combin. **3** (1994), no. 3, 291–305, DOI 10.1023/A:1022463918288. MR1285498

[11] M. Kashiwara, T. Miwa, and E. Stern, *Decomposition of q-deformed Fock spaces*, Selecta Math. (N.S.) **1** (1995), no. 4, 787–805, DOI 10.1007/BF01587910. MR1383585

[12] I. G. Macdonald, *Symmetric functions and Hall polynomials*, 2nd ed., Oxford Classic Texts in the Physical Sciences, The Clarendon Press, Oxford University Press, New York, 2015. With contribution by A. V. Zelevinsky and a foreword by Richard Stanley; Reprint of the 2008 paperback edition [MR1354144]. MR3443860

[13] Kei Miki, *Toroidal braid group action and an automorphism of toroidal algebra* $U_q(\mathrm{sl}_{n+1,\mathrm{tor}})$ $(n \geq 2)$, Lett. Math. Phys. **47** (1999), no. 4, 365–378, DOI 10.1023/A:1007556926350. MR1693755

[14] Kei Miki, *Representations of quantum toroidal algebra* $U_q(\mathrm{sl}_{n+1,\mathrm{tor}})$ $(n \geq 2)$, J. Math. Phys. **41** (2000), no. 10, 7079–7098, DOI 10.1063/1.1287436. MR1781425

[15] Kentaro Nagao, *K-theory of quiver varieties, q-Fock space and nonsymmetric Macdonald polynomials*, Osaka J. Math. **46** (2009), no. 3, 877–907. MR2583334

[16] Andrei Negut, *The shuffle algebra revisited*, Int. Math. Res. Not. IMRN **22** (2014), 6242–6275, DOI 10.1093/imrn/rnt156. MR3283004

[17] Andrei Negut, *Quantum Algebras and Cyclic Quiver Varieties*, ProQuest LLC, Ann Arbor, MI, 2015, Thesis (Ph.D.)–Columbia University. MR3358156

[18] Andrei Negut, *Quantum toroidal and shuffle algebras*, Adv. Math. **372** (2020), 107288, 60. MR4128572

[19] Daniel Orr and Mark Shimozono, *Wreath Macdonald polynomials, a survey*, arXiv preprint `arXiv:2308.12166` (2023).

[20] Daniel Orr, Mark Shimozono, and Joshua Jeishing Wen, *Wreath Macdonald operators*, Sém. Lothar. Combin. **89B** (2023), Art. 51, 12. MR4659559

[21] Yoshihisa Saito, *Quantum toroidal algebras and their vertex representations*, Publ. Res. Inst. Math. Sci. **34** (1998), no. 2, 155–177, DOI 10.2977/prims/1195144759. MR1617066

[22] Olivier Schiffmann and Eric Vasserot, *The elliptic Hall algebra and the K-theory of the Hilbert scheme of $\mathbb{A}^2$*, Duke Math. J. **162** (2013), no. 2, 279–366, DOI 10.1215/00127094-1961849. MR3018956

[23] Alexander Tsymbaliuk, *Several realizations of Fock modules for toroidal $\ddot{U}_{q,d}(\mathfrak{sl}_n)$*, Algebr. Represent. Theory **22** (2019), no. 1, 177–209, DOI 10.1007/s10468-017-9761-5. MR3908899

[24] Joshua Jeishing Wen, *Wreath Macdonald polynomials as eigenstates*, arXiv preprint arXiv:1904.05015 (2019).

DEPARTMENT OF MATHEMATICS, NORTHEASTERN UNIVERSITY, BOSTON, MASSACHUSETTS
Email address: j.wen@northeastern.edu

Contemporary Mathematics
Volume **815**, 2025
https://doi.org/10.1090/conm/815/16316

New formulas for Macdonald polynomials via the multispecies exclusion and zero range processes

Olya Mandelshtam

ABSTRACT. We describe some recently discovered connections between one-dimensional interacting particle models and Macdonald polynomials. The first such model is the *multispecies asymmetric simple exclusion process (ASEP)* on a ring, linked to the symmetric Macdonald polynomial $P_\lambda(X;q,t)$ through its partition function. Through this connection, a new formula was found for P_λ by generalizing *multiline queues*, which were introduced by Martin in 2018 to compute stationary probabilities of the ASEP. The second particle model is the *multispecies totally asymmetric zero range process (TAZRP)* on a ring, which was very recently found to have an analogous connection to the modified Macdonald polynomial $\widetilde{H}_\lambda(X;q,t)$ through its partition function. This discovery coincided with a new formula for $\widetilde{H}_\lambda$, this time in terms of tableaux with a *queue inversion* statistic, which also compute stationary probabilities of the TAZRP. We explain the plethystic relationship between multiline queues and queue inversion tableaux, and along the way, derive a new formula for P_λ using the queue inversion statistic. This plethystic correspondence is closely related to fusion in the setting of integrable systems.

1. Introduction

The goal of this article is to describe some recent developments that explore the relationships between interacting particle models and Macdonald polynomials. Over the last couple of decades, a growing body of research has been devoted to studying these relationships, revealing remarkable connections between particle systems and symmetric functions. Much of this research has been focused on discovering new combinatorial objects and positive formulas with the goal of gaining a deeper understanding of the two areas. This article will focus on the role of two particle models which have been found to play a significant role in recent years: the *asymmetric simple exclusion process* (ASEP) and the *zero range process* (ZRP).

The *symmetric Macdonald polynomials* $P_\lambda(X;q,t)$, introduced by Macdonald [**Mac88**], are a multivariate family of orthogonal polynomials in the variables $X = \{x_1, x_2, \ldots\}$, characterized by certain triangularity and orthogonality axioms. They are indexed by partitions, and their coefficients are rational functions in the parameters q and t. They notably generalize the Schur functions, Hall–Littlewood

---

2020 *Mathematics Subject Classification.* Primary 05E05; Secondary 60G10.

The author was partially supported by NSERC grant RGPIN-2021- 02568 and NSF grant DMS-1953891.

polynomials, Jack polynomials, and q-Whittaker functions. The *modified Macdonald polynomials* $\widetilde{\mathrm{H}}_\lambda(X;q,t)$ are a transformed version of the P_λ's, obtained through the formal operation of plethysm applied to a scalar multiple of the P_λ's. The $\widetilde{\mathrm{H}}_\lambda$'s famously enjoy many "nice" properties such as being Schur positive, and appear in a variety of formulas with positive expansions in various bases. There has been much attention devoted to combinatorial study of the P_λ's, the $\widetilde{\mathrm{H}}_\lambda$'s, and related functions, including the discovery of various combinatorial objects that lead to positive formulas for these polynomials.

The initiation of the study of one-dimensional non-equilibrium exactly solvable particle models is historically attributed to Spitzer in [**Spi70**], where he introduced both the ASEP and the ZRP. In general, exclusion processes are a class of continuous time interacting particle models on a one-dimensional lattice, such that there is at most one particle per site (hence, exclusion) – with the ASEP as the most famous model in this class. On the other hand, zero range processes are in a sense defined to have the opposite property: any site can contain any number of particles. For both models, transitions are defined by particles hopping between adjacent sites. We consider *multispecies* versions of both types of models, by assigning labels in $\mathbb{N}$ to the particles, so that particles with higher labels are considered stronger, and particles of the same label are indistinguishable. See Sections 2.2 and 2.3 for the complete definitions.

The ASEP enjoys rich combinatorial structure, which in many cases has been exploited to obtain combinatorial formulas for its exact stationary distributions. Another enticing aspect for algebraic combinatorics has been its intriguing connection to Macdonald polynomials. The first such link was found for the ASEP with open boundaries, whose partition function was discovered to be closely linked to Askey–Wilson polynomials in the single-species case [**Sas99, USW04**] and Koornwinder polynomials (Macdonald polynomials of type BC) in the multispecies case [**CW18, Can17, CGdGW16**]. Remarkably, a similar connection was also discovered for classical (type A) Macdonald polynomials in [**CdGW15**], who found that the partition function of the multispecies ASEP on a circular lattice is a specialization of $P_\lambda(X;q,t)$ at $x_1 = \cdots = x_n = q = 1$, with stationary probabilities appearing as certain nonsymmetric Macdonald polynomials. This connection was made explicit through two different approaches: via *multiline queues*, which are probabilistic objects that project to the dynamics of the ASEP, in [**CMW19**], and via vertex models, which are a related object coming from Yang–Baxter integrability, in [**BW19**]. The latter was generalized in several works (see [**ABW21**] and the references therein) to colored vertex models to obtain a number of analogous formulas for various families of symmetric and nonsymmetric functions related to Macdonald polynomials, including the modified Macdonald polynomial $\widetilde{\mathrm{H}}_\lambda$ in [**GW20**]. Independently, a new multiline-queue-esque formula for $\widetilde{\mathrm{H}}_\lambda$ was found in [**AMM20**], using a *queue inversion* statistic on tableaux. It was then a natural question to find a particle process analogous to the ASEP whose partition function corresponds to $\widetilde{\mathrm{H}}_\lambda$. In [**AMM22**], such a process was found to be the multispecies totally asymmetric zero range process on a circle (mTAZRP).

The purpose of the present article is to tie together the "P_λ triangle" and the "$\widetilde{\mathrm{H}}_\lambda$ triangle" of Figure 1. In Section 3, we describe the multiline queue formula that connects the multispecies ASEP on a circle to P_λ. In Section 4 we describe the mTAZRP and a tableau formula that illustrates the connection to $\widetilde{\mathrm{H}}_\lambda$. In Section 5 we give a tableaux characterization of the multiline queue formula and show how the the two formulas are tied together through plethysm. This analysis brings us to a new formula and a new conjectural compact formula for $P_\lambda(X;q,t)$, in terms of the queue inversion statistic.

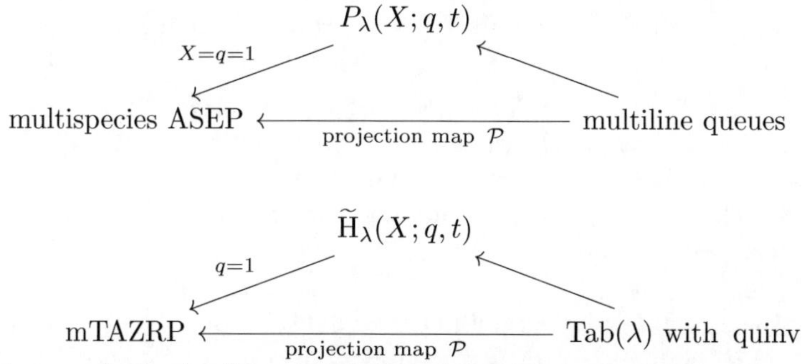

FIGURE 1. In the "P_λ triangle", multiline queues are the combinatorial object interpolating between probabilities of the ASEP on a circle and the classical Macdonald polynomials. In the "$\widetilde{\mathrm{H}}_\lambda$ triangle", fillings of tableaux Tab(λ) with the *queue inversion* statistic are the combinatorial object interpolating between the modified Macdonald polynomials and the multispecies TAZRP on a circle.

2. Notation and definitions

2.1. Combinatorial statistics and tableaux. A *composition* is a sequence $\alpha = (\alpha_1, \dots, \alpha_k)$, of k non negative integers, which are called its *parts*. If some parts are 0, it is called a *weak composition*. The number of nonzero parts is given by $\ell(\alpha)$, and $|\alpha| = \sum_i \alpha_i$ is the sum of the parts. Denote by $\lambda(\alpha)$ and $\mathrm{inc}(\alpha)$ the compositions obtained by rearranging the parts of α in weakly decreasing and increasing order, respectively. For example, if $\alpha = (0,2,0,2,1,3)$ then $\mathrm{inc}(\alpha) = (0,0,1,2,2,3)$, $\lambda(\alpha) = (3,2,2,1,0,0)$, and $\ell(\alpha) = 4$.

If λ is a weakly decreasing composition, we call it a partition, and we denote it as $\lambda \vdash n$ where $n := |\lambda|$. We may equivalently use the *frequency notation* to describe a partition λ by giving the multiplicity of its parts. Let $m_i(\lambda)$ be the number of parts of size i in λ; then we may write λ as $\langle 1^{m_1(\lambda)} 2^{m_2(\lambda)} \dots \rangle$. For example, $\lambda = (4,3,3,3,1,1)$ may be written as $\langle 1^2 2^0 3^3 4^1 \rangle$. If each part of λ appears at most once, we call λ a *strict* partition.

Given a partition $\lambda = (\lambda_1, \dots, \lambda_n)$, its *diagram* $\mathrm{dg}(\lambda)$ is a configuration of bottom justified columns with λ_i cells in the i'th column from left to right. Make note that our definition of $\mathrm{dg}(\lambda)$ corresponds to the Ferrers graph of the conjugate partition λ'.

The columns of $\mathrm{dg}(\lambda)$ are labeled from left to right, and the rows from bottom to top. The notation $u = (r, j)$ refers to the cell in row r and column j (again, departing from the convention of Cartesian coordinates). We denote by $\mathrm{South}(u) = \mathrm{South}(r,j)$ the cell $(r-1, j)$ directly below u, if it exists; if $r = 1$ and $\mathrm{South}(u)$ doesn't exist, we set the convention $\mathrm{South}(u) = \infty$. See Figure 2 for the diagram of the partition $(3,3,3,2,2,1)$.

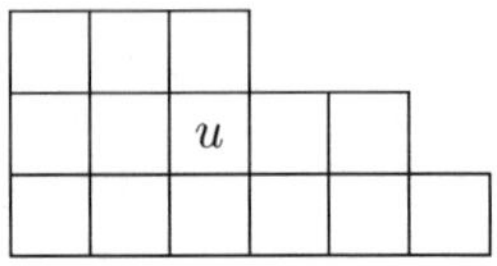

FIGURE 2. We show $\mathrm{dg}(\lambda)$ for $\lambda = (3,3,3,2,2,1)$. The cell $u = (2,3)$ has $\mathrm{leg}(u) = 1$ and $\widehat{\mathrm{arm}}(u) = 3$ (fromDefinition 5.2).

The *leg* of a cell $(r,j) \in \mathrm{dg}(\lambda)$, denoted by $\mathrm{leg}(u) = \mathrm{leg}(r,j)$, is equal to $\lambda_j - r$: the number of cells above in the same column as u.

A *filling* $\sigma : \mathrm{dg}(\lambda) \to \mathbb{Z}_+$ is an assignment of positive integers to the cells of $\mathrm{dg}(\lambda)$. For a cell $u \in \mathrm{dg}(\lambda)$, $\sigma(u)$ denotes the entry in the cell u. The content of a filling σ is recorded as a monomial in the variables $X = x_1, x_2, \ldots,$ and is denoted by $x^\sigma := \prod_{u \in \mathrm{dg}(\lambda)} x_{\sigma(u)}$.

Define the *set of descents* of a filling σ of $\mathrm{dg}(\lambda)$ to be

$$\mathrm{Des}(\sigma) = \{u \in \mathrm{dg}(\lambda) \ : \ \sigma(u) > \sigma(\mathrm{South}(u))\}.$$

The *major index* $\mathrm{maj}(\sigma)$ is then defined as:

$$\mathrm{maj}(\sigma) = \sum_{s \in \mathrm{Des}(\sigma)} (\mathrm{leg}(s) + 1).$$

We also define the notion of attacking cells and non-attacking fillings in our setting. (Our definition is the reverse of that of [**HHL04**], so as to be compatible with the queue inversion statistic of Definition 4.1.)

DEFINITION 2.1. Two cells $x = (r_1, i)$ and $y = (r_2, j)$ with $i < j$ are considered a *quinv-attacking pair* if $r_1 = r_2$ or $r_2 = r_1 - 1$. That is, the cells are in one of the two configurations:

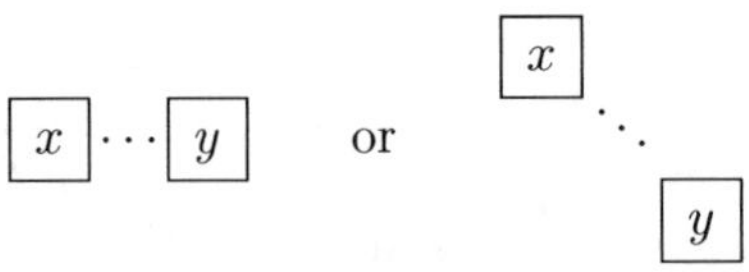

A filling σ is called *quinv-non-attacking* (or simply non-attacking if the context is clear) if $\sigma(x) \neq \sigma(y)$ for any quinv-attacking pair of cells x and y.

2.2. Multispecies ASEP. The ASEP was originally introduced as a model for protein synthesis [**MGP68**] and soon after was brought into the mathematical physics sphere by Spitzer in the late 1960s [**Spi70**] as a canonical example of an exactly solvable non-equilibrium process that exhibits boundary-induced phase transitions. Since the initiation of the field, the ASEP has been widely studied by physicists and mathematicians, and has since found connections and applications to many areas including random matrix theory, total positivity on the Grassmanian, formation of shocks, molecular transport, and others.

The ASEP has a number of interesting variations, many of which are exactly solvable; the version we shall focus on in this article is a continuous time multispecies ASEP on a circular lattice with n sites (i.e. with periodic boundary conditions) and a hopping parameter $0 \leq t \leq 1$. In this model, particles are labeled by

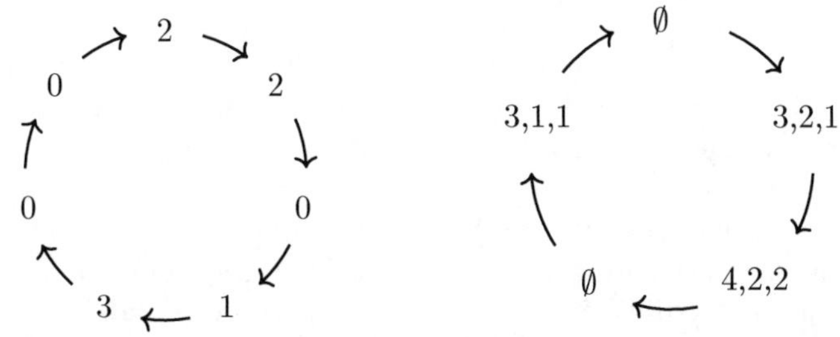

FIGURE 3. On the left is a state of the $\mathcal{A}((3,2,2,1,0,0,0),n)$. Read clockwise starting from the topmost site, the state is $(2,2,0,1,3,0,0)$. On the right is a state of the TAZRP of type $(4,3,3,2,2,2,1,1,1)$ on 5 sites. The state is $(\cdot|321|422|\cdot|311)$ (also read clockwise starting from the topmost site.) The arrows show the direction of hopping.

positive integers where the labels represent different "species"; our convention is that particles with higher labels are considered stronger. Moreover, for ease of notation, we shall consider vacancies as particles with the smallest label 0. As the boundary is closed, particles are conserved, and so the state space is determined by a positive integer n and a partition $\lambda = (\lambda_1, \ldots, \lambda_n)$, so that the j'th part of λ corresponds to a particle of species λ_j, for each j. The state space of this process is enumerated by permutations of the parts of λ, and is denoted by $\mathcal{A}(\lambda, n)$. See Figure 3 for an example of a state of the ASEP on $n = 7$ sites with particles of types $3, 2, 2, 1$.

The multispecies ASEP is a continuous time Markov chain on $\mathcal{A}(\lambda, n)$ with the following dynamics. For any pair of adjacent sites $j, j+1$ (considered cyclically modulo n, so that sites n and 1 are adjacent) their respective contents a, b can swap to become b, a, such that the rate of swapping when $a > b$ is t times the rate when $a < b$. (Particles with the same label are indistinguishable, so if $a = b$, no transition occurs.)

EXAMPLE 2.2. For example, $\mathcal{A}((3,3,1),4)$ consists of the following 12 states:

$$(3,3,1,0),(3,1,3,0),(1,3,3,0),(3,3,0,1),(3,1,0,3),(1,3,0,3),$$
$$(3,0,3,1),(3,0,1,3),(1,0,3,3),(0,3,3,1),(0,3,1,3),(0,1,3,3).$$

Examples of some transitions are:

- The jumps from $(3,1,3,0)$ are to $(1,3,3,0)$ and $(3,1,0,3)$ with rate t and to $(0,1,3,3)$ and $(3,3,1,0)$ with rate 1.
- The jumps from $(0,3,3,1)$ are to $(1,3,3,0)$ with rate t and to $(3,0,3,1)$ and $(0,3,1,3)$ with rate 1.

REMARK 2.3. Although we describe the multispecies ASEP as a continuous time process, it it also natural to consider it in discrete time: at each time step, select a site with probability $1/n$, and then let the hopping rate t dictate whether or not a hop occurs from that site. The discrete and continuous time processes will have the same stationary distribution.

2.3. Multispecies TAZRP. The zero range process (ZRP) is a typical example of a one-dimensional non-equilibrium process that exhibits condensation transitions. A ZRP can generally be defined on any graph (with possibly some additional data), where every site may contain an arbitrary number of particles. Particles hop from one site to the other in continuous time with a rate that depends solely on the content of the origin site, whence the name 'zero range'. A key property of this process, and one of the many reasons the ZRP is of great interest to physicists, is that the stationary distribution is a product measure; see [**EH05**] for a review.

Like the ASEP, physicists and mathematicians have studied a range of different versions of the ZRP; the one we consider in this paper is called the *multispecies totally asymmetric zero-range process* (mTAZRP), which had been introduced in [**Tak15**], and is a specialization of a more general multispecies ZRP that is described in [**KOW17**].

For positive integer n and a partition $\lambda = (\lambda_1, \ldots, \lambda_k)$, define $\mathcal{T}(\lambda, n)$ to be the set of arrangements of the k particles labeled by the parts of λ amongst the n sites. $\mathcal{T}(\lambda, n)$ can be described as the set of multiset compositions of length n that rearrange to λ: a state $w = (w_1, \ldots, w_n) \in \mathcal{T}(\lambda, n)$ is a sequence of (possibly empty) multisets such that the union of all the parts $\biguplus_{j=1}^n w_j$ is equal to the multiset λ.

The mTAZRP is a continuous-time Markov process on $\mathcal{T}(\lambda, n)$ with a global parameter $t \geq 0$ and site-dependent parameters $x_1, \ldots, x_n$. Each particle is equipped with an exponential clock, such that when a clock goes off, a particle jumps to an adjacent site (a particle at site j jumps to site $j+1$ cyclically, so that a particle jumping out of site n enters site 1). Define the transition rates as follows. For each $j \in \{1, \ldots, n\}$ and each species $r \geq 1$, if site j contains c_r particles of type r and d_r particles of type larger than r, then the total rate of jumps of particles of type r from site j is $x_j^{-1} t^{d_r} \sum_{i=0}^{c_r - 1} t^i$.

EXAMPLE 2.4. $\mathcal{T}((3,1,1),3)$ consists of the following 18 states:

$$(311|\cdot|\cdot), (31|1|\cdot), (31|\cdot|1), (3|11|\cdot), (3|1|1), (3|\cdot|11), (11|3|\cdot), (11|\cdot|3), (1|31|\cdot),$$
$$(1|1|3), (1|3|1), (1|\cdot|31), (\cdot|311|\cdot), (\cdot|31|1), (\cdot|3|11), (\cdot|11|3), (\cdot|1|31), (\cdot|\cdot|311).$$

Examples of transitions of the mTAZRP on $\mathcal{T}((3,1,1),3)$ are:

- The jumps from $(\cdot|311|\cdot)$ are to $(\cdot|11|3)$ with rate x_2^{-1}, and to $(\cdot|31|1)$ with rate $x_2^{-1}(t + t^2)$;
- The jumps from $(\cdot|1|31)$ are to $(3|1|1)$ with rate x_3^{-1}, to $(1|1|3)$ with rate $x_3^{-1}t$, and to $(\cdot|\cdot|311)$ with rate x_2^{-1}.

3. The ASEP and multiline queues

The concept of using multiline queues as a tool to study the ASEP was famously discovered by Ferrari and Martin in [**FM07**] to compute probabilities for the *totally* asymmetric simple exclusion process (TASEP), which is the $t = 0$ case. More recently, Martin generalized the result for the general t setting in [**Mar20**]. In this article, we shall define multiline queues with some slight modifications with respect to Martin's definition.

DEFINITION 3.1 (Ball system). Fix a positive integer n and a partition λ with $\ell(\lambda) \leq n$. A *ball system* B of size λ, n is an $L \times n$, where $L := \lambda_1$, with rows labeled 1 through L from bottom to top and columns labeled 1 through n from left to right. The array is a cylinder, so that column n is adjacent to column 1. Each of the Ln positions in the array is either empty or occupied by a ball, such that row j contains exactly λ'_j balls. We can represent a ball system by an L-tuple of subsets of n where the j'th subset corresponds to the set of columns that contain balls in row j. Denote the set of ball systems of size λ, n by $\mathcal{B}(\lambda, n)$. Then

$$\mathcal{B}(\lambda, n) = \{(b_1, \ldots, b_L) : b_j \subseteq \{1, \ldots, n\} \text{ and } |b_j| = \lambda'_j \text{ for } 1 \leq j \leq L\}.$$

DEFINITION 3.2 (Pairing). Let $B \in \mathcal{B}(\lambda, n)$ be a ball system. A *pairing* of B is an algorithm by which balls are labeled and connected by strands from higher rows to lower rows according to the following rules.

- Every ball in a row $r > 1$ of B is matched to a unique ball in row $r - 1$. Two balls are connected by the shortest strand (looking weakly to the right and wrapping if necessary around the cylinder) to connect the two balls. We abuse the definition slightly by also calling every such strand a "pairing".
- The label of each ball is equal to the length of the strand containing it. (In other words, a strand originating at row r has all balls contained in it labeled "r".)
- If a ball with label a is directly above a ball with label a in the same column, then necessarily, the two balls are connected by a strand. Such pairings are called *trivial pairings.*
- If a ball with label a is directly above a ball with label b in the same column, then $a < b$.

If a strand is pairing balls from row r to row $r - 1$, the ball in row r is called the *departure site* and the ball in row $r - 1$ is called the *arrival site.*

REMARK 3.3. The above definition of a pairing corresponds to the multiline queues defined in [**CMW19**], and is a slight variation on the multiline queues introduced by Martin in [**Mar20**]: in Martin's definition, the third bullet point is missing. Our definition results in a slightly different (and generally smaller) set of multiline queues corresponding to any given size. We revisit this difference in the discussion of Section 5.2.

See Figure 4 for an example of a ball system of size $(4, 4, 3, 1, 1)$ and $n = 6$, and see Figure 5 for an example of a pairing on that ball system.

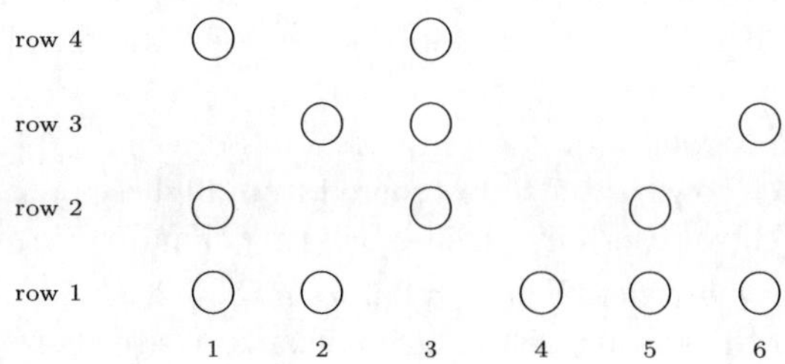

FIGURE 4. A ball system of type $\lambda = (4, 4, 3, 1, 1)$ and $n = 6$, represented as the tuple $(\{1, 2, 4, 5, 6\}, \{1, 3, 5\}, \{2, 3, 6\}, \{1, 3\})$.

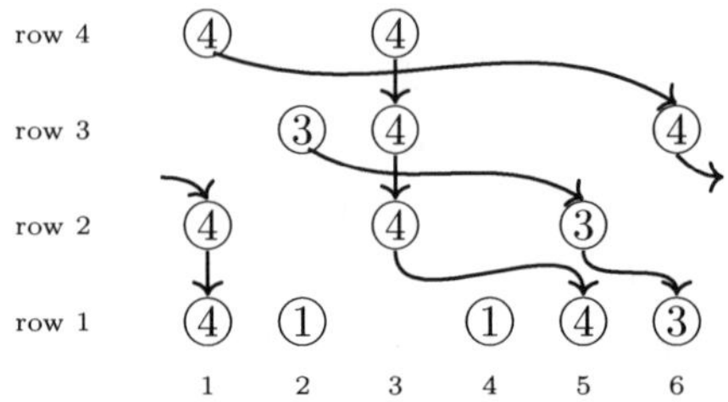

FIGURE 5. A multiline queue M of size $\lambda = (4,4,3,1,1)$ with $n = 6$, projecting to the state $\mathcal{P}(M) = (4,1,0,1,4,3) \in \mathcal{A}(\lambda, n)$.

Note that a pairing uniquely determines the labeling, but not vice versa– it is possible to have two different pairings corresponding to a single labeling of a ball system, for example:

DEFINITION 3.4 (Multiline queue). A *multiline queue* of size λ, n is a pairing on a ball system in $\mathcal{B}(\lambda, n)$, and the set of them is denoted by $\mathrm{MLQ}(\lambda, n)$. The *state* of a multiline queue is the sequence of integers $\mathcal{P}(M) = (\alpha_1, \ldots, \alpha_n)$ read off the labels of the balls in the bottom row (from left to right), with the vacancies read as 0's. The state of a multiline queue can be thought of as the *projection* of the multiline diagram onto its bottom row, and we denote this projection map by $\mathcal{P} : \mathrm{MLQ}(\lambda, n) \to \mathcal{A}(\lambda, n)$. Denote the set of multiline queues corresponding to state $\alpha = (\alpha_1, \ldots, \alpha_n)$ by $\mathrm{MLQ}(\alpha)$. (It is immediate from the definitions that if $M \in \mathrm{MLQ}(\lambda, n)$ and $\alpha = \mathcal{P}(M)$, then $\lambda(\alpha) = \lambda$.)

We now define a weight $\mathrm{wt}(M)$ associated to each multiline queue M, as a function of the variables X and parameters q, t. For each M, $\mathrm{wt}(M)$ is a monomial in X multiplied by a rational function in q, t.

DEFINITION 3.5. Let M be a multiline queue of size (λ, n) corresponding to the ball system $B = (b_1, \ldots, b_L)$ where $L := \lambda_1$. Define x^M to be the monomial $\prod_{r=1}^{L} \prod_{j \in b_r} x_j$. In other words, the power of x_j is equal to the number of balls in column j of M. For example, the multiline queue in Figure 5 has $x^M = x_1^3 x_2^2 x_3^3 x_4 x_5^2 x_6^2$.

We also define the *q, t-weight* of M by associating a weight to each pairing strand p. We shall consider the strands being created one at a time, pairing from the departure site to the arrival site. The pairing order must satisfy the following:

- strands are ordered by rows from top to bottom;
- within each row, strands are ordered from highest to lowest label;
- within each row, for any given label, the trivial pairings must come first.

Let π be a fixed pairing order. For each ball u in rows $2, \ldots, L$, define $d_\pi(u)$ to be the position in π of the pairing with u as its departure site. For each ball u in rows $1, \ldots, L-1$ with label ≥ 2, define $a_\pi(u)$ to be the position in π of the pairing with u as its arrival site. Call $\mathrm{dep}_\pi(M)$ the list of values of $d_\pi(u)$ ranging over all balls u in rows $2, \ldots, L$, read row by row from top to bottom and from left to right within each row. For example, in Figure 5, there is only one possible ordering, given by $\mathrm{dep}_\pi(M) = (2, 1, 5, 3, 4, 6, 7, 8)$.

We require that if the labelings of two multiline queues M_1 and M_2 coincide with respective pairing orders π_1 and π_2, then the orders of their departures are equal: $\mathrm{dep}_{\pi_1}(M_1) = \mathrm{dep}_{\pi_2}(M_2)$.

DEFINITION 3.6. Let $M = (b_1, \ldots, b_L)$ be a multiline queue with pairing order π; $\mathrm{wt}_\pi(M)$ depends on this order. Let p be a pairing of balls $u = (r, i)$ from row r to $v = (r-1, j)$ with the label ℓ. Define $\mathrm{leg}(p) := \ell - r$. If p is a trivial pairing, it has weight 1. Otherwise,

$$\mathrm{wt}_\pi(p)(q,t) = \frac{t^{\mathrm{skip}(p)} q^{\xi(p)(\mathrm{leg}(p)+1)}(1-t)}{1 - t^{\mathrm{free}(p)} q^{\mathrm{leg}(p)+1}},$$

where $\xi(p) = 1$ if p wraps around the cylinder (i.e. $i > j$) and 0 otherwise. $\mathrm{skip}(p)$ is the number of balls x in row $r-1$ *passed over* by p, whose arrival time $a_\pi(x)$ comes after the departure time $d_\pi(u)$ of p. $\mathrm{free}(p)$ is the total number of balls in row $r-1$ whose arrival time comes after $d_\pi(u) - 1$. Formally,

$$\begin{aligned} \mathrm{free}(p) &= |\{x \in b_{r-1} : d_\pi(u) \leq a_\pi(x)\}|, \quad \text{and} \\ \mathrm{skip}(p) &= |\{x = (r-1, k) \in b_{r-1} : d_\pi(u) < a_\pi(x) \text{ and } i < k < j \pmod{n}\}|, \end{aligned}$$

Now we define $\mathrm{wt}_\pi(M) = \prod_p \mathrm{wt}_\pi(p)$ to be the product of the weights over all the pairings in M. See Example 3.10 for a computation of the weight of a multiline queue.

Somewhat surprisingly, the total weight over all possible multiline queues with the same labeling is independent of the chosen pairing order! This is due to the following lemma.

LEMMA 3.7 ([**CMW19**, Lemma 2.1]). *Consider a ball system with a labeling that is compatible with some pairing. Let π_1 and π_2 be two fixed orderings corresponding to that labeling. Let $\mathcal{O}$ be the set of possible pairings corresponding to that labeling. Then,*

$$\sum_{M \in \mathcal{O}} \mathrm{wt}_{\pi_1}(M) = \sum_{M \in \mathcal{O}} \mathrm{wt}_{\pi_2}(M).$$

That is, the total weight of all possible pairings for a fixed labeling is independent of the order in which we pair those balls.

This allows us to fix a canonical ordering π (say, ordering balls from left to right when there is a choice), and drop the subscript when we refer to the weight $\mathrm{wt}(M)$ of a multiline queue. Thus we can present the definition of $\mathrm{wt}(M)$ as follows. Call $\mathrm{skip}(M) = \sum_p \mathrm{skip}(p)$ where the sum is over all non-trivial pairings p (according to a fixed pairing order), and $\mathrm{maj}(M) = \sum_{p:\xi(p)=1} \mathrm{leg}(p) + 1$. Then we may write

$$\mathrm{wt}(M) = q^{\mathrm{maj}(M)} t^{\mathrm{skip}(M)} \prod_{p \text{ non-trivial}} \frac{1-t}{1 - q^{\mathrm{leg}(p)+1} t^{\mathrm{free}(p)}}. \tag{3.1}$$

REMARK 3.8. Our definition of $\mathrm{wt}(M)$ is based on [**CMW19**], and is a generalization with the added parameters X, q of a slight modification to the weight function on multiline queues defined in [**Mar20**]. Specifically, the difference in the definitions is our rule for trivial pairings.

For a (weak) composition α and $n = \ell(\alpha)$, define

$$F_\alpha(x_1,\dots,x_n;q,t) = \sum_{M\in \mathrm{MLQ}(\alpha)} x^M \,\mathrm{wt}(M) \tag{3.2}$$

and for a partition λ with $\ell(\lambda)\le n$, define

$$\begin{aligned} Z_\lambda(X;q,t) = Z_\lambda(x_1,\dots,x_n;q,t) &= \sum_{M\in \mathrm{MLQ}(\lambda,n)} x^M \,\mathrm{wt}(M) \\ &= \sum_{\alpha:\lambda(\alpha)=\lambda} F_\alpha(X;q,t). \end{aligned} \tag{3.3}$$

Notably, it is shown in [**CMW19**] that $F_\alpha(X;q,t)$ is precisely the *permuted basement Macdonald polynomial* $E^{\nu}_{\mathrm{inc}(\alpha)}(X;q,t)$, where ν is the longest permutation such that $\nu\circ\alpha = \mathrm{inc}(\alpha)$ and $\mathrm{inc}(\alpha)$ is the composition obtained by rearranging α to be weakly increasing. Combining with the result of [**CdGW15**], it follows that $F_\alpha(X;q,t)$ interpolates between the permuted basement Macdonald polynomials and the stationary probabilities of the multispecies ASEP.

THEOREM 3.9 ([**CMW19**, Theorem 1.11]). *Let λ be a partition and n a positive integer. Let α be a composition such that $\lambda(\alpha)=\lambda$. The stationary probability of the state $\alpha\in\mathcal{A}(\lambda,n)$ is given by*

$$\Pr(\alpha) = \frac{1}{\mathcal{Z}_\lambda(t)} F_\alpha(x_1=\dots=x_n=1;1,t),$$

where

$$\mathcal{Z}_\lambda(t) = Z_\lambda(x_1=\dots=x_n=1;1,t).$$

Moreover, the symmetric Macdonald polynomial is equal to $Z_\lambda(X;q,t)$, which can be written as

$$P_\lambda(X;q,t) = \sum_{M\in\mathrm{MLQ}(\lambda,n)} x^M q^{\mathrm{maj}(M)} t^{\mathrm{skip}(M)} \prod_{p \text{ non-trivial}} \frac{1-t}{1-q^{\mathrm{leg}(p)+1}t^{\mathrm{free}(p)}}. \tag{3.4}$$

EXAMPLE 3.10. We compute $\mathrm{wt}(M)$ for the multiline queue M in Figure 5. Fix a pairing order $(p_1,\dots,p_8)$ and let $\mathrm{dep}(M) = (2,1,5,3,4,6,7,8)$ be the corresponding order of departures when scanning the rows from top to bottom and from left to right within each row. Thus p_1 is the pairing from $(4,3)$ to $(3,3)$ which is $\mathrm{dep}(M)_2$, p_2 is the pairing from $(4,1)$ to $(3,6)$ which is $\mathrm{dep}(M)_1$, etc.

- p_1,p_3,p_6 are trivial, so $\mathrm{wt}(p_1)=\mathrm{wt}(p_3)=\mathrm{wt}(p_6)=1$.
- p_2 skips over $(3,2)$, $\mathrm{leg}(p_2)=0$, and $\mathrm{free}(p_2)=2$, so $\mathrm{wt}(p_2)=\frac{t(1-t)}{1-qt^2}$.
- p_4 wraps around the cylinder so $\xi(p_4)=1$, $\mathrm{leg}(p_4)=2$, $\mathrm{skip}(p_4)=0$, $\mathrm{free}(p_4)=2$, so $\mathrm{wt}(p_4)=\frac{q^2(1-t)}{1-q^2t^2}$.
- Finally, $\mathrm{wt}(p_5)=\frac{1-t}{1-qt}$, $\mathrm{wt}(p_7)=\frac{t(1-t)}{1-q^3t^4}$, and $\mathrm{wt}(p_5)=\frac{1-t}{1-q^2t^3}$.

Thus

$$\mathrm{wt}(M) = \frac{q^2t^2(1-t)^5}{(1-q^3t^4)(1-q^2t^3)(1-q^2t^2)(1-qt^2)(1-qt)}.$$

3.1. The integral form J_λ. As described in [**CHM+22**, Section 4], we get for free a multiline queue formula for $J_\lambda(X;q,t)$, the *integral form* of the Macdonald polynomial P_λ. The integral form is defined as $J_\lambda(X;q,t) = \Pi_\lambda(q,t)P_\lambda(X;q,t)$, where

$$\Pi(\lambda) = \prod_{x\in M}(1 - q^{\mathrm{leg}(x)}t^{\mathrm{free}(x)+1}).$$

Looking at the contribution to $\Pi(\lambda)$ from cells x for which $\mathrm{leg}(x) = 0$, we get a factor of $\prod_{i\geq 1}(t;t)_{m_i}$, where $(t;t)_k = (1-t)(1-t^2)\cdots(1-t^k)$ and $\lambda = \langle 1^{m_1}2^{m_2}\cdots\rangle$ (i.e. for $i \geq 1$, m_i is the number of parts of length i in λ). We shall write this term as $(1-t)^{\ell(\lambda)}\,\mathrm{perm}(\lambda)$, as it is precisely the t-analog of the number of ways of permuting the columns of the same height in $\mathrm{dg}(\lambda)$. We will revisit the statistic perm in Definition 4.10.

The multiline queue formula for P_λ can be modified to describe J_λ as follows. From Definition 3.6, we observe that if every pairing of a multiline queue of size λ, n is non-trivial, then the product of the denominators $(1 - q^{\ell-r+1}t^{\mathrm{free}(p)})$ over all pairings p precisely equals $\Pi(\lambda)(1-t)^{-\ell(\lambda)}\,\mathrm{perm}(\lambda)^{-1}$. Thus we only need to consider those pairings which do not contribute a factor in the denominator to the product, namely the trivial pairings. Then,

(3.5)

$$J_\lambda(X;q,t) = (1-t)^{\ell(\lambda)}\,\mathrm{perm}(\lambda) \sum_{M\in \mathrm{MLQ}(\lambda)} x^M q^{\mathrm{maj}(M)} t^{\mathrm{skip}(M)} \times \prod_{p \text{ non-trivial}} (1-t) \times \prod_{p \text{ trivial}} (1 - q^{\mathrm{leg}(p)+1}t^{\mathrm{free}(p)}).$$

3.2. Multiline queues as tableaux. There is a natural mapping from a multiline queue of size (λ, n) to a filling of $\mathrm{dg}(\lambda)$ with integers $\{1,\ldots,n\}$. We correspond each strand of length ℓ to a column of length ℓ in $\mathrm{dg}(\lambda)$ by filling the j'th box of that column (from bottom to top) with the site number of the ball in the strand at row j. We name this queue-tableau map Π (at this stage, it is only well-defined when λ is a strict partition). We then impose an ordering on columns of the same height by requiring that the entries at row 1 are in increasing order: this makes Π a bijection. See Figure 6 for an example.

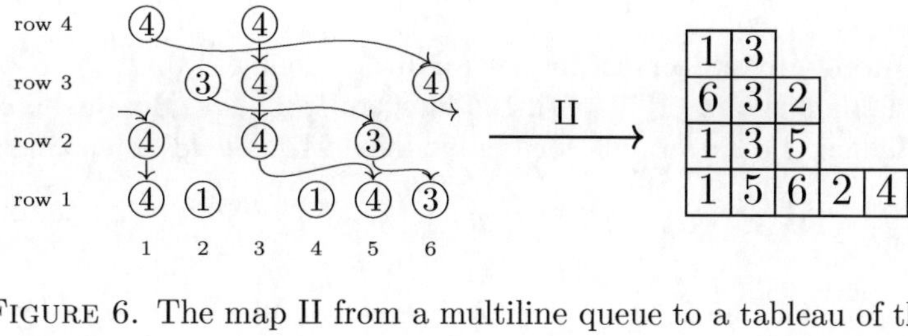

FIGURE 6. The map Π from a multiline queue to a tableau of the same size.

We observe a few things about how MLQ rules and statistics translate to rules and statistics on tableaux via this map. Let $M \in \mathrm{MLQ}(\lambda, n)$.

i. At each row r of the MLQ, there is at most one ball per site, so each entry in the set $\{1, \ldots, n\}$ can appear at most once in each row of the tableau.
ii. If a ball of label i is above a ball of label j in the MLQ, then either $i < j$, or $i = j$ and they are in the same strand. This means that if the same entry appears in rows r and $r-1$ of the tableau, i.e. $\sigma(r,k) = \sigma(r-1,m)$, then $k > m$ and $\lambda_k < \lambda_m$ (the entry below is to the southwest in a larger column), or they are in the same column ($k = m$). In particular, they can never be an *attacking pair* (see Definition 2.1).
iii. If a strand p from a ball (r,i) to a ball $(r-1,j)$ is wrapping, then $i > j$, corresponding to a *descent* in the filling coming from the pair of cells $\boxed{\substack{i\\j}}$, and $\mathrm{leg}(p) = \mathrm{leg}(u)$ where u is the cell in the tableau containing the i. Thus $\mathrm{maj}(M) = \mathrm{maj}(\amalg(M))$.
iv. free(p) and skip(p) correspond to certain "arm" and "coinversion" statistics, which have a technical definition that can be found in [**CMW19**, Section 5]. However, in Conjecture 5.6 we will see a simplifying modification.
v. We can naturally adapt the projection map $\mathcal{P}$ for fillings by passing through $\amalg$: reading off the bottom row of a filling σ, for each j, the cell $(1,j)$ corresponds to a particle of type λ_j at site $\sigma(1,j)$ of the ASEP of size λ, n.

We observe that these tableaux are suspiciously similar to *permuted basement tableaux*, introduced by Ferreira in [**Fer11**] and further studied by Alexandersson in [**Ale19**]. To be precise, if we reflect the tableaux across the y-axis, we *almost* obtain the permuted basement tableaux corresponding to $E^{\nu}_{\mathrm{inc}(\lambda)}$ for some permutation ν. This is, of course, to be expected, since $F_\alpha(X;q,t) = E^{\nu}_{\mathrm{inc}(\alpha)}$ where $\nu \circ \alpha = \mathrm{inc}(\alpha)$ from (3.2). See [**CMW19**, Remark 5.7] for details.

4. The modified Macdonald polynomials $\widetilde{\mathrm{H}}_\lambda(X;q,t)$

The modified Macdonald polynomials $\widetilde{\mathrm{H}}_\lambda(X;q,t)$ were originally introduced by Garsia–Haiman in [**GH96**], and defined using plethystic notation in terms of the J_λ's (the integral form of P_λ defined in Section 3.1) as

$$\widetilde{\mathrm{H}}_\lambda(X;q,t) = t^{n(\lambda)} J_\lambda\left[\frac{X}{1-t^{-1}};q,t^{-1}\right] \tag{4.1}$$

where $n(\lambda) = \sum_i \binom{\lambda'_i}{2}$.

We postpone delving into the details of this formula until Section 5, and instead turn our focus to the combinatorics of the monomial expansion of $\widetilde{\mathrm{H}}_\lambda$.

4.1. Queue inversion tableaux formula. Define $\mathrm{Tab}(\lambda, n)$ to be the set of fillings of the diagram $\mathrm{dg}(\lambda)$ with the integers $1, \ldots, n$. Recall the celebrated Haglund–Haiman–Loehr formula for the modified Macdonald polynomials:

$$\widetilde{\mathrm{H}}_\lambda(x_1,\ldots,x_n;q,t) = \sum_{\sigma \in \mathrm{Tab}(\lambda,n)} x^\sigma q^{\mathrm{maj}(\sigma)} t^{\mathrm{inv}(\sigma)}, \tag{4.2}$$

Here maj is from Section 2.1, and inv is a classical statistic introduced by Haglund–Haiman–Loehr in [**HHL04**].

In [**AMM20**], an analogous formula was given for $\widetilde{\mathrm{H}}_\lambda$ over the same set of fillings, but with the *queue inversion* statistic "quinv" instead of "inv", the motivation for which is discussed in Section 5. We describe this statistic. For comparison to the inv statistic, please refer to [**HHL04**].

DEFINITION 4.1 (Queue inversions). Given a diagram $\mathrm{dg}(\lambda)$, an *L-triple* consists of either

- three cells $x = (r+1, i)$, $y = (r, i)$ and $z = (r, j)$ with $i < j$; or
- two cells $y = (r, i)$ and $z = (r, j)$ with $i < j$ and $\lambda_i = r$ (in which case the triple is called a *degenerate triple*).

We shall refer to L-triples by the trio of cells (x, y, z), in the orientation described. Let $a = \sigma(x)$, $b = \sigma(y)$, and $c = \sigma(z)$ be the contents of the cells of an L-triple, so the triple is in one of the configurations below:

$$\begin{array}{|c|ccc} \hline a & & & \\ \hline b & \cdots & \boxed{c} & \\ \hline \end{array}, \qquad \text{or} \qquad \begin{array}{ccc} \emptyset & & \\ \boxed{b} & \cdots & \boxed{c} \end{array}$$

We say that a triple (x, y, z) is a *queue inversion (quinv) triple* if its entries are oriented counterclockwise when the entries are read in increasing order, with ties being broken with respect to top-to-bottom, right-to-left reading order: in other words, if

$$a \leq b < c, \text{ or } b < c < a, \text{ or } c < a \leq b.$$

If the triple is degenerate with content b, c, it is a quinv triple if and only if $b < c$.

The *weight* of a filling σ is then $x^\sigma t^{\mathrm{quinv}(\sigma)} q^{\mathrm{maj}(\sigma)}$, where $\mathrm{quinv}(\sigma)$ is the total number of quinv triples in σ, and maj is the classical major index statistic. See Figure 7 for an example of a tableau and its weight. Then we have the following.

THEOREM 4.2 ([**AMM20**, Theorem 2.6]). *Let λ be a partition. The modified Macdonald polynomial can be written as*

$$\widetilde{\mathrm{H}}_\lambda(X; q, t) = \sum_{\sigma \in \mathrm{Tab}(\lambda)} x^\sigma t^{\mathrm{quinv}(\sigma)} q^{\mathrm{maj}(\sigma)}. \tag{4.3}$$

The proof of this result in [**AMM20**] is in the spirit of the original Haglund–Haiman–Loehr proof of (4.1): using LLT polynomials and super-symmetric fillings to show the formula satisfies the axioms that characterize the modified Macdonald polynomials. In Section 5, modulo some technical details, we sketch a separate proof via plethysm by comparing tableaux with the quinv statistic to multiline queues.

3	2					
1	3	1	3	3		
1	1	2	1	2	3	3

FIGURE 7. A tableau of type $\lambda = (3, 3, 2, 2, 2, 1, 1)$ and $n = 3$. The weight of this filling is $x_1^5 x_2^3 x_3^6 q^5 t^{12}$.

REMARK 4.3. Although (4.3) appears very similar to (4.2), with the statistics inv and quinv being based on triples in the seemingly symmetric configurations $\begin{array}{cc} \boxed{x} & \boxed{z} \\ \boxed{y} & \end{array}$ and $\begin{array}{cc} \boxed{x} & \\ \boxed{y} & \boxed{z} \end{array}$ respectively, finding an explicit bijection between the two remains an open problem.[1]

4.2. Queue inversion tableaux and probabilities of the mTAZRP. The intended application for the fillings $\mathrm{Tab}(\lambda, n)$ with the quinv weight was originally to obtain formulas for probabilities of the mTAZRP of size (λ, n). Recall the projection map $\mathcal{P} : \mathrm{MLQ}(\lambda, n) \to \mathcal{A}(\lambda, n)$ from Definition 3.4. We generalize this map to a map $\mathcal{P} : \mathrm{Tab}(\lambda, n) \to \mathcal{T}(\lambda, n)$, to which we give the same name. (Implicitly, we are passing through the queue-tableau map Π.)

Let $\sigma \in \mathrm{Tab}(\lambda, n)$. Define $\mathcal{P}(\sigma) = (w_1, \ldots, w_n) \in \mathcal{T}(\lambda, n)$, where $w_j = \{\lambda_s : 1 \leq s \leq \ell(\lambda), \sigma(1, s) = j\}$. In other words, each site w_j contains the multiset of column lengths of $\mathrm{dg}(\lambda)$ such that the entry in row 1 of those columns is equal to j.

EXAMPLE 4.4. For the filling in Figure 7, $\mathcal{P}(\sigma) = (332|22|11) \in \mathcal{T}(3^2 2^3 1^2, 3)$.

Now, we obtain a formula for the stationary probabilities of the mTAZRP via this projection map, as an analog to Theorem 3.9.

THEOREM 4.5 ([**AMM22**, Theorem 1.1]). *Let $w \in \mathcal{T}(\lambda, n)$ be a state of the mTAZRP. Its stationary probability is given by*

$$\Pr(w) = \frac{1}{\widetilde{\mathcal{Z}}_{\lambda,n}} \sum_{\substack{\sigma \in \mathrm{Tab}(\lambda,n) \\ \mathcal{P}(\sigma) = w}} x^{\sigma} t^{\mathrm{quinv}(\sigma)},$$

where $\widetilde{\mathcal{Z}}_{\lambda,n}$ is the combinatorial partition function of the mTAZRP:

$$\widetilde{\mathcal{Z}}_{\lambda,n} = \sum_{\sigma \in \mathrm{Tab}(\lambda,n)} x^{\sigma} t^{\mathrm{quinv}(\sigma)}.$$

This result is proved in [**AMM22**] by explicitly constructing a Markov chain on $\mathrm{Tab}(\lambda, n)$ whose stationary distribution is given by $\mathrm{wt}(\sigma)(q = 1) = x^{\sigma} t^{\mathrm{quinv}(\sigma)}$, and which lumps to (projects onto) the mTAZRP Markov chain via the projection map.

EXAMPLE 4.6. We show all six tableaux and their weights $x^{\sigma} t^{\mathrm{quinv}(\sigma)}$ corresponding to $w = (\cdot|21|1) \in \mathcal{T}((2, 1, 1), 3)$:

$$\begin{array}{cccccc}
\begin{array}{|c|cc}\hline 1 & \multicolumn{2}{c}{} \\ \hline 2 & \multicolumn{1}{|c|}{3} & \multicolumn{1}{c|}{2} \\ \hline\end{array} &
\begin{array}{|c|cc}\hline 2 & \multicolumn{2}{c}{} \\ \hline 2 & \multicolumn{1}{|c|}{3} & \multicolumn{1}{c|}{2} \\ \hline\end{array} &
\begin{array}{|c|cc}\hline 3 & \multicolumn{2}{c}{} \\ \hline 2 & \multicolumn{1}{|c|}{3} & \multicolumn{1}{c|}{2} \\ \hline\end{array} &
\begin{array}{|c|cc}\hline 1 & \multicolumn{2}{c}{} \\ \hline 2 & \multicolumn{1}{|c|}{2} & \multicolumn{1}{c|}{3} \\ \hline\end{array} &
\begin{array}{|c|cc}\hline 2 & \multicolumn{2}{c}{} \\ \hline 2 & \multicolumn{1}{|c|}{2} & \multicolumn{1}{c|}{3} \\ \hline\end{array} &
\begin{array}{|c|cc}\hline 3 & \multicolumn{2}{c}{} \\ \hline 2 & \multicolumn{1}{|c|}{2} & \multicolumn{1}{c|}{3} \\ \hline\end{array} \\
x_1x_2^2x_3t & x_2^3x_3t & x_2^2x_3^2 & x_1x_2^2x_3t^2 & x_2^3x_3t^2 & x_2^2x_3^2t
\end{array}$$

Thus $\widetilde{\mathcal{Z}}_{(2,1,1),3} \Pr(\cdot|21|1) = x_2^2 x_3 (1 + t)(x_1 t + x_2 t + x_3)$.

[1]Since the writing of this article, Loehr gave bijective proofs of certain identities relating inv to quinv in [**Loe22**].

4.3. A compact quinv **formula for $\widetilde{\mathrm{H}}_\lambda$.** In [**CHM$^+$22**], a compact version of the Haglund–Haiman–Loehr formula was given for $\widetilde{\mathrm{H}}_\lambda$ when λ is not a strict partition, by summing over fillings with a canonical ordering on columns of equal height:

THEOREM 4.7 ([**CHM$^+$22**]). *The modified Macdonald polynomial can be written as*

$$\widetilde{\mathrm{H}}_\lambda(X;q,t) = \sum_{\substack{\sigma\in\mathrm{Tab}(\lambda)\\ \sigma \text{ inv-sorted}}} \mathrm{perm}(\sigma)x^\sigma q^{\mathrm{maj}(\sigma)}t^{\mathrm{inv}(\sigma)}, \tag{4.4}$$

where a filling is inv sorted *if its columns of equal height are arranged in the unique order such that the filling has minimal* inv. perm(σ) *is defined in Definition* 4.10.

It turns out that an analogous compact formula exists for Theorem 4.2.

DEFINITION 4.8. We say $\sigma \in \mathrm{Tab}(\lambda, n)$ is *quinv sorted* if for any pair of columns of equal height, the topmost pair of differing entries is not contained in a queue inversion. More precisely, for a pair of unequal columns A, B of equal height ℓ with entries $a_\ell, \ldots, a_1$ and $b_\ell, \ldots, b_1$ respectively from top to bottom, let $1 \le r \le \ell$ be maximal such that $a_r \neq b_r$. Then, we say $A > B$ if the cells corresponding to a_{r+1}, a_r, b_r form a quinv triple (as before, if $r = \ell$, by our convention we set $a_{r+1} = 0$). A filling is *quinv-sorted* if for every pair of columns of equal height A, B, if $A < B$, then A is to the left of B. See Example 4.9 for the ordering and Example 4.15 for a quinv-sorted filling.

EXAMPLE 4.9. The quinv-ordering for the following set of columns is:

$$\begin{array}{|c|}\hline 4\\\hline 2\\\hline 3\\\hline\end{array} < \begin{array}{|c|}\hline 4\\\hline 1\\\hline 2\\\hline\end{array} < \begin{array}{|c|}\hline 4\\\hline 1\\\hline 1\\\hline\end{array} < \begin{array}{|c|}\hline 3\\\hline 2\\\hline 3\\\hline\end{array} < \begin{array}{|c|}\hline 2\\\hline 2\\\hline 3\\\hline\end{array} < \begin{array}{|c|}\hline 1\\\hline 3\\\hline 2\\\hline\end{array}$$

Define the t-integer $[k]_t = \frac{1-t^k}{1-t} = 1 + t + \cdots + t^{k-1}$, the t-factorial

$$[m]_t! = [m]_t[m-1]_t \cdots [2]_t,$$

and the t-multinomial coefficient for $m = m_1 + \cdots + m_k$:

$$\begin{bmatrix} n \\ m_1, \ldots, m_k \end{bmatrix} = \frac{[n]_t!}{[m_1]_t![m_2]_t! \cdots [m_k]_t!}.$$

DEFINITION 4.10. Let $\sigma^{(i)}$ be a filling of a $m \times i$ rectangular diagram (m columns of length i). Suppose $\sigma^{(i)}$ has k distinct columns with multiplicities $m_1, \ldots, m_k$ (whence $m = m_1 + \cdots + m_k$). We define the statistic perm$(\sigma^{(i)})$ to be the inversion generating function of a word in the letters $\langle 1^{m_1} \cdots k^{m_k} \rangle$ (in frequency notation):

$$\mathrm{perm}(\sigma^{(i)}) = \begin{bmatrix} m \\ m_1, \ldots, m_k \end{bmatrix}_t.$$

Then, for a filling $\sigma \in \mathrm{Tab}(\lambda, n)$, let $\sigma^{(i)}$ be the (possibly empty) rectangular block consisting of the columns of height i of σ. Define

$$\mathrm{perm}(\sigma) = \prod_{i=1}^{\lambda_1} \mathrm{perm}(\sigma^{(i)}).$$

Described in another way, $\mathrm{perm}(\sigma)$ is simply the t-analog of the number of distinct ways of permuting the columns of σ within the shape $\mathrm{dg}(\lambda)$.

We note that if all columns of σ are distinct, then $\mathrm{perm}(\sigma)$ is equal to $\mathrm{perm}(\lambda)$ as defined in Section 3.1.

EXAMPLE 4.11. Consider the tableau σ in Figure 7. $\mathrm{perm}(\sigma^{(1)}) = 1$ since the two columns of height 1 are equal, $\mathrm{perm}(\sigma^{(2)}) = [3]_t[2]_t$, and $\mathrm{perm}(\sigma^{(1)}) = [2]_t$. Thus $\mathrm{perm}(\sigma) = (1+t+t^2)(1+t)^2$.

THEOREM 4.12. *The modified Macdonald polynomial can be written as*

$$\widetilde{\mathrm{H}}_\lambda(X;q,t) = \sum_{\substack{\sigma \in \mathrm{Tab}(\lambda) \\ \sigma \text{ quinv-sorted}}} \mathrm{perm}(\sigma) x^\sigma q^{\mathrm{maj}(\sigma)} t^{\mathrm{quinv}(\sigma)}. \tag{4.5}$$

EXAMPLE 4.13. As an example, we compute the following coefficient using six fillings with (4.5) instead of the ten needed with (4.3).

$$[x_1^3 x_2^2] H_{(2,2,1)}(X;q,t) = 1 + t + q + 2t^2 + 2qt + q^2 + t^3 + qt^2.$$

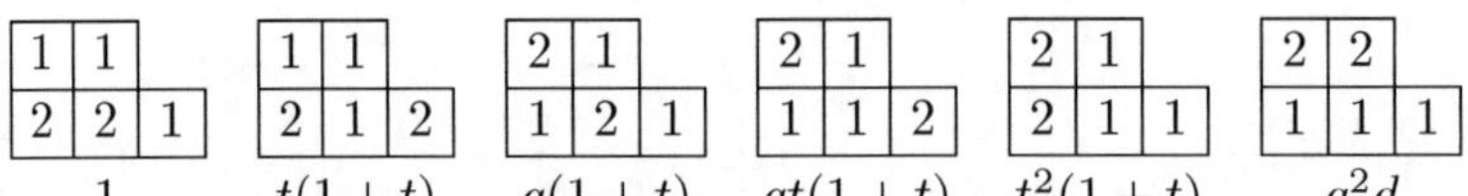

$1 \quad t(1+t) \quad q(1+t) \quad qt(1+t) \quad t^2(1+t) \quad q^2 d$

The proof of Theorem 4.12 follows the strategy of [**CHM**$^+$**22**]. We introduce entry-swapping operators $\{\tau_j\}$ to generate the entire set $\mathrm{Tab}(\lambda, n)$ from the set of quinv-sorted tableaux, so that each sorted tableau σ generates a set of fillings with total weight precisely equal to $\mathrm{wt}(\sigma)\,\mathrm{perm}(\sigma)$.

DEFINITION 4.14 (The operators τ_j). Let $\sigma \in \mathrm{Tab}(\lambda, n)$, and suppose $\lambda_j = \lambda_{j+1} = k$. We will define an operator τ_j which exchanges contents of certain cells between columns j and $j+1$. If the fillings of columns j and $j+1$ are equal in σ, then $\tau_j(\sigma) = \sigma$. Otherwise, write $\sigma(r, j) = a_r$ and $\sigma(r, j+1) = b_r$ for $r = 1, \ldots, k$, and let ℓ be maximal such that $a_\ell \neq b_\ell$. τ_j swaps the pair a_ℓ and b_ℓ in columns $j, j+1$ in row ℓ, and iteratively, if the pair in columns $j, j+1$ in row $i \leq \ell$ is swapped and this makes a difference to whether the triple $\{(i,j), (i-1,j), (i-1,j+1)\}$ is a quinv triple, then τ_j swaps the pair in columns $j, j+1$ of row $i-1$ also. The swapping terminates either when $i = 1$, or when the swap at row i doesn't change whether the triple $((i, j), (i-1, j), (i-1, j+1))$ is a quinv triple. Then $\tau_j(\sigma)$ is the tableau obtained by performing these iterative swaps in columns $j, j+1$, while keeping all other entries unchanged. See the diagram below, where the swapping terminates at row h. See Example 4.16.

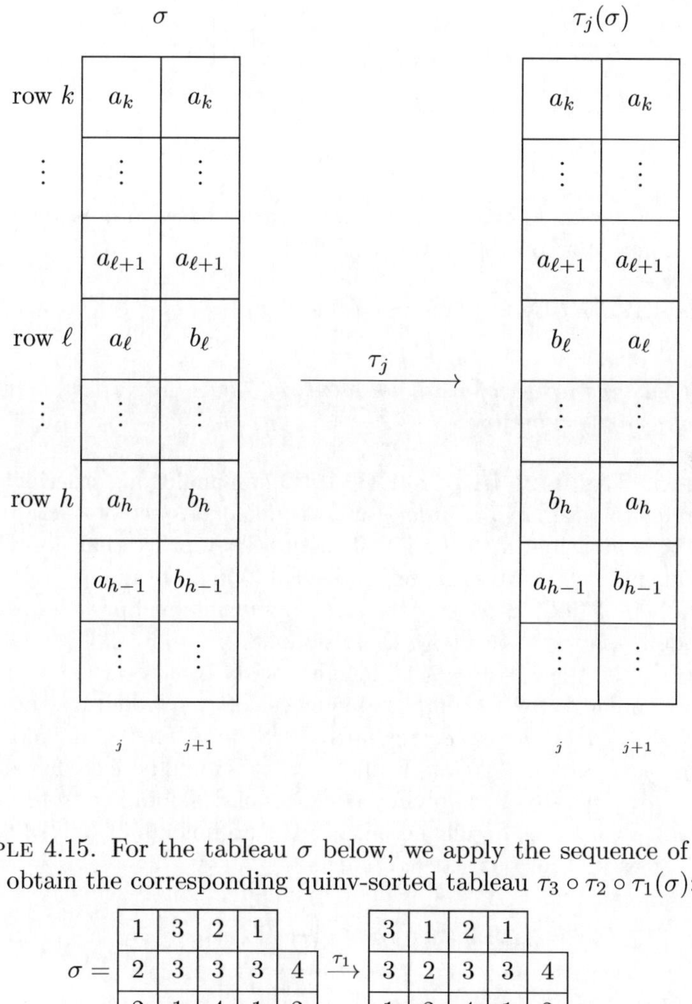

EXAMPLE 4.15. For the tableau σ below, we apply the sequence of operators τ_1, τ_2, τ_3 to obtain the corresponding quinv-sorted tableau $\tau_3 \circ \tau_2 \circ \tau_1(\sigma)$:

$\sigma =$

1	3	2	1	
2	3	3	3	4
2	1	4	1	2

$\xrightarrow{\tau_1}$

3	1	2	1	
3	2	3	3	4
1	2	4	1	2

$\xrightarrow{\tau_2}$

3	2	1	1	
3	2	3	3	4
1	2	4	1	2

$\xrightarrow{\tau_3}$

3	2	1	1	
3	2	3	3	4
1	2	1	4	2

EXAMPLE 4.16. Suppose σ has columns $j, j+1$ as shown below. Then $k = 6$, $\ell = 5$, and $h = 3$, since swapping the 2 with the 3 at row 3 changes the triple at rows $h, h-1$ from $\begin{smallmatrix} 2 & \\ 3 & 4 \end{smallmatrix}$ to $\begin{smallmatrix} 3 & \\ 3 & 4 \end{smallmatrix}$, which is a quinv triple in both cases. Thus applying the operator τ_j gives the following.

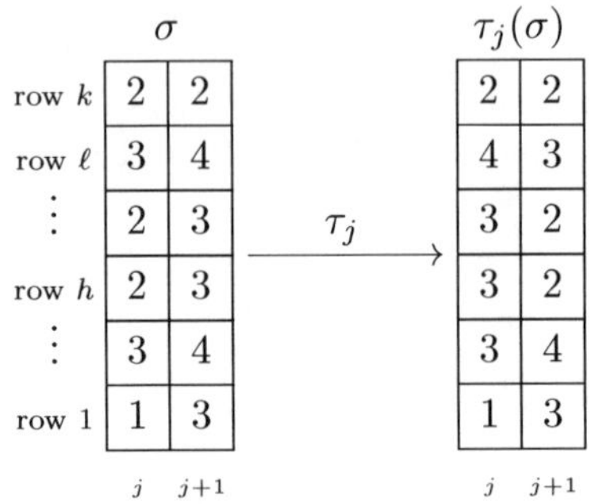

The main function of the τ_j's is to swap entries so as to preserve the maj and change the quinv in a controlled way, as follows.

LEMMA 4.17 ([**AMM20**, Lemmas 7.5 and 7.6]). *Let* $\sigma \in \mathrm{Tab}(\lambda, n)$ *with* $\lambda_j = \lambda_{j+1}$.

- *If columns* $j, j+1$ *of* σ *are not identical, then* $\mathrm{quinv}(\tau_j(\sigma)) = \mathrm{quinv}(\sigma) \pm 1$.
- $\mathrm{maj}(\tau_j(\sigma)) = \mathrm{maj}(\sigma)$.

PROOF OF THEOREM 4.12. In [**CHM**$^+$**22**], an analogous ordering on columns (we shall refer to it as an *inv-order*) and an analogous set of operators $\{\widehat{\tau}_j\}$ are defined to be compatible with the inv statistic. They prove that applying the $\widehat{\tau}_j$'s to the set of inv-sorted tableaux generates $\mathrm{Tab}(\lambda, n)$, obtaining (4.4), which is a compact version of (4.2). Precisely the same arguments will prove (4.5) for the τ_j's applied to quinv-sorted tableaux of Definition 4.8.

Unfortunately, the τ_j's do not in general satisfy braid relations. Thus we need to choose a canonical way to compose sequences of τ_j's, which for the $\widehat{\tau}_j$'s is done using *positive distinguished subexpressions* (PDS) in [**CHM**$^+$**22**, Section 3.2]. Let σ be a quinv-sorted filling. We apply the same reasoning here to define a family of tableaux $\{\tau_j\}\sigma$, generated by applying sequences of τ_j's obtained from a suitable set of PDS to σ. We omit all details here, since the arguments in [**CHM**$^+$**22**, Section 3.3] work just as well for our setting. Then,

$$\mathrm{Tab}(\lambda, n) = \biguplus_{\substack{\sigma \in \mathrm{Tab}(\lambda,n) \\ \sigma \text{ inv-sorted}}} \{\tau_j\}\sigma. \tag{4.6}$$

Let $\mathrm{quinv}' = \tau_{i_k} \circ \cdots \circ \tau_{i_1}(\sigma) \in \{\tau_j\}\sigma$, where $s_{i_k} \cdots s_{i_1}$ is a PDS. By Lemma 4.17, we deduce that $\mathrm{quinv}(\sigma') = t^k \,\mathrm{quinv}(\sigma)$. (Morally, this is permuting the columns of σ by the permutation $s_{i_k} \cdots s_{i_1}$.) The length (i.e. inversion) generating function of the set of PDS compatible with σ is precisely equal to $\mathrm{perm}(\sigma)$. Thus we obtain

$$\sum_{\sigma' \in \{\tau_j\}\sigma} x^{\sigma'} q^{\mathrm{maj}(\sigma')} t^{\mathrm{quinv}(\sigma')} = x^{\sigma} q^{\mathrm{maj}(\sigma)} t^{\mathrm{quinv}(\sigma)} \,\mathrm{perm}(\sigma).$$

Combined with (4.6), this proves the result. $\square$

5. On the plethystic relationship between the quinv tableau formula and multiline queues

We begin with a discussion of the modified Macdonald polynomials $\widetilde{\mathrm{H}}_\lambda(X;q,t)$ and the intuition behind the combinatorial interpretation of the plethystic substitution leading to the introduction of the *queue inversion* (quinv) statistic and the discovery of the connection with the mTAZRP. Recall the definition of $\widetilde{\mathrm{H}}_\lambda$ in (4.1) and the multiline queue formula (3.5) for $J_\lambda(X;q,t)$. For a symmetric function f, the plethystic notation $f[X/(1-t^1)]$ represents the substitution $p_k(x) \to p_k(x)/(1-t^k)$ into the power-sum expansion of f (where $p_k(x) = \sum_i x^k$ is the k'th power sum symmetric function). In (4.1), this translates to the formal substitution of the monomials $X = x_1, \ldots, x_n$ by the monomials $X/(1-t^{-1}) = x_1, x_1t^{-1}, x_1t^{-2}, \ldots,$ $x_2, x_2t^{-1}, x_2t^{-2}, \ldots, x_n, x_nt^{-1}, x_nt^{-2}, \ldots$ into $J_\lambda(X;q,t^{-1})$. Expanding (4.1) as

$$\widetilde{\mathrm{H}}_\lambda(X;q,t) = t^{n(\lambda)} J_\lambda(x_1, x_1t^{-1}, x_1t^{-2}, \ldots, x_2, x_2t^{-1}, x_2t^{-2}, \ldots; q, t^{-1})$$

and using (3.5), we may interpret this polynomial as a generating function over infinite multiline queues. Specifically, we consider multiline queues with a countable number of columns labeled by x_1, x_1t^{-1}, x_1t^{-2}, $\ldots$, x_2, x_2t^{-1}, x_2t^{-2}, $\ldots$, so that any ball in a column labeled by x_it^{-j} contributes x_it^{-j} to the weight of the configuration. To get back to a finite object, we group the infinite multiline queues into a finite set of equivalence classes, by "forgetting" the powers of t labeling each column while keeping the same pairings, and associating an appropriate weight to each pairing to account for the compression and the lost factors of t^{-j}. This results in a "multi-capacity multiline queue", which we define to be a *multiline diagram* in which multiple balls can occupy any given site.[2]

Although it becomes cumbersome to evaluate the compression factor directly, this notion of a multi-capacity queue led the authors of [**CHM**$^+$**22**] to a new combinatorial formula for $\widetilde{\mathrm{H}}_\lambda(X;q,t)$ through the skip statistic of multiline queues, which was reformulated as the queue inversion statistic (the formula was subsequently proved in [**AMM20**]). To be precise, the queue inversion statistic is essentially the negation of the "skip" statistic in the multiline queues, and this relationship in many ways parallels the transformation from Haglund–Haiman–Loehr nonattacking fillings and coinversions for computing P_λ and J_λ to unrestricted fillings and inversions for computing $\widetilde{\mathrm{H}}_\lambda$. We note that [**AMM20**] focused on tableaux rather than multiline diagrams, so obtaining analogous formulas in terms of multiline diagrams is yet to be done in a rigorous manner.

REMARK 5.1. In fact, it was shown in [**GW20**] that the plethystic substitution to obtain $\widetilde{\mathrm{H}}_\lambda$ from J_λ corresponds to a known concept in physics called *fusion*, introduced by Kulish–Reshetikhin–Sklyanin [**KRS81**]. In physics, a *fermionic* particle system is one having the property of "exclusion", where any site can contain at most one particle, whereas a *bosonic* particle system is one without this restriction. With this terminology, the ASEP is fermionic and the TAZRP is bosonic. The compression we describe above of an infinite family of fermionic objects (multiline queues with countably many columns for P_λ) into a finite family of bosonic objects (multiline diagrams for $\widetilde{\mathrm{H}}_\lambda$) is an example of fusion. Using matrix products, [**GW20**] express the fused weights of the fermionic lattice model for J_λ from [**CGdGW16**]

[2] One could also call this a "fused multiline queue", or "bosonic multiline queue" following Remark 5.1.

as infinite sums that end up being positive polynomials in all parameters. We have not found a way to explicitly connect our results to theirs, but it would be very interesting to understand this relationship.

5.1. Using superization to obtain a new formula for P_λ from (4.3). In [**HHL04**, Section 8] a formula for $J_\lambda(X;q,t)$ is derived from the extension of (4.2) to super-fillings. Their formula reduces to the formula [**HHL04**, Proposition 8.1] for $J_\lambda(X;q,t)$ as a sum over non-attacking fillings of $\mathrm{dg}(\lambda)$ (see Definition 2.1) using the coinversion statistic. We use the same strategy here to derive a quinv analog, which will correspond to the "compression factor" described above.

For a given $n \in \mathbb{N}$, define the alphabet $\mathcal{A} = \{1,\bar{1},2,\bar{2},\dots,n,\bar{n}\}$. A *super-filling* of $\mathrm{dg}(\lambda)$ is a function $\sigma : \mathrm{dg}(\lambda) \to \mathcal{A}$, and is denoted by $\widetilde{\mathrm{Tab}}(\lambda,n)$. For $\sigma \in \widetilde{\mathrm{Tab}}(\lambda,n)$, define $p(\sigma)$ and $m(\sigma)$ to be the numbers of positive and negative entries in σ, respectively, and define $|\sigma| \in \mathrm{Tab}(\lambda,n)$ to be the filling whose entries are the absolute values of the entries in σ. Fix the total ordering $<$ on $\mathcal{A}\cup\{0\}$ to be $\{0<1<\bar{1}<2<\bar{2}<\cdots<n<\bar{n}\}$. For $a,b\in\mathcal{A}\cup\{0\}$ we introduce the notation

$$I(a,b) = \begin{cases} a>b & a\neq b \\ 0 & a\in\mathbb{Z}_+ \\ 1 & a\in\mathbb{Z}_-. \end{cases}$$

to generalize the notion of "greater than" on $\mathbb{Z}_+$. We can now extend the definition of a descent to a super-filling: $x\in\mathrm{dg}(\lambda)\in\mathrm{Des}(\sigma)$ if $I(\sigma(x),\sigma(\mathrm{South}(x)))=1$. We also extend the definition of a quinv triple: an L-triple (x,y,z) is a quinv triple if exactly *one* of the following is true:

$$\{I(\sigma(x),\sigma(y))=1, I(\sigma(z),\sigma(y))=0, I(\sigma(x),\sigma(z))=0\}.$$

When $\sigma=|\sigma|$, we recover the original definitions of Des and quinv.

Using [**HHL04**, Proposition 4.3], the following tableaux formula is derived in [**AMM20**, Section 3]:

$$\widetilde{\mathrm{H}}_\lambda[X(t-1);q,t] = \sum_{\sigma\in\widetilde{\mathrm{Tab}}(\lambda,n)} (-1)^{m(\sigma)}x^{|\sigma|}q^{\mathrm{maj}(\sigma)}t^{p(\sigma)+\mathrm{quinv}(\sigma)}. \tag{5.1}$$

Rearranging (4.1), one obtains an expression for $J_\lambda(X;q,t)$:

$$\widetilde{\mathrm{H}}_\lambda(X;q,t^{-1}) = t^{-n(\lambda)}J_\lambda\Big[\frac{X}{1-t};q,t\Big],$$

$$t^{n(\lambda)}\widetilde{\mathrm{H}}_\lambda[X(1-t);q,t^{-1}] = J_\lambda\Big[X;q,t\Big],$$

$$t^{n(\lambda)+n}\widetilde{\mathrm{H}}_\lambda[X(t^{-1}-1);q,t^{-1}] = J_\lambda(X;q,t). \tag{5.2}$$

We invoke [**AMM20**, Theorem 5.3] to write (5.1) as a sum over quinv-non-attacking fillings (see Definition 2.1):

$$J_\lambda(X;q,t) = t^{n(\lambda)+n}\sum_{\substack{\sigma\in\widetilde{\mathrm{Tab}}(\lambda,n)\\ |\sigma|\ \text{quinv-non-attacking}}} (-1)^{m(\sigma)}x^{|\sigma|}q^{\mathrm{maj}(\sigma)}t^{-p(\sigma)-\mathrm{quinv}(\sigma)}. \tag{5.3}$$

To reduce (5.3) to quinv-non-attacking fillings of $\mathrm{Tab}(\lambda,n)$ in the positive alphabet, we follow the strategy in [**HHL04**, Section 8] . For each $\tau\in\mathrm{Tab}(\lambda,n)$, we consider the set of fillings $\sigma\in\widetilde{\mathrm{Tab}}(\lambda,n)$ such that $\tau=|\sigma|$. We observe that for any entries $a,b\in\mathcal{A}$, if $|a|\neq|b|$, then $I(a,b)=I(|a|,|b|)$. Moreover, in any L-triple

(x,y,z), either $\tau(x) \neq \tau(y) \neq \tau(z)$, or $\tau(x) = \tau(y) \neq \tau(z)$ (where $y = \mathrm{South}(x)$) by the non-attacking condition. So, to obtain the difference $\mathrm{quinv}(\sigma) - \mathrm{quinv}(\tau)$ and $\mathrm{maj}(\sigma) - \mathrm{maj}(\tau)$, we need only to consider descents and L-triples that involve a cell $u \in \mathrm{dg}(\lambda)$ such that $\tau(u) = \tau(\mathrm{South}(u))$.

DEFINITION 5.2. For a cell $u \in \mathrm{dg}(\lambda)$ in a row $r > 1$, define the statistic $\widehat{\mathrm{arm}}(u)$ to be the number of cells to the right of u in the row below. In other words, $\widehat{\mathrm{arm}}(r,c) = |\{(r-1,j) \in \mathrm{dg}(\lambda) : j > c\}|$, which is also equal to the leg of $\mathrm{South}(u)$ in $\mathrm{dg}(\lambda')$. See Figure 2 for an example of $\widehat{\mathrm{arm}}$.

Each $u \in \mathrm{dg}(\lambda)$ such that $\tau(u) \neq \tau(\mathrm{South}(u))$ brings a factor of $(t^{-1} - 1)$ for $\sigma(u) \in \mathbb{Z}_+$ and $\sigma(u) \in \mathbb{Z}_-$, respectively (as $\mathrm{quinv}(\sigma) = \mathrm{quinv}(\tau)$ and $\mathrm{maj}(\sigma) = \mathrm{maj}(\tau)$ for such u). Now consider $u \in \mathrm{dg}(\lambda)$ such that $\tau(u) = \tau(\mathrm{South}(u))$.

- If $\sigma(u) \in \mathbb{Z}_+$, then $I(\sigma(u), \sigma(\mathrm{South}(u))) = I(\tau(u), \tau(\mathrm{South}(u))) = 0$ and so $\mathrm{maj}(\sigma) = \mathrm{maj}(\tau)$. Moreover, every L-triple $(u, \mathrm{South}(u), z)$ is a quinv triple in σ and in τ. Thus every such u contributes a t^{-1}.
- If $\sigma(u) \in \mathbb{Z}_-$, then $I(\sigma(u), \sigma(\mathrm{South}(u))) = 1 - I(\tau(u), \tau(\mathrm{South}(u))) = 1$ and thus $(u, \mathrm{South}(u), z)$ is never a quinv triple in σ for any z, while it is one in τ; the number of choices for the cell z is equal to $\widehat{\mathrm{arm}}(u)$. Moreover, this u is a descent in σ, but not one in τ. Thus, every such u contributes $(-\widehat{\mathrm{arm}}(u))$ to the difference $\mathrm{quinv}(\sigma) - \mathrm{quinv}(\tau)$, and $\mathrm{leg}(u) + 1$ to the difference $\mathrm{maj}(\sigma) - \mathrm{maj}(\tau)$.

We conclude that the total contribution for every u such that $\tau(u) = \tau(\mathrm{South}(u))$ from all choices for σ is $(t^{-1} - q^{\mathrm{leg}(u)+1} t^{\widehat{\mathrm{arm}}(u)})$.

Thus we obtain the following formula, after distributing the t^n to the contribution for each cell $u \in \mathrm{dg}(\lambda)$, where (5.5) follows from (5.2). Define $\mathrm{coquinv}(\sigma) := n(\lambda) - \mathrm{quinv}(\sigma)$.

THEOREM 5.3. *Fix a partition* λ. *Then*

$$J_\lambda(X;q,t) = \sum_{\substack{\tau \in \mathrm{Tab}(\lambda) \\ \tau \text{ quinv-non-attacking}}} x^\tau q^{\mathrm{maj}(\tau)} t^{\mathrm{coquinv}(\tau)} \prod_{\substack{u \in \mathrm{dg}(\lambda) \\ \tau(u) \neq \tau(\mathrm{South}(u))}} (1-t) \times \prod_{\substack{u \in \mathrm{dg}(\lambda) \\ \tau(u) = \tau(\mathrm{South}(u))}} (1 - q^{\mathrm{leg}(u)+1} t^{\widehat{\mathrm{arm}}(u)+1}). \tag{5.4}$$

and

$$P_\lambda(X;q,t) = \frac{1}{\mathrm{perm}(\lambda)} \sum_{\substack{\tau \in \mathrm{Tab}(\lambda) \\ \tau \text{ quinv-non-attacking}}} x^\tau q^{\mathrm{maj}(\tau)} t^{\mathrm{coquinv}(\tau)} \times \prod_{\substack{u \in \mathrm{dg}(\lambda) \\ \tau(u) \neq \tau(\mathrm{South}(u))}} \frac{1-t}{1 - q^{\mathrm{leg}(u)+1} t^{\widehat{\mathrm{arm}}(u)+1}}, \tag{5.5}$$

where both sums are over quinv-non-attacking tableaux.

We have ended up with formulas for $J_\lambda(X;q,t)$ and $P_\lambda(X;q,t)$ as sums over quinv-non-attacking tableaux in terms of the statistics coquinv and $\widehat{\mathrm{arm}}$, a simplification from the more technical Haglund–Haiman–Loehr (HHL) formula which involves "Type A" and "Type B" triples (see [**HHL04**] for the definitions).

EXAMPLE 5.4. We compute the coefficients of m_{22} and m_{211} for $P_{(2,2)}(X;q,t)$ using (5.5) to get $m_{22}+\frac{(1+q)(1-t)}{1-qt}m_{211}$ after dividing by $\mathrm{perm}((2,2))=(1+t)$.

$$\underset{1}{\begin{array}{|c|c|}\hline 1&2\\\hline 1&2\\\hline\end{array}}\quad \underset{t}{\begin{array}{|c|c|}\hline 2&1\\\hline 2&1\\\hline\end{array}}\quad\Bigg|\quad \underset{\frac{(1-t)}{1-qt}}{\begin{array}{|c|c|}\hline 1&2\\\hline 1&3\\\hline\end{array}}\quad \underset{\frac{t(1-t)}{1-qt^2}}{\begin{array}{|c|c|}\hline 2&1\\\hline 3&1\\\hline\end{array}}\quad \underset{\frac{q(1-t)}{1-qt}}{\begin{array}{|c|c|}\hline 1&3\\\hline 1&2\\\hline\end{array}}\quad \underset{\frac{qt^2(1-t)}{1-qt^2}}{\begin{array}{|c|c|}\hline 3&1\\\hline 2&1\\\hline\end{array}}\quad \underset{\frac{qt(1-t)^2}{(1-qt)(1-qt^2)}}{\begin{array}{|c|c|}\hline 3&1\\\hline 1&2\\\hline\end{array}}\quad \underset{\frac{qt^2(1-t)^2}{(1-qt)(1-qt^2)}}{\begin{array}{|c|c|}\hline 2&1\\\hline 1&3\\\hline\end{array}}$$

REMARK 5.5. Notice the extra factor of $\mathrm{perm}(\lambda)$ in (5.5) that is not present in (3.4). This is due to the fact that there is no prescribed order on the columns of a filling in (5.5), the way there is in a multiline queue, but instead, each permutation of columns of the same height corresponds to a distinct tableau. Thus the formula (5.5), just like the original HHL formula for P_λ, enumerates $\mathrm{perm}(\lambda)(t=1)$ times more tableaux than the multiline queue formula (3.4)! Asymptotically, this is the same factor that compares the number of terms in the compact formulas (4.5) and (4.4) for $\widetilde{\mathrm{H}}_\lambda$ to the number of terms in the original HHL formula (4.2).

In particular, we conjecture a *new* compact formula for P_λ summing over quinv-sorted tableaux and using the coquinv statistic.

CONJECTURE 5.6. *The Macdonald polynomial $P_\lambda(X;q,t)$ is given by*

(5.6)
$$P_\lambda(X;q,t)=\sum_{\substack{\tau\in\mathrm{Tab}(\lambda)\\ \tau\ \textit{quinv-non-attacking}\\ \tau\ \textit{quinv-sorted}}} x^\tau q^{\mathrm{maj}(\tau)}t^{\mathrm{coquinv}(\tau)}\prod_{\substack{u\in\mathrm{dg}(\lambda)\\ \tau(u)\neq\tau(\mathrm{South}(u))}}\frac{1-t}{1-q^{\mathrm{leg}(u)+1}t^{\widehat{\mathrm{arm}}(u)+1}}.$$

EXAMPLE 5.7. We compute $P_{(2,2)}(X;q,t)$ using (5.6) to get

$$P_{22}(X;q,t)=m_{22}+\frac{(1+q)(1-t)}{1-qt}m_{211}+\frac{(2+t+3q+q^2+3qt+2q^2t)(1-t)^2}{(1-qt)(1-qt^2)}m_{1^4}.$$

$$\underset{1}{\begin{array}{|c|c|}\hline 1&2\\\hline 1&2\\\hline\end{array}}\quad\Bigg|\quad \underset{\frac{(1-t)}{1-qt}}{\begin{array}{|c|c|}\hline 1&2\\\hline 1&3\\\hline\end{array}}\quad \underset{\frac{q(1-t)}{1-qt}}{\begin{array}{|c|c|}\hline 1&3\\\hline 1&2\\\hline\end{array}}\quad\Bigg|\quad \underset{1}{\begin{array}{|c|c|}\hline 1&2\\\hline 3&4\\\hline\end{array}}\quad \underset{t}{\begin{array}{|c|c|}\hline 1&2\\\hline 4&3\\\hline\end{array}}\quad \underset{1}{\begin{array}{|c|c|}\hline 1&3\\\hline 2&4\\\hline\end{array}}\quad \underset{qt}{\begin{array}{|c|c|}\hline 1&3\\\hline 4&2\\\hline\end{array}}\quad \underset{q}{\begin{array}{|c|c|}\hline 1&4\\\hline 2&3\\\hline\end{array}}\quad \underset{qt}{\begin{array}{|c|c|}\hline 1&4\\\hline 3&2\\\hline\end{array}}$$

$$\underset{qt}{\begin{array}{|c|c|}\hline 2&3\\\hline 1&4\\\hline\end{array}}\quad \underset{q}{\begin{array}{|c|c|}\hline 2&3\\\hline 4&1\\\hline\end{array}}\quad \underset{q^2t}{\begin{array}{|c|c|}\hline 2&4\\\hline 1&3\\\hline\end{array}}\quad \underset{q}{\begin{array}{|c|c|}\hline 2&4\\\hline 3&1\\\hline\end{array}}\quad \underset{q^2}{\begin{array}{|c|c|}\hline 3&4\\\hline 1&2\\\hline\end{array}}\quad \underset{q^2t}{\begin{array}{|c|c|}\hline 3&4\\\hline 2&1\\\hline\end{array}}$$

(To avoid clutter, the common factor of $\frac{(1-t)^2}{(1-qt)(1-qt^2)}$ was left out from the weights of the fillings for m_{1111}.)

One aesthetically pleasing property of this formula is its symmetry. The orientation of L-triples agrees with the orientation of the Ferrers diagram. Moreover, for each cell $u\in\mathrm{dg}(\lambda)$ in a row $r>1$, the sum $\mathrm{leg}(u)+1+\widehat{\mathrm{arm}}(u)+1$ is equal to the hook length of $\mathrm{South}(u)$. This symmetry is curious in light of the mysterious q,t-symmetry of the Macdonald polynomial.

We also note that this formula coincides with Lenart's formula for P_λ in [**Len09**] (as well as with (3.4)) when λ is a strict partition (has no repeating parts).

REMARK 5.8. During the preparation of this article for publication, the author proved the conjecture in [**Man24**] using probabilistic column swapping operators that generalize the τ_i's defined in Definition 4.14.

5.2. Comparison to multiline queues. Observe that quinv-sorted non-attacking tableaux appear remarkably similar to the multiline queues defined in [**CMW19**] after passing through the queue-tableau map Π. In fact, the difference comes exclusively from our rule about trivial pairings: within each row of a multiline queue, a pairing order as defined in Definition 3.5 is required to place trivial pairings first within each class of labels. Thus trivially-paired balls cannot count for the skip or free statistics. Moreover, in a quinv non-attacking tableau, the same content can only appear in adjacent rows if it is in the same column, or in two columns of different height. On the other hand, in the multiline queues of Martin in [**Mar20**], the trivial-pairing condition is replaced by the condition "if there is a free ball directly below a departure site, then the pairing is forced to be trivial". In fact, Martin's multiline queues in [**Mar20**] are in bijection with quinv-non-attacking quinv sorted tableaux via the queue-tableau map Π, which also preserves the associated statistics.

References

[ABW21] Amol Aggarwal, Alexei Borodin, and Michael Wheeler, *Colored fermionic vertex models and symmetric functions*, Comm. Amer. Math. Soc. **3** (2023), 400–630, DOI 10.1090/cams/24. MR4628347

[Ale19] Per Alexandersson, *Non-symmetric Macdonald polynomials and Demazure-Lusztig operators*, Sém. Lothar. Combin. **76** (2016), Art. B76d, 27. MR4063906

[AMM20] Arvind Ayyer, Olya Mandelshtam, and James B. Martin, *Modified Macdonald polynomials and the multispecies zero-range process: I*, Algebr. Comb. **6** (2023), no. 1, 243–284, DOI 10.5802/alco.248. MR4552715

[AMM22] Arvind Ayyer, Olya Mandelshtam, and James B. Martin, *Modified Macdonald polynomials and the multispecies zero range process: II*, Math. Z. **308** (2024), no. 31.

[BW19] Alexei Borodin and Michael Wheeler, *Nonsymmetric Macdonald polynomials via integrable vertex models*, Trans. Amer. Math. Soc. **375** (2022), no. 12, 8353–8397, DOI 10.1090/tran/8309. MR4504641

[Can17] Luigi Cantini, *Asymmetric simple exclusion process with open boundaries and Koornwinder polynomials*, Ann. Henri Poincaré **18** (2017), no. 4, 1121–1151, DOI 10.1007/s00023-016-0540-3. MR3626299

[CdGW15] Luigi Cantini, Jan de Gier, and Michael Wheeler, *Matrix product formula for Macdonald polynomials*, J. Phys. A **48** (2015), no. 38, 384001, 25, DOI 10.1088/1751-8113/48/38/384001. MR3400909

[CGdGW16] Luigi Cantini, Alexandr Garbali, Jan de Gier, and Michael Wheeler, *Koornwinder polynomials and the stationary multi-species asymmetric exclusion process with open boundaries*, J. Phys. A **49** (2016), no. 44, 444002, 23, DOI 10.1088/1751-8113/49/44/444002. MR3562836

[CHM+22] Sylvie Corteel, Jim Haglund, Olya Mandelshtam, Sarah Mason, and Lauren Williams, *Compact formulas for Macdonald polynomials and quasisymmetric Macdonald polynomials*, Selecta Math. (N.S.) **28** (2022), no. 2, Paper No. 32, 33, DOI 10.1007/s00029-021-00721-7. MR4359567

[CMW19] Sylvie Corteel, Olya Mandelshtam, and Lauren Williams, *From multiline queues to Macdonald polynomials via the exclusion process*, Amer. J. Math. **144** (2022), no. 2, 395–436, DOI 10.1353/ajm.2022.0007. MR4401508

[CW18] Sylvie Corteel and Lauren K. Williams, *Macdonald-Koornwinder moments and the two-species exclusion process*, Selecta Math. (N.S.) **24** (2018), no. 3, 2275–2317, DOI 10.1007/s00029-017-0375-x. MR3816505

[EH05] M. R. Evans and T. Hanney, *Nonequilibrium statistical mechanics of the zero-range process and related models*, J. Phys. A **38** (2005), no. 19, R195–R240, DOI 10.1088/0305-4470/38/19/R01. MR2145800

[Fer11] Jeffrey Paul Ferreira, *Row-strict Quasisymmetric Schur Functions, Characterizations of Demazure Atoms, and Permuted Basement Nonsymmetric Macdonald Polynomials*, ProQuest LLC, Ann Arbor, MI, 2011. Thesis (Ph.D.)–University of California, Davis. MR3022565

[FM07] Pablo A. Ferrari and James B. Martin, *Stationary distributions of multi-type totally asymmetric exclusion processes*, Ann. Probab. **35** (2007), no. 3, 807–832, DOI 10.1214/009117906000000944. MR2319708

[GH96] A. M. Garsia and M. Haiman, *Some natural bigraded S_n-modules and q,t-Kostka coefficients*, Electron. J. Combin. **3** (1996), no. 2, Research Paper 24, approx. 60, DOI 10.37236/1282. The Foata Festschrift. MR1392509

[GW20] Alexandr Garbali and Michael Wheeler, *Modified Macdonald polynomials and integrability*, Comm. Math. Phys. **374** (2020), no. 3, 1809–1876, DOI 10.1007/s00220-020-03680-w. MR4076089

[HHL04] J. Haglund, M. Haiman, and N. Loehr, *A combinatorial formula for Macdonald polynomials*, J. Amer. Math. Soc. **18** (2005), no. 3, 735–761, DOI 10.1090/S0894-0347-05-00485-6. MR2138143

[KOW17] Atsuo Kuniba, Masato Okado, and Satoshi Watanabe, *Integrable structure of multispecies zero range process*, SIGMA Symmetry Integrability Geom. Methods Appl. **13** (2017), Paper No. 044, 29, DOI 10.3842/SIGMA.2017.044. MR3663589

[KRS81] P. P. Kulish, N. Yu. Reshetikhin, and E. K. Sklyanin, *Yang-Baxter equations and representation theory. I*, Lett. Math. Phys. **5** (1981), no. 5, 393–403, DOI 10.1007/BF02285311. MR649704

[Len09] Cristian Lenart, *On combinatorial formulas for Macdonald polynomials*, Adv. Math. **220** (2009), no. 1, 324–340, DOI 10.1016/j.aim.2008.09.007. MR2462843

[Loe22] Nicholas A. Loehr, *Bijective proofs of some coinversion identities related to Macdonald polynomials*, 2022, Arxiv preprint `arXiv:2210.11286`.

[Mac88] Ian Macdonald, *A new class of symmetric functions*, Sém. Lothar. Combin **20** (1988).

[Man24] Olya Mandelshtam, *A compact formula for the symmetric Macdonald polynomias*, 2024, Arxiv preprint `arXiv:2401:17223`.

[Mar20] James B. Martin, *Stationary distributions of the multi-type ASEP*, Electron. J. Probab. **25** (2020), Paper No. 43, 41, DOI 10.1214/20-ejp421. MR4089793

[MGP68] J. Macdonald, J. Gibbs, and A. Pipkin, *Kinetics of biopolymerization on nucleic acid templates*, Biopolymers **6** (1968).

[Sas99] Tomohiro Sasamoto, *One-dimensional partially asymmetric simple exclusion process with open boundaries: orthogonal polynomials approach*, J. Phys. A **32** (1999), no. 41, 7109–7131, DOI 10.1088/0305-4470/32/41/306. MR1732546

[Spi70] Frank Spitzer, *Interaction of Markov processes*, Advances in Math. **5** (1970), 246–290, DOI 10.1016/0001-8708(70)90034-4. MR268959

[Tak15] Yoshihiro Takeyama, *Algebraic construction of multi-species q-boson system*, Arxiv preprint `arXiv:1507.02033`, 2015.

[USW04] Masaru Uchiyama, Tomohiro Sasamoto, and Miki Wadati, *Asymmetric simple exclusion process with open boundaries and Askey-Wilson polynomials*, J. Phys. A **37** (2004), no. 18, 4985–5002, DOI 10.1088/0305-4470/37/18/006. MR2065218

UNIVERSITY OF WATERLOO, WATERLOO, ONTARIO, CANADA
Email address: `omandels@uwaterloo.ca`

Contemporary Mathematics
Volume **815**, 2025
https://doi.org/10.1090/conm/815/16317

Some combinatorial aspects of Theta and Delta operators

Marino Romero

ABSTRACT. Theta operators are essential in the statement and proof of the Compositional Delta Conjecture, and they are also used in describing the conjectural Frobenius characteristic for the S_n-coinvariants of the polynomial ring with two commuting and two anticommuting sets of variables. We will look at how two of the most recent advancements in the theory of Macdonald polynomials, and related symmetric function operators, describe Theta operators. The first is in regards to the proof of the Delta Conjecture, by D'Adderio and Mellit, and the five-term relation of Garsia and Mellit. The second is with regards to the recent advances by Blasiak, Haiman, Morse, Pun, and Seelinger, regarding connections between the Schiffmann Algebra, the Shuffle Algebra, and LLT polynomials. We end by describing new results regarding Theta and Delta operators, and we present some open questions whose answers may lay hidden between the connections of these advancements.

1. Introduction

We have two main goals. We will begin by laying out some connections between two important developments regarding eigenoperators of the modified Macdonald basis. One development is with regards to the proof of the Compositional Delta Conjecture [**14**], its relation to Theta operators [**13**], and the five-term relation of Garsia and Mellit [**19**]. The second important advance is in terms of understanding the Schiffmann algebra (or Elliptic Hall Algebra of Burban and Schiffmann) [**9**], its connection to the Shuffle Algebra [**32**], a proof of the Extended Delta Conjecture [**8**], and ultimately a series of combinatorial constructions which follow from these amazing relations [**7**],[**6**], [**5**]. Our second goal is to give some examples of the new combinatorial aspects surrounding these operators. It may very well be the case that the connection between the two solutions of the Delta Conjecture and Shuffle Conjecture are key in understanding these new questions.

In this introduction, we will give some background and some representation theoretic motivations for studying these operators. Then, we will go through many symmetric function identities and basic constructs related to the aforementioned recent advances. In the second half, we will look at some of the combinatorial constructs that appear in studying Theta operators.

2020 *Mathematics Subject Classification.* 05E05, 05E10, 20C30, 33D52.

The author was partially supported by the NSF Mathematical Sciences Postdoctoral Research Fellowship DMS-1902731.

1.1. Representation theory background and the space of Diagonal Harmonics. Given an action of the symmetric group S_n on a graded module $M = \bigoplus_\alpha M_\alpha$, we get a graded character $\chi^M = \sum_\alpha q^\alpha \chi^{M_\alpha}$. The Frobenius map, $\mathcal{F}$, sends a graded character to a symmetric function with coefficients in $\mathbb{C}[q^\alpha]$: the conjugacy class of type $\lambda \vdash n$, C_λ, is sent to the power sum symmetric function p_λ/z_λ, where $z_\lambda = n!/|C_\lambda|$ is given by $1^{m_1}m_1!2^{m_2}m_2!\cdots$, m_i being the multiplicity of i in λ. One of the most important aspects of this map is that if we have the expansion

$$\mathcal{F}\left(\chi^M\right) = \sum_{\lambda\vdash n} n_\lambda(q)s_\lambda$$

into the Schur function basis, then the coefficient of q^α in $n_\lambda(q)$ gives the multiplicity of the irreducible χ^λ in M_α. Remarkably, one can know that $\mathcal{F}(\chi^M) = F$, yet not know the Schur expansion of F. One way to then understand F is to use a combinatorial model in order to produce a symmetric function G equal to F.

These connections are some of the most important aspects concerning Macdonald polynomials and their related symmetric function operators. Modified Macdonald polynomials are a prime example of this. Along with introducing the original Macdonald polynomials [**31**], Macdonald found that when expanded in terms of a particular basis, the coefficients all appeared to be in $\mathbb{N}[q,t]$. He conjectures that not only are they positive, but the terms in these polynomials are given by a pair of statistics on standard tableaux. In an effort to give this positivity a representation theoretical interpretation, Garsia and Haiman introduced the modified version of Macdonald's polynomials, $\{\widetilde{H}_\mu[X;q,t]\}$, and they defined what is now referred to as the Garsia-Haiman module

$$M_\mu = \{P(\partial_{X_n}, \partial_{Y_n})\Delta_\mu : P \in \mathbb{C}[X_n, Y_n]\},$$

given by the linear span of derivatives of an alternant Δ_μ in two sets of variables $X_n = x_1, \ldots, x_n$ and $Y_n = y_1, \ldots, y_n$ [**16**]. The action of $\sigma \in S_n$ is given by permuting the indices of the variables: $\sigma x_i = x_{\sigma_i}$ and $\sigma y_i = y_{\sigma_i}$. This is a bigraded module. Each graded component of bidegree r, s corresponds to the polynomials that are of degree r in X_n and of degree s in Y_n. Using a t to record degrees in X_n and q to record degrees in Y_n, they conjecture that

$$\mathcal{F}(\chi^{M_\mu}) = \widetilde{H}_\mu[X;q,t].$$

This means that the modified q,t-Kostka, $\widetilde{K}_{\lambda,\mu}(q,t)$, appearing in the expansion

$$\widetilde{H}_\mu[X;q,t] = \sum_{\lambda\vdash n} \widetilde{K}_{\lambda,\mu}(q,t)s_\lambda,$$

are counting multiplicities of irreducible representations. Therefore they are polynomials in $\mathbb{N}[q,t]$. This was proved by Haiman with some amazing connections to the geometry of the Hilbert scheme of points in the plane $\mathbb{C}^2$, and more specifically the isospectral Hilbert scheme [**26**]. However, it remains open to find a precise, combinatorial description for the q,t-Kostka.

We must also mention here that there is a beautiful combinatorial formula for the modified Macdonald basis in terms of monomials due to Haglund, Haiman, and Loehr [**22**]. It gives explicit combinatorial objects and statistics for computing $M_{\lambda,\mu}$ in the expansion

$$\widetilde{H}_\mu = \sum_\lambda M_{\lambda,\mu} m_\lambda.$$

The question still remains to find the Schur expansion of this symmetric function.

The space of diagonal coinvariants can be defined as

$$\mathcal{A}^{(2,0)} = \frac{\mathbb{C}[X_n, Y_n]}{\left(\mathbb{C}[X_n, Y_n]_+^{S_n}\right)},$$

where $\left(\mathbb{C}[X_n, Y_n]_+^{S_n}\right)$ is the ideal generated by symmetric polynomials with no constant term. By looking at the orthogonal complement of the ideal, one can equivalently study the space of Diagonal Harmonics

$$\mathrm{DH}_n = \left\{ P \in \mathbb{C}[X_n, Y_n] : \sum_i \partial_{x_i}^r \partial_{y_i}^s P = 0 \right\}.$$

It turns out that all the Garsia-Haiman modules live in this space. Garsia and Haiman then conjectured [**17**] that

$$\mathcal{F}(DH_n) = \sum_{\mu \vdash n} \frac{MB_\mu \Pi_\mu}{w_\mu} T_\mu \widetilde{H}_\mu = \overline{\nabla} e_n.$$

These factors will be defined in the following section. Haiman proved this equality [**27**], yet the Schur expansion is unknown. The next best result is then given by the Shuffle Conjecture [**23**] and Compositional Shuffle Conjecture [**24**], which gives a monomial expansion of $\overline{\nabla} e_n$. This was first proved by Carlsson and Mellit [**10**]. Instead of writing the combinatorial details of the Shuffle Conjecture, we would like to instead point out a combinatorial expansion when $t = 1$ that can be easily overlooked.

Recall that a parking function can be defined in the following two ways. First, a Dyck path in the $n \times n$ square can be drawn as a sequence of North and East segments from $(0, 0)$ to (n, n) such that all the steps are always weakly above the line $y = x$. Now place a label from $\{1, \ldots, n\}$ next to each North step so that every number is used precisely once, and consecutive North steps have increasing labels from bottom to top. This gives us an element of PF_n. For instance, Figure 1 depicts a parking function in the 8×8 square. Instead of drawing a main diagonal which may be touched by the Dyck path, we instead drew a staircase which the Dyck path is never allowed to touch.

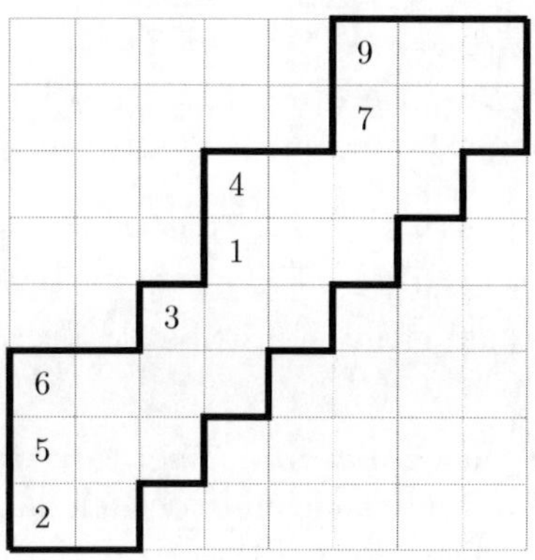

FIGURE 1. A parking function $P \in \mathrm{PF}_n$ with area 10.

Given $P \in \mathrm{PF}_n$, let $c_i(P)$ be the column in which i is placed. For the example in Figure 1, we get

$$(c_1(P), \ldots, c_n(P)) = (4, 1, 3, 4, 1, 1, 6, 6).$$

This gives the second definition of a parking function: any sequence $(c_1, \ldots, c_n)$ for which when we write it in increasing order, $(a_1, \ldots, a_n)$, we have $a_i \leq i$. Given a parking function in sequence form, $(c_1, \ldots, c_n)$, we can go back to a parking function drawn on a Dyck path D. The number of full lattice cells between D and the main diagonal is denoted by area(D); and we can define the area of a parking function to be the area of its underlying Dyck path.

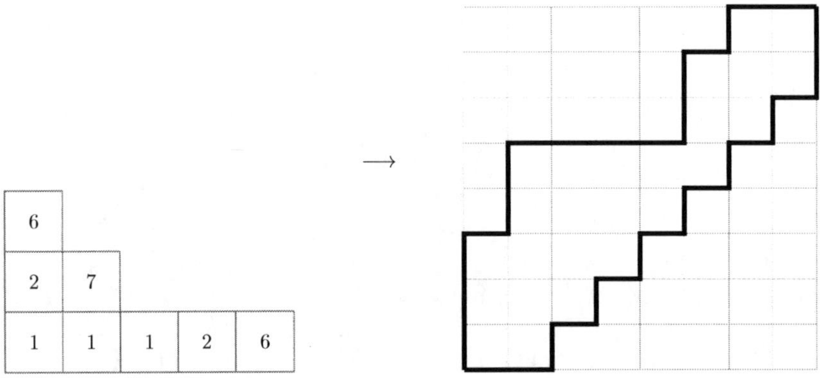

FIGURE 2. An element of $\mathrm{PS}_{(5,2,1)}$ with area 10 and the Dyck path corresponding to the entries inside the tableaux.

Let PS_λ denote the set of all semistandard Young tableaux of shape λ whose entries, when read in any order, give a parking function. For $T \in \mathrm{PS}_\lambda$, we let area($T$) be the area of any parking function one gets by reading the entries in T in some order. See Figure 2 for an example. Then

THEOREM 1.1 ([**33**]).

$$\overline{\nabla} e_n \Big|_{t=1} = \sum_{\lambda \vdash n} s_\lambda \sum_{T \in \mathrm{PS}_{\lambda'}} q^{\mathrm{area}(T)}.$$

Since $\overline{\nabla} e_n$ is the Frobenius characteristic of a module, we know this expression has to be q, t-Schur positive. This means that there must exist some statistic, stat, such that

$$\overline{\nabla} e_n = \sum_{\lambda \vdash n} s_\lambda \sum_{T \in \mathrm{PS}_{\lambda'}} q^{\mathrm{area}(T)} t^{\mathrm{stat}(T)}.$$

Finding such a statistic would give the distribution of irreducibles in the space of Diagonal Harmonics.

1.2. Generalized coinvariants and Theta operators. Before moving on to the proper definitions behind the symmetric functions here, we would like to first give one of the main motivations of this work. Theta operators were defined in [**13**] in order to give the Delta Conjecture of Haglund, Remmel, and Wilson [**25**] a compositional refinement, which was ultimately proved in [**14**]. During this time,

Zabrocki gave a conjectured module for the representation theoretical side of $\Delta_{e_k} e_n$. This was extended even further, by D'Adderio, Iraci, and Vanden Wyngaerd.

CONJECTURE 1.2 (D'Adderio, Iraci, Vanden Wyngaerd [**13**]). *Let $\Gamma_n = \gamma_1, \ldots, \gamma_n$ and $T_n = \tau_1, \ldots, \tau_n$ be two sets of anticommuting variables, so that $\gamma_i\gamma_j = -\gamma_i\gamma_j$ and $\tau_i\tau_j = -\tau_i\tau_j$. Then for*

$$\mathcal{A}^{(2,2)} = \frac{\mathbb{C}[X_n, Y_n, \Gamma_n, T_n]}{\left(\mathbb{C}[X_n, Y_n, \Gamma_n, T_n]_{+}^{S_n}\right)}$$

we have

$$\mathcal{F}(A^{(2,2)}) = \sum_{i+j<n} u^i v^j \Theta_i \Theta_j \overline{\nabla} e_{n-(i+j)}.$$

Here, u records the degrees in Γ_n and v records degrees in T_n. So far, this has been confirmed for $q = t = 0$ [**29**] and for the case when $u = v = 0$, which is the Diagonal Harmonics case. Attached to this conjecture there should be some natural combinatorial model; one is yet to be given.

Some work has been done to extend these sorts of results to generalized coinvariant rings. Conjecture 8.3 in [**13**] gives a method using Theta operators for computing lower degree terms in the Frobenius Characteristic of the coinvariants with two commuting and any number of anticommuting variables. In [**2**], it is conjectured that in order to compute the Frobenius characteristic of some general coinvariant ring, with say r sets of commuting variables and s sets of anticommuting variables, then it is enough to know the case where there are infinitely many sets of commuting variables.

We now go on to define the symmetric functions involved in these expressions.

2. Symmetric function basics

As a reference for introducing symmetric functions and the original Macdonald polynomials, we have [**31**]. We should note that a lot of the plethystic machinery we will use in this section was first developed in [**4**],[**18**], and [**20**].

The space of symmetric functions has many families of bases. The usual classical ones are the power, homogeneous, elementary, monomial, forgotten, and Schur bases, denoted respectively by $\{p_\lambda\}$, $\{h_\lambda\}$, $\{e_\lambda\}$, $\{m_\lambda\}$, $\{f_\lambda\}$, and $\{s_\lambda\}$. They come with the Hall inner product, defined by any of the following pairs of bases:

$$\begin{aligned} \chi(\lambda = \mu) \qquad &= \langle p_\lambda, p_\mu / z_\mu \rangle \qquad &= \langle h_\lambda, m_\mu \rangle \\ &= \langle e_\lambda, f_\mu \rangle \qquad &= \langle s_\lambda, s_\mu \rangle, \end{aligned}$$

where $\chi(A)$ gives 1 if the statement A is true and 0 otherwise. Therefore, any symmetric function F can be expressed in terms of the monomial or power bases in the following way:

$$F = \sum_\lambda \langle F, h_\lambda \rangle \, m_\lambda = \sum_\lambda \langle F, p_\lambda / z_\lambda \rangle \, p_\lambda$$

One can also define the adjoint $f^\perp$ of a symmetric function f to be the operator which for all g and h gives

$$\langle f^\perp g, h \rangle = \langle g, fh \rangle.$$

The ring of symmetric functions also comes with an involution ω given by one of the following equalities:

$$\omega(p_\lambda) = (-1)^{|\lambda|-\ell(\lambda)}p_\lambda, \qquad \omega(h_\lambda) = e_\lambda, \qquad \omega(m_\lambda) = f_\lambda, \qquad \omega(s_\lambda) = s_{\lambda'}.$$

We recall a few basic properties of plethystic substitution. For any expression $E(t_1, t_2, \dots)$, we let $p_k[E(t_1, t_2, \dots)] = E(t_1^k, t_2^k, \dots)$. Then extend this substitution to all symmetric functions by setting $p_\lambda[E] = p_{\lambda_1}[E] \cdots p_{\lambda_{\ell(\lambda)}}[E]$ and

$$F[E] = \sum_\lambda \langle F, p_\lambda / z_\lambda \rangle \, p_\lambda[E].$$

If we let $X = x_1 + x_2 + \cdots$, then $p_k[X] = x_1^k + x_2^k + \cdots$ is the usual power sum symmetric function. We should note that if a is a constant, then

$$p_k[-aX] = -ax_1^k - ax_2^k - \cdots = -ap_k[X].$$

This means that variable substitution does not commute with plethystic substitution, and in particular, there are two negative signs. Let ϵ be a variable within the plethystic bracket that equals -1 outside the bracket. Then

$$p_k[-\epsilon X] = -\epsilon^k p_k[X] = (-1)^{k-1} p_k[X] = \omega(p_k)[X],$$

It follows that $\omega(F)[E] = F[-\epsilon E]$, converting the involution ω into an algebraic operation.

Starting from the Hall inner product, we can get new inner products by picking an expression Y and setting

$$\langle F[X], G[X] \rangle_Y = \langle F[X], G[XY] \rangle.$$

The "star" scalar product may be seen as the following transformation:

$$\langle F, G \rangle_* = \langle F, G \rangle_{-\epsilon M} = \langle F, G[-\epsilon XM] \rangle,$$

where the factor $M = (1-q)(1-t)$ becomes so prevalent that it is best to give it a symbol. We let $F^* = {}^*(F) = F[X/M]$ and $F^{\overline{*}} = F[XM]$. Then the star scalar product can be written as

$$\langle F, G \rangle_* = \langle F, \omega G^{\overline{*}} \rangle.$$

Lastly, the modified star scalar product is given by

$$\langle F, G \rangle_\bullet = \langle F, G \rangle_{-M} = \langle F, G[-XM] \rangle.$$

The plethystic exponential is the generating function of homogeneous symmetric functions, given by

$$\mathrm{Exp}[X] = \exp\left(\sum_{k>0} p_k[X]/k \right) = \sum_{n \geq 0} h_n[X].$$

The Cauchy identity states that for any two dual bases, $\{u_\lambda\}_\lambda$ and $\{v_\lambda\}_\lambda$ under the Hall Scalar product, we have

$$\mathrm{Exp}[XY] = \sum_\lambda u_\lambda[X] v_\lambda[Y].$$

To introduce the modified Macdonald basis, we need to define a few combinatorial factors. In what follows, we draw partitions in the French convention. A cell $c \in \mu$ is also denoted by the coordinates (i, j) of the lower-left corner of c. These coordinates (i, j) are also called the coarm and coleg respectively. The cells strictly East of c are called the arm of c, and the cells strictly North of c are called the

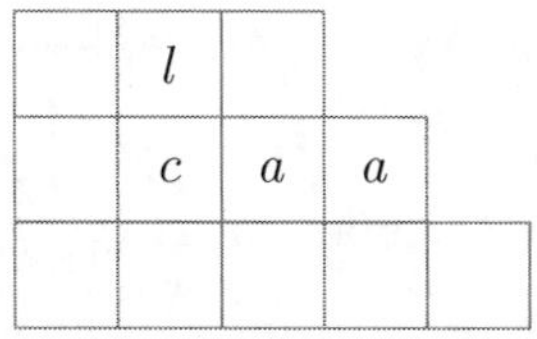

FIGURE 3. The cell $c = (1,1)$ in $\mu = (5,4,3)$. The cells in the arm of c are labeled "a," and the cells in the leg are labeled "l."

leg of c. The lengths of the arm and leg of c in μ are denoted by $a_\mu(c)$ and $l_\mu(c)$, respectively. Figure 3 gives an example of a partition, a cell, and the collection of cells in the arm or leg of c.

Let

$$B_\mu = \sum_{(i,j)\in\mu} q^i t^j, \qquad T_\mu = \prod_{(i,j)\in\mu} q^i t^j, \qquad \Pi_\mu = \prod_{(i,j)\in\mu/(1)} 1 - q^i t^j, \text{ and}$$

$$w_\mu = \prod_{c\in\mu} (q^{a_\mu(c)} - t^{l_\mu(c)+1})(t^{l_\mu(c)} - q^{a_\mu(c)+1}).$$

Here, we will list some of the important properties of the modified Macdonald basis $\{\widetilde{H}_\mu\}_\mu$. Different subsets of this list can be taken as the defining properties of the modified Macdonald basis; however, we make no effort to define them in this way. We will simply say that they can be uniquely determined by the triangularity relations and one of the linear functional relations.

a) Orthogonality relations: $\langle \widetilde{H}_\lambda, \widetilde{H}_\mu \rangle_* = w_\mu \, \chi(\lambda = \mu)$.
b) Linear functional relations: $\widetilde{H}_\mu[1] = 1$ and $\widetilde{H}_\mu[-\epsilon] = T_\mu$.
c) Triangularity relations: $\widetilde{H}_\mu = \sum_{\lambda \geq \mu'} A_{\lambda,\mu} s_\lambda \left[\frac{X}{1-t}\right]$ and $\widetilde{H}_\mu = \sum_{\lambda\geq\mu} B_{\lambda,\mu} s_\lambda \left[\frac{X}{1-q}\right]$.
d) Macdonald reciprocity: $\widetilde{H}_\mu[MB_\lambda]/\Pi_\mu = \widetilde{H}_\lambda[MB_\mu]/\Pi_\lambda$.
e) The "flip" relation: $\widetilde{H}_\mu[X;t,q] = \widetilde{H}_{\mu'}[X;q,t]$.

Using the Cauchy formula and Macdonald reciprocity, it is not difficult to see that

$$e_n = \sum_{\mu\vdash n} \frac{MB_\mu \Pi_\mu}{w_\mu} \widetilde{H}_\mu = \sum_{\mu\vdash n} \frac{\langle \widetilde{H}_\mu, e_n\rangle_*}{w_\mu} \widetilde{H}_\mu.$$

This can be compared with

$$e_n^* = \sum_{\mu\vdash n} \frac{\langle \widetilde{H}_\mu, e_n^*\rangle_*}{w_\mu} \widetilde{H}_\mu = \sum_{\mu\vdash n} \frac{\langle \widetilde{H}_\mu, h_n\rangle}{w_\mu} \widetilde{H}_\mu = \sum_{\mu\vdash n} \frac{1}{w_\mu} \widetilde{H}_\mu$$

We will see in Section 2.3 that the connection between these two expressions gives rise to a wide variety of symmetric functions with remarkable combinatorial structure. To this end, we will start by recalling some eigenoperators of the modified Macdonald basis: For any symmetric function F, set

$$\Delta_F \widetilde{H}_\mu = F[B_\mu]\widetilde{H}_\mu, \qquad \nabla \widetilde{H}_\mu = (-1)^{|\mu|} T_\mu \widetilde{H}_\mu, \qquad \text{and} \qquad \Pi\widetilde{H}_\mu = \Pi_\mu \widetilde{H}_\mu.$$

The Delta operator was first introduced in [**4**], while this nabla operator differs by a sign from the version which was originally defined by Bergeron and Garsia in [**3**].

We will call the original version the "normalized" nabla operator and set

$$\overline{\nabla}\widetilde{H}_\mu = T_\mu \widetilde{H}_\mu.$$

It is often useful to look at the operator $\Delta_{\mathrm{Exp}[-vX]} = \Delta_v$, and note that

$$\Delta_v \widetilde{H}_\mu = \Pi_\mu(v)\widetilde{H}_\mu \qquad \text{with} \qquad \Pi_\mu(v) = \prod_{(i,j)\in\mu} 1 - vq^it^j.$$

Note here that if μ is nonempty, then $\Pi_\mu(v)/(1-v)$ specializes to Π_μ at $v=1$.

There is also a modified version of Δ, defined by setting

$$\Delta'_{f[X]} = \Delta_{f[X-1/M]}.$$

It then follows that

$$\begin{aligned}\Delta'_v &= \Delta_{\mathrm{Exp}[-v(X-1/M)]} = \Delta_{\mathrm{Exp}[v/M]}\Delta_{\mathrm{Exp}[-vX]}\\ &= \mathrm{Exp}\left[\frac{v}{M}\right]\Delta_v.\end{aligned}$$

One should be careful: this is not the same Delta prime operator used in [**25**].

2.1. Plethystic operators. The translation and multiplication operators, $\mathcal{T}_Y$ and $\mathcal{P}_Y$ respectively, are two fundamental plethystic operations which form invaluable tools for finding symmetric function operator identities. They are defined by setting

$$\mathcal{T}_Y F[X] = F[X+Y] \qquad \text{and} \qquad \mathcal{P}_Y F[X] = \mathrm{Exp}[XY]F[X].$$

Alternatively, these operators are equivalent to

$$\mathcal{T}_Y = \sum_\lambda s_\lambda[Y]s_\lambda^\perp \qquad \text{and} \qquad \mathcal{P}_Y = \sum_\lambda s_\lambda[Y]\underline{s_\lambda}[X],$$

where $\underline{f}[X]$ is the operation of multiplication by $f[X]$. This, for instance, gives the following specializations:

$$\mathcal{T}_u = \sum_{k\geq 0} u^k h_k^\perp \qquad \text{and} \qquad \mathcal{P}_{-\frac{u}{M}} = \sum_{k\geq 0}(-u)^k \underline{e}_k^*.$$

One of the fundamental relations between these operators and Δ_v is the following special case of Garsia-Mellit's five-term relation:

THEOREM 2.1 (Garsia,Mellit [**19**]).

$$\nabla^{-1}\mathcal{T}_{uv}\nabla = \Delta_v^{-1}\mathcal{T}_u\Delta_v\mathcal{T}_{-u} \qquad \textit{or} \qquad \nabla\mathcal{P}_{-\frac{uv}{M}}\nabla^{-1} = \mathcal{P}_{\frac{u}{M}}\Delta_v\mathcal{P}_{-\frac{u}{M}}\Delta_v^{-1}.$$

We now define

$$\widetilde{\Theta}(z,v) = \Delta_v\mathcal{P}_{-\frac{z}{M}}\Delta_v^{-1} \qquad \text{and} \qquad \widetilde{\Theta}(z,v)^{-1} = \Delta_v\mathcal{P}_{\frac{z}{M}}\Delta_v^{-1}.$$

It was shown in [**15**] that these operators satisfy the following identity:

THEOREM 2.2.

$$\widetilde{\Theta}(z,v)^{-1}\mathcal{T}_u\widetilde{\Theta}(z,v) = \mathrm{Exp}\left[\frac{uz(v-1)}{M}\right]\Delta_{uzv}\mathcal{T}_u.$$

This theorem has an amazing number of consequences. In particular, it gives identities regarding Theta operators. We are now ready to connect these identities back to the operators described in the introduction.

DEFINITION 2.3. If F is homogeneous of degree n and G is homogeneous of degree m, then we define the Theta operator of [**13**] by setting

$$\Theta_F G = \begin{cases} 0 & m = 0 \text{ and } n > 0 \\ \Pi \underline{F}^* \Pi^{-1} G & \text{otherwise.} \end{cases}$$

The special case when $F = e_k$ has been denoted by $\Theta_k = \Theta_{e_k}$.

Then one finds that

$$\widetilde{\Theta}(z,1) = \sum_{k \geq 0} (-z)^k \Theta_k.$$

We can then set $\Theta_z = \Theta_{\mathrm{Exp}[-zX]}$ and find that $\widetilde{\Theta}(z,1) = \Theta_z$. If we specialize Theorem 2.2 to when $v = 1$, we get

$$\Theta_z^{-1} \mathcal{T}_u \Theta_z \mathcal{T}_{-u} = \Delta_{uz}.$$

This is an important identity that relates the Theta operators (of the compositional Delta conjecture and generalized coinvariant algebras) with translation and Delta operators (generators from the Schiffmann algebra, or elliptic Hall algebra).

2.2. The Schiffmann algebra and the five-term relation. For any expression $A(q,t)$ in q and t, we let $\overline{A}(q,t) = A(1/q, 1/t)$. This operation carries over to delta operators as follows:

$$\overline{\Delta}_F \widetilde{H}_\mu = \overline{F\left[B_\mu\right]} \widetilde{H}_\mu \qquad \text{and} \qquad \overline{\Delta}'_F \widetilde{H}_\mu = \overline{F\left[B_\mu - 1/M\right]} \widetilde{H}_\mu$$

DEFINITION 2.4. For any symmetric function F, we define

$$F\left[X^{1,0}\right] = \Delta'_{\omega F}, \qquad F\left[X^{-1,0}\right] = \Delta_{(\omega F)\overline{[1/M - X]}} = \overline{\Delta'}_{\overline{F}[-X]},$$
$$F\left[X^{0,1}\right] = \underline{F}[-X/M], \text{ and} \qquad F[X^{0,-1}] = F^\perp.$$

These operators can be related to terms appearing in the five-term relation as follows:

$$T_{0,1} = \mathcal{P}_{-\frac{u}{M}} = \mathrm{Exp}[uX^{0,1}], \qquad T_{0,-1} = \mathcal{T}_u = \mathrm{Exp}[uX^{0,-1}],$$
$$T_{1,0} = \Delta'_v = \mathrm{Exp}[\epsilon v X^{1,0}], \qquad T_{-1,0} = \overline{\Delta}'_v = \mathrm{Exp}[\epsilon v X^{-1,0}].$$

There are two important actions on these linear operators that we should now point out. The first is defined on any operator L acting on symmetric functions by setting $\mathcal{N}(L) = \nabla L \nabla^{-1}$. We will also need to consider conjugating by the normalized nabla operator: $\overline{\mathcal{N}}(L) = \overline{\nabla} L \overline{\nabla}^{-1}$. The second can be defined as the unique operator $\mathcal{S}^{-1}(L)$ for which we have the commutation $\mathcal{T}_{\frac{-1}{M}} \mathcal{T}_1 L = \mathcal{S}^{-1}(L) \mathcal{T}_1 \mathcal{P}_{\frac{-1}{M}}$ as operators on symmetric functions. Shortly, we will describe in further detail that this map gives $\mathcal{S}^{-1}(p_1[-MX^{1,a}]) = p_1[-MX^{1,a-1}]$. The $T_{m,n}$ come with a u and v variable. So we add the condition that $\mathcal{N}^{-1}$ acts on the variables by sending (u,v) to (uv, v). Also let $\mathcal{S}^{-1}$ send (u,v) to (u, uv).

It turns out that we must have the following relations amongst the $T_{m,n}$:

$$\mathcal{N}^{-1}(T_{m,n}) = T_{m,m+n} \qquad \text{and} \qquad \mathcal{S}^{-1}(T_{m,n}) = T_{m+n,n} \qquad (\text{ for } n > 0),$$

with $\mathcal{S}^{-1}(T_{1,0}) = \mathrm{Exp}[u/M] T_{1,0}$. The five-term relation, in its general form, can be stated now as follows.

THEOREM 2.5 (Garsia,Mellit [**19**]). *For $m, n, m', n' \geq 0$ satisfying $mn' - m'n = 1$, we have*

$$T_{m,n}T_{m',n'} = T_{m',n'}T_{m+m',n+n'}T_{m,n}.$$

There is then a unique way of building $T_{m,n}$ using this identity and the operators $\mathcal{N}$ and $\mathcal{S}$, starting from $T_{1,0}$ and $T_{0,1}$. For instance,

$$T_{1,1} = T_{0,1}^{-1}T_{1,0}T_{0,1}T_{1,0}^{-1} = \mathcal{N}^{-1}(T_{1,0}) = \mathcal{S}^{-1}(T_{0,1}).$$

These series, when expanded in terms of u and v give

$$T_{m,n} = \sum_{k\geq 0} u^{mk}v^{nk}R_{mk,nk},$$

where $R_{mk,nk}$ are thus generated by the collection of $R_{k,0}$ and $R_{0,k}$.

As is done in [**7**] or [**9**], we can extend from $X^{1,0}$ and $X^{0,1}$ to a family $\{X^{m,n}\}$ of operators. Let $\Lambda[X^{m,n}]$ be the space of symmetric functions evaluated on all $X^{m,n}$ for an integer pair (m, n). We will not include the relations between them, which can essentially be found in [**9**] after translating to the conventions used in [**7**]. We should know however that $\overline{\mathcal{N}}$ acts on this space as follows.

$$\overline{\nabla}(F[X^{m,n}])\overline{\nabla}^{-1} = F[X^{m+n,n}],$$

and the operators $p_1[-MX^{1,k}] = (-1)^k D_k$ give the operators generated by

$$D(z) = \mathcal{P}_{-z}\mathcal{T}_{\frac{M}{z}} = \sum_{k\in\mathbb{Z}} z^k D_k.$$

One of the relations between Theta operators and the D_k operators is the following relation conjectured in [**13**] and proved in [**35**].

$$[\Theta_{e_k}, p_1[-MX^{1,1}]] = \sum_{i=1}^{k} p_1[-MX^{1,i+1}]\Theta_{e_{k-i}},$$

or simply

$$[\Theta_{e_k}, p_1[X^{1,1}]] = \sum_{i=1}^{k} p_1[X^{1,i+1}]\Theta_{e_{k-i}}.$$

More importantly, this relation follows from the following operator identity:

$$-u\widetilde{\Theta}(u,v)p_1[-MX^{1,1}]\widetilde{\Theta}(u,v)^{-1} = \sum_{k>0}(-u)^k p_1[-MX^{1,k}]$$

In the language of the Schiffmann algebra, this says

$$T_{1,0}T_{0,1}T_{1,0}^{-1}p_1[-MX^{1,1}]T_{1,0}^{-1}T_{0,1}^{-1}T_{1,0} = \sum_{k>0}(-u)^{k-1}p_1[-MX^{1,k}].$$

Furthermore, we see precisely how Theta operators live in the Schiffmann algebra:

THEOREM 2.6. *We have*

$$\Theta_{e_k} = h_k[X^{1,1}] + \sum_{r+s\leq k-1}(-1)^{k-r}h_r[X^{1,0}]h_k[X^{0,1}]e_s[X^{1,0}].$$

PROOF. The operator $(-1)^k\Theta_{e_k}$ is found by setting $v = 1$ in the following expression:

$$\begin{aligned}\widetilde{\Theta}(u,v)\Big|_{u^k} = T_{1,0}T_{0,1}T_{1,0}^{-1}\Big|_{u^k} &= \Delta'_v \mathcal{P}_{\frac{-u}{M}} \Delta'^{-1}_v\Big|_{u^k} \\ &= \Delta'_v h_k[X^{0,1}]\Delta'^{-1}_v \\ &= \sum_{r\geq 0}(-1)^r h_r[vX^{1,0}]h_k[X^{0,1}]\sum_{s\geq 0} e_s[vX^{1,0}].\end{aligned}$$

The last equality appears as a series. However it is a polynomial in v. This polynomiality is the adjoint version of Proposition 2.2 in [**19**]. For any homogeneous symmetric function F of degree k, the operator

$$\Delta'_v(\omega F)[X^{0,1}]\Delta'^{-1}_v = \Delta_v(\omega F)[X^{0,1}]\Delta_v^{-1}$$

acts on the modified Macdonald basis as follows:

$$\begin{aligned}\Delta_v \underline{F}^* \Delta_v^{-1}\widetilde{H}_\mu &= \sum_{\nu\supset\mu}\widetilde{H}_\mu \frac{\langle F^*\widetilde{H}_\mu, \widetilde{H}_\nu\rangle_*}{w_\mu}\prod_{(i,j)\in\nu}(1-vq^it^j)\prod_{(i,j)\in\mu}\frac{1}{1-vq^it^j} \\ &= \sum_{\nu\supset\mu}\widetilde{H}_\mu \frac{\langle F^*\widetilde{H}_\mu, \widetilde{H}_\nu\rangle_*}{w_\mu}\prod_{(i,j)\in\nu/\mu}(1-vq^it^j).\end{aligned}$$

By the Macdonald Pieri rules, the sums run only over partitions $\nu \vdash |\mu| + k$ containing μ, meaning the two products cancel and leave a product over cells in ν but not μ. We also see that the highest degree in v is $|\nu| - |\mu| = k$, and in degree k we have that the products contribute a factor of

$$(-1)^k \prod_{(i,j)\in\nu/\mu} q^it^j = (-1)^k T_\nu/T_\mu.$$

Therefore, the degree k term of $\Delta_v F[X^{0,1}]\Delta_v^{-1}$ is given by

$$\nabla F[X^{0,1}]\nabla^{-1} = (-1)^k\overline{\mathcal{N}}(F[X^{0,1}]) = (-1)^k F[X^{1,1}].$$

In all, we have shown that

$$(-1)^k\Theta_{e_k} = (-1)^k h_k[X^{1,1}] + \sum_{r+s\leq k-1}(-1)^r h_r[X^{1,0}]h_k[X^{0,1}]e_s[X^{1,0}],$$

or in other terms

$$\Theta_{e_k} = h_k[X^{1,1}] + \sum_{r+s\leq k-1} e_r[-X^{1,0}]e_k[-X^{0,1}]e_s[X^{1,0}], \qquad \square$$

In the end, it would be worthwhile to investigate the deeper connections between the $X^{m,n}$, the $R_{m,n}$, and Theta operators.

2.3. A new operator to consider. In [**30**], a new symmetric function operator was considered. Recall that the elementary symmetric function has the form

$$e_n = \sum_{\mu\vdash n} MB_\mu\Pi_\mu\frac{\widetilde{H}_\mu}{w_\mu} = M\Delta_{e_1}\Pi\sum_{\mu\vdash n}\frac{\widetilde{H}_\mu}{w_\mu} = M\Delta_{e_1}\Pi e_n^*$$

It will then be convenient to introduce the following operator:

DEFINITION 2.7. For any symmetric function F, we set

$$\Xi F = M\Delta_{e_1}\Pi F^*.$$

With this, the above expansion of e_n is actually stating that

$$\Xi e_n = e_n.$$

Thus, e_n is an eigenfunction of Ξ with eigenvalue 1.

Say we want to study the general behavior of Theta operators, and in particular, we would like to know how expressions of the form $\Theta_F G$ expand in terms of some classical basis (like the elementary, Schur, or monomial bases). Since this expression is linear in both F and G, we can solve this problem by picking F and G to be from a particular basis. Choosing $F = e_\lambda$ and $G = \Pi e^*_\mu$, we find that

$$\Theta_{e_\lambda} \Pi e^*_\mu = \Pi^*(e_\lambda e_\mu).$$

In order to produce positivities, we then include a normalization given by $M\Delta_{e_1}$. In the end, we are looking at Ξ applied to an elementary symmetric function. This means that giving expansions for symmetric functions of the form Ξe_λ is already enough to determine the general behavior of Theta operators.

3. Combinatorial aspects

We will now look at some of the combinatorial aspects associated with Theta, Delta and Xi operators. One of the most important underlying structures we will see are parallelogram polyominoes. A parallelogram polyomino is a pair of lattice paths (P, Q) consisting of North and East unit steps such that P is strictly above Q except at the beginning $(0,0)$ and end (m,n). We will then say that $(P,Q) \in \mathbf{P}^{m,n}$. For instance, in Figure 4 we have a polyomino (R,S) in the 12 by 8 rectangle: We can then represent P as a sequence $N^{a_1}E^{b_1}\cdots N^{a_r}E^{b_r}$ with a_1, b_1, $\ldots$, a_r, b_r all nonzero. And similarly we can represent Q as a sequence $E^{c_1}N^{d_1}\cdots E^{c_s}, N^{d_s}$ with c_1, d_1, $\ldots$, c_s, d_s all nonzero. Define $\lambda(P,Q) = (a_1, \ldots, a_r)$ to be the composition whose parts are the lengths of the vertical segments in P; and define $\mu(P,Q) = (c_1 - 1, c_2, \ldots, c_s)$ to be the composition whose parts are the lengths of the horizontal segments in Q. Note here that the first step in Q is forced and always ignored when finding the composition for Q. For instance, in Figure 4, the top path is given by $R = N^3E^1N^1E^5N^4E^6$. This gives the composition $\lambda(R,S) = (3,1,4)$. We also have $S = E^2N^1E^2N^2E^3N^1E^2N^1E^1N^1E^2N^2$. Therefore $\mu(R,S) = (1,2,3,2,1,2)$. If $\lambda(P,Q)$ rearranges to the partition λ and $\mu(P,Q)$ rearranges to the partition μ,

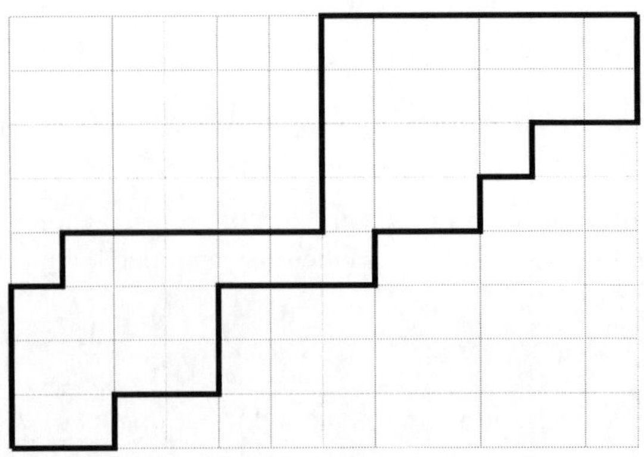

FIGURE 4. The polyomino $(R,S) \in \mathbf{P}^{12,8}$ with area $35 - (12 + 8 - 1)$.

then we will write $(P,Q) \in \mathbf{P}_{\lambda,\mu}$. We call a cell in between P and Q an area cell if it does not touch Q, and denote the total number of area cells by $\mathrm{area}(P,Q)$.

There is an amazing relation between polyominoes and Delta eigenoperators, and this connection is yet to be fully understood. It can be found in [**28**] and [**8**].

THEOREM 3.1. *Given a positive integer n and any partition λ,*

$$\Delta_{m_\lambda} e_n \Big|_{t=1} = \sum_{\mu \vdash n} e_\mu \sum_{P \in \mathbf{P}_{\lambda,\mu}} q^{\mathrm{area}(P)}.$$

This symmetric function $\Delta_{m_\lambda} e_n$ is not Schur positive, yet remarkably, we get an e-positivity when setting either t or q to 1. There is an even further remarkable positivity which we refer to as the e-positivity phenomenon.

DEFINITION 3.2. We say that a symmetric function $F[X;q,t]$ exhibits the e-positivity phenomenon if $F[X;1+q,t]$ is e-positive. This is to say that

$$F[X;1+q,t] = \sum_{\mu} c_\mu(q,t) e_\mu$$

for some $c_\mu(q,t) \in \mathbb{N}[q,t]$.

In particular, it is conjectured in [**34**] that $\Delta_{m_\lambda} e_n$ exhibits the e-positivity phenomenon. This conjecture would imply that there is a statistic $\mathrm{stat}(P,S)$, depending on a polyomino P and subset $S \subseteq \mathrm{Area}(P)$ of the area cells of P, for which

$$\Delta_{m_\lambda} e_n = \sum_{\mu \vdash n} e_\mu \sum_{P \in \mathbf{P}_{\lambda,\mu}} \sum_{S \subseteq \mathrm{Area}(P)} q^{|S|} t^{\mathrm{stat}(P,S)}.$$

We should mention that the special case when $\lambda = (1^k)$ has now been proved since this corresponds to the Delta Conjecture. We will now describe this connection.

A Dyck Path $D \in D_n$ is a polyomino (P,Q) whose bottom path is given by $Q = E(EN)^n$. This means that P corresponds to a lattice path consisting of North and East unit steps from $(0,0)$ to (n,n) which never go below the main diagonal $y = x$. Given such a Dyck path D, we construct the set of labeled Dyck paths supported by D, $\mathrm{LD}(D)$ by labelling North steps in D with positive integers so that columns are increasing. Reading the labeling from bottom to top, we have a word $w = (w_1, \ldots, w_n)$, and our labeled Dyck path can be denoted by $P = (D,w)$. Let $x^P = x_{w_1} \cdots w_{w_n}$. Each letter w_i corresponds to a North segment that starts on the diagonal $y = x + d_i$. We set

$$\mathrm{dinv}(D,w) = \Big| \{(i<j) : w_i < w_j \wedge d_i = d_j \text{ or } w_i > w_j \wedge d_i = d_j + 1\} \Big|.$$

The vertical strip LLT polynomial associated to D can be written as

$$LLT_D[X;t] = \sum_{P \in \mathrm{LD}(D)} t^{\mathrm{dinv}(P)} x^P$$

Given a Dyck path D, let $a_i(D)$ be the number of complete, unit lattice cells between the ith North step and the main diagonal $y = x$. Then define the Haglund factor by setting

$$H_D(z) = (1+z) \prod_{a_{i+1} = a_i + 1} \left(1 + \frac{z}{q^{a_{i+1}}}\right).$$

Then the rise version of the Delta Conjecture can be stated as

$$\sum_{k=0}^{n} z^{n-k}\Delta_{e_k} e_n = \sum_{D\in D_n} H_D(z) q^{\text{area}(D)} LLT[X;t].$$

Note that the Shuffle conjecture is given by the coefficient of z^0 on both sides. D'Adderio showed in [**11**] that vertical strip LLT polynomials exhibit the e-positivity phenomenon, and a specific combinatorial formula is given in [**1**]. This implies that $\Delta_{e_k} e_n$, and in particular $\overline{\nabla} e_n$, exhibits the e-positivity phenomenon.

It is unclear if there is a vertical strip LLT expansion of $\Delta_{m_\lambda} e_n$. In fact, since $\Delta_{m_\lambda} e_n$ is not even Schur positive, it cannot be a positive sum of vertical strip LLT polynomials. However, we should note that an expression can be found using the methods of Blasiak, Haiman, Morse, Pun and Seelinger. Furthermore, the generalized q,t-Catalan numbers in [**21**] can be viewed to give a refinement of $\langle \Delta_{m_\lambda} e_n, e_n\rangle$. There are still many open questions regarding this case.

We now look at a more general setting in which these symmetric functions appear.

3.1. A variation on parking functions. Here, we will introduce the set of γ-parking functions of type λ, denoted PF^γ_λ. A γ-parking function of height n is constructed by the following steps:

(1) Select a rearrangement $(a_1,\dots,a_n)$ of $\gamma_1, \dots, \gamma_{\ell(\gamma)}, 0^{n-\ell(\gamma)}$. Let $Q = EE^{a_1+1}NE^{a_2+1}N\cdots E^{a_n+1}N$.
(2) Choose a path P from $(0,0)$ to $(n+|\gamma|+1, n)$ that stays strictly above Q except at the beginning and end points.
(3) Place the numbers $1^{m_1(\lambda)}, 2^{m_2(\lambda)}, \dots, n^{m_n(\lambda)}$ to the right of North steps in P so that columns are strictly increasing from bottom to top. Let w be the word you get by reading the entries from bottom to top.

If the multiplicity of i in w is λ_i, then we will say that w is of type λ, and $(P,Q,w)\in \text{PF}^\gamma_\lambda$. A γ-parking function is said to be lattice if w is a lattice word, i.e. the number of occurrences of i in any prefix $w_1, \dots, w_r$ is at least the number of $i+1$'s in the prefix. The set of lattice γ-parking functions of type λ will then be denoted by $\text{LPF}^\gamma_\lambda$ Lastly, the area of a γ-parking function is the same as the area of its underlying polyomino.

DEFINITION 3.3. The η-composition of a γ-parking function is found as follows: Suppose we are given $A = (P,Q,w)\in \text{PF}^\gamma_\lambda$, and suppose that $P = N^{a_1}E^{b_1}\cdots N^{a_k}E^{b_k}$ with all a_i and b_i nonzero. Start with the composition $\lambda(P) = (a_1, a_2, \dots, a_k)$. Two North segments N^{a_i} and $N^{a_{i+1}}$ are adjacent if $b_i = 1$. Scan from right to left for the first i such that the ith North segment is adjacent to the next North segment. If $w_{a_1+\cdots+a_i+1} < w_{a_1+\cdots+a_i}$, in other words $a_1+\cdots+a_i \notin \text{Asc}(w)$, then we join the two segments together to make (a_i, a_{i+1}) into $(a_i + a_{i+1})$. Now continue inductively to the left of i. Once we reach a_1, we will end with a composition which we denote by $\eta(A)$.

Two adjacent North segments are combined if the labels in the columns do not form an ascent as we move from one column to the next. For example, in Figure 5 we have a γ-parking function for when $\gamma = (2,1,1,1)$ and $\lambda = (2,2,2,1)$. If we go up the first column and jump to the second, we see that the cell with a 4 is corner-to-corner with a cell labeled with a $2 < 4$. This is not an ascent, and so we

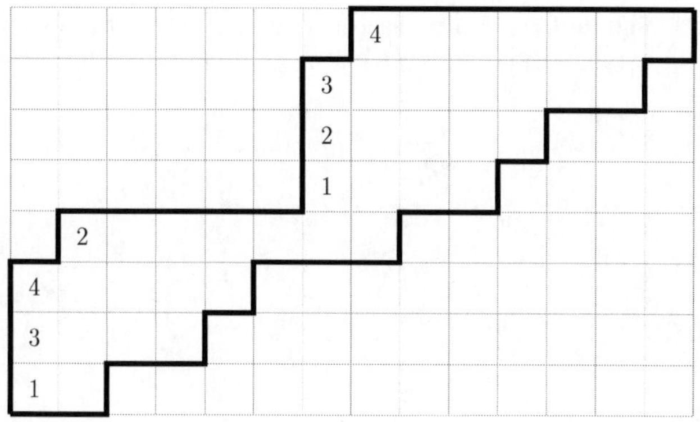

FIGURE 5. A γ-parking function $\mathbf{P}^{(2,1,1,1)}_{(2,2,2,1)}$ with area 21.

include the 2 in the same compositional part as the 4. This gives a part of size 4. We then continue to the next vertical segment which is a column filled with $1, 2, 3$ followed by an adjacent column with only a 4. Since 3 to 4 is an ascent, we keep these two segments separate. All together, we have $\eta(P) = (4, 3, 1)$.

The main results in **[30]** regarding γ-parking functions and Ξ say that

THEOREM 3.4. *For any partition γ and any partition λ, we have*

$$\Delta_{m_\gamma} \Xi e_\lambda \Big|_{t=1} = \sum_{P \in \mathrm{PF}^\gamma_\lambda} q^{\mathrm{area}(P)} e_{\eta(P)}.$$

Similarly, there is a an expression for Schur functions and lattice γ-parking functions:

THEOREM 3.5. *For any partition γ and any partition λ, we have*

$$\Delta_{m_\gamma} \Xi s_{\lambda'} \Big|_{t=1} = \sum_{P \in \mathrm{LPF}^\gamma_\lambda} q^{\mathrm{area}(P)} e_{\eta(P)}.$$

We should note that the original notion of a parking function is in our case an element of $\mathrm{PF}^{\emptyset}_{1^n}$. This means that the bottom path must form a staircase, the top path forms a Dyck path, and the numbers $1, \ldots, n$ are used as labels for the top path, where each number appears precisely once. We have

$$\Xi e_{1^n} \Big|_{t=1} = \sum_{P \in \mathrm{PF}^{\emptyset}_{1^n}} q^{\mathrm{area}(P)} e_{\eta(P)}.$$

On the other hand, the case of the Shuffle Conjecture is given by

$$\Delta_{m_{1^n}} \Xi e_n \Big|_{t=1} = \sum_{P \in \mathrm{PF}^{1^n}_n} q^{\mathrm{area}(P)} e_{\eta(P)}.$$

Note here that the bottom path of a (1^n)-parking function is not a staircase, but rather has the form $EE^2NE^2N \cdots E^2N$. This is not the typical Dyck path picture we have seen in the previous sections. Surprisingly, the original, combinatorial statement of the Shuffle Conjecture is best viewed through the following equality: Recall from the properties of the modified Macdonald basis, we have $\langle \widetilde{H}_\mu, e_{|\mu|}\rangle = T_\mu$. It follows that

$$h_n^* = \sum_{\mu \vdash n} \langle h_n^*, \widetilde{H}_\mu \rangle \frac{\widetilde{H}_\mu}{w_\mu} = \sum_{\mu \vdash n} T_\mu \frac{\widetilde{H}_\mu}{w_\mu}.$$

Therefore

PROPOSITION 3.6.

$$\Xi h_n = \overline{\nabla} e_n.$$

In this setting, the Dyck path picture is more apparent:

$$\Xi s_n \Big|_{t=1} = \sum_{P \in \mathrm{LPF}_n^{\emptyset}} q^{\mathrm{area}(P)} e_{\eta(P)}.$$

The elements $(P, Q, w) \in \mathrm{LPF}_n^{\emptyset}$ can be described in the following way: Q is a staircase of the form $E(EN)^n$, P then must correspond to a Dyck path, and $w = 1\,2 \cdots n$. This means that every i is an Ascent of w. Thus, in this case, we have $\eta(P, Q, w) = \lambda(P, Q)$.

3.2. Tiered trees and Theta operators. One of the most recent results regarding Theta operators, relates symmetric functions of the form $\Theta_{e_\lambda} e_1$ with tiered trees. A rooted tiered tree is a tree $T = (V, E)$ consisting of vertices V and edges E with a level function $\mathrm{lv} : V \to \mathbb{N}$ and a labeling $w : V \to \mathbb{N}_+$ such that the following hold:

(a) If $\{i, j\} \in E$, then $\mathrm{lv}(i) \neq \mathrm{lv}(j)$.
(b) If $\{i, j\} \in E$ and $\mathrm{lv}(i) < \mathrm{lv}(j)$, then $w(i) < w(j)$.
(c) If $p(i) = p(j)$ and $\mathrm{lv}(i) = \mathrm{lv}(j)$, then $w(i) \neq w(j)$. Here $p(i)$ denotes the parent of i, which is the unique neighbor of i closest to the root.
(d) The root r is the only vertex at level 0 ($\mathrm{lv}^{-1}(0) = \{r\}$).

A rooted α-tree is one in which $|\mathrm{lv}^{-1}(i)| = \alpha_i$, so that α gives the number of vertices at level i. The collection of all rooted, tiered α-trees will be denoted by $\mathrm{RTT}(\alpha)$; and the collection of all rooted, tiered α-trees where the root is the unique vertex labeled with 0 will be denoted by $\mathrm{RTT}_0(\alpha)$. See Figure 6 for an example.

Recall that j is a descendant of i if there is a $k > 0$ for which $p^k(j) = i$. We will say that two vertices i and j are compatible if $\mathrm{lv}(i) < \mathrm{lv}(j)$ and $w(i) < w(j)$, or if the reverse inequalities hold. This means that absent of all other vertices, one is able to connect i and j without violating the labelling rules (both levels and labels increase, or they both decrease).

DEFINITION 3.7. An inversion of T is a pair (i, j) of non-root vertices for which the following hold:

(1) j is a descendant of i,
(2) j is compatible with $p(i)$, and
(3) either $w(j) < w(i)$ or $w(j) = w(i) \wedge \mathrm{lv}(j) > \mathrm{lv}(i)$.

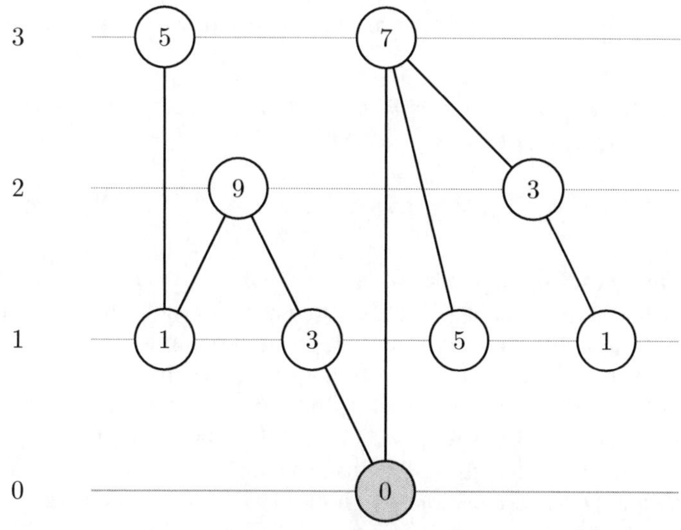

FIGURE 6. An element of $\mathrm{RTT}_0(4,2,2)$ with 6 inversions.

CONJECTURE 3.8 (The Theta Trees Conjecture[**12**]). *For any composition* α,

$$\Theta_{e_\alpha} e_1\Big|_{t=1} = \sum_{T\in \mathrm{RTT}(\alpha)} q^{\mathrm{inv}(T)} x^T.$$

Also, one has

$$\Xi e_\alpha\Big|_{t=1} = \sum_{\mathrm{RTT}_0(\alpha)} q^{\mathrm{inv}(T)} x^T.$$

There are two main comments to make about this conjecture. The first is that the case of $\Theta_{e_1^n} e_1\Big|_{t=1}$ was shown to hold [**12**]. In order to get this connection, it was essential to view this symmetric function as giving Kac polynomials for dandelion quivers: the coefficient of m_μ is the number of absolutely indecomoposable representations over $\mathbb{F}_q$ with dimension vector depending on μ. There is then the question of whether general Theta operator expressions can be interpreted using the representations of certain quivers.

The second main comment to make is that in [**30**], as described in the previous portions of this section, we have an elementary symmetric function for $\Xi e_\lambda\Big|_{t=1}$. Since the change of basis from elementary symmetric functions to monomials is easy to understand, it is expected that we should be able to solve this portion of the Theta Trees Conjecture. However, it has remained unclear precisely how this can be done.

4. Concluding Remarks

The combinatorial expansions achieved in [**8**] make use of Neguţ's elements [**32**] and a connection between the shuffle algebra, Schiffmann's algebra, and LLT polynomials. Though we made no effort to make such connections to Theta operators in this article, we think this may have a vital role in properly understanding and solving some of these open questions.

Many positivities have yet to be fully understood. For instance, from [**30**], we have the following conjecture:

CONJECTURE 4.1. *For any partitions* γ *and* λ, $\Delta_{m_\gamma}\Xi s^*_\lambda$ *exhibits the e-positivity phenomenon. This means there is a statistic* stat *for which*

$$\Delta_{m_\gamma}\Xi s^*_\lambda\Big|_{q\to 1+q} = \sum_{P\in \mathrm{LPF}^\gamma_\lambda,}\ \sum_{S\subseteq \mathrm{Area}(P)} q^{|S|}t^{\mathrm{stat}(P,S)}e_\eta(P).$$

It is our hope that by understanding how Theta operators lie in the Schiffmann and shuffle algebras, we will get a better understanding of what stat should be in this conjecture. It would also be interesting to understand the symmetric function side of Conjecture 1.2 and give a combinatorial description for the conjectural distribution of irreducible representations in $\mathcal{A}^{(2,2)}$. Lastly, we saw that the case $\alpha = (1^n)$ of Conjecture 3.8 was proved by relating $\Theta_{e_1^n}e_1$ to the Kac polynomial for dandelion quivers. Perhaps this is just the surface as to how Theta operators are related to quiver representation. We hope that these questions and connections spark further interest in the study of Theta operators.

References

[1] P. Alexandersson and R. Sulzgruber, *A combinatorial expansion of vertical-strip LLT polynomials in the basis of elementary symmetric functions*, Adv. Math. **400** (2022), Paper No. 108256, 58, DOI 10.1016/j.aim.2022.108256. MR4385139

[2] F. Bergeron, *The Bosonic-Fermionic Diagonal Coinvariant Modules Conjecture*, Preprint, `arXiv:2005.00924`, 2020.

[3] F. Bergeron and A. M. Garsia, *Science fiction and Macdonald's polynomials*, Algebraic methods and q-special functions (Montréal, QC, 1996), CRM Proc. Lecture Notes, vol. 22, Amer. Math. Soc., Providence, RI, 1999, pp. 1–52, DOI 10.1090/crmp/022/01. MR1726826

[4] F. Bergeron, A. M. Garsia, M. Haiman, and G. Tesler, *Identities and positivity conjectures for some remarkable operators in the theory of symmetric functions*, Methods Appl. Anal. **6** (1999), no. 3, 363–420, DOI 10.4310/MAA.1999.v6.n3.a7. Dedicated to Richard A. Askey on the occasion of his 65th birthday, Part III. MR1803316

[5] J. Blasiak, M. Haiman, J. Morse, A. Pun, and G. H. Seelinger, *Dens, nests, and the Loehr-Warrington conjecture*, Preprint, `arXiv:2112.07070`, 2021.

[6] J. Blasiak, M. Haiman, J. Morse, A. Pun, and G. H. Seelinger, *LLT polynomials in the Schiffmann algebra*, J. Reine Angew. Math. **811** (2024), 93–133, DOI 10.1515/crelle-2024-0012. MR4751118

[7] J. Blasiak, M. Haiman, J. Morse, A. Pun, and G. H. Seelinger, *A shuffle theorem for paths under any line*, Forum Math. Pi **11** (2023), Paper No. e5, 38, DOI 10.1017/fmp.2023.4. MR4553914

[8] J. Blasiak, M. Haiman, J. Morse, A. Pun, and G. H. Seelinger, *A proof of the extended delta conjecture*, Forum Math. Pi **11** (2023), Paper No. e6, 28, DOI 10.1017/fmp.2023.3. MR4553915

[9] I. Burban and O. Schiffmann, *On the Hall algebra of an elliptic curve, I*, Duke Math. J. **161** (2012), no. 7, 1171–1231, DOI 10.1215/00127094-1593263. MR2922373

[10] E. Carlsson and A. Mellit, *A proof of the shuffle conjecture*, J. Amer. Math. Soc. **31** (2018), no. 3, 661–697, DOI 10.1090/jams/893. MR3787405

[11] M. D'Adderio, *e-positivity of vertical strip LLT polynomials*, J. Combin. Theory Ser. A **172** (2020), 105212, 15, DOI 10.1016/j.jcta.2020.105212. MR4054520

[12] M. D'Adderio, A. Iraci, Y. Le Borgne, M. Romero, and A. Vanden Wyngaerd, *Tiered trees and Theta operators*, Int. Math. Res. Not. IMRN **24** (2023), 20748–20783, DOI 10.1093/imrn/rnac258. MR4681271

[13] M. D'Adderio, A. Iraci, and A. Vanden Wyngaerd, *Theta operators, refined delta conjectures, and coinvariants*, Adv. Math. **376** (2021), Paper No. 107447, 59, DOI 10.1016/j.aim.2020.107447. MR4178919

[14] M. D'Adderio and A. Mellit, *A proof of the compositional delta conjecture*, Adv. Math. **402** (2022), Paper No. 108342, 17, DOI 10.1016/j.aim.2022.108342. MR4401822

[15] M. D'Adderio and M. Romero, *New identities for theta operators*, Trans. Amer. Math. Soc. **376** (2023), no. 8, 5775–5807, DOI 10.1090/tran/8911. MR4630759

[16] A. M. Garsia and M. Haiman, *A graded representation model for Macdonald's polynomials*, Proc. Nat. Acad. Sci. U.S.A. **90** (1993), no. 8, 3607–3610, DOI 10.1073/pnas.90.8.3607. MR1214091

[17] A. M. Garsia and M. Haiman, *A remarkable q,t-Catalan sequence and q-Lagrange inversion*, J. Algebr. Comb. **5** (1996), no. 3, 191–244, DOI 10.1023/A:1022476211638. MR1394305

[18] A. M. Garsia, M. Haiman, and G. Tesler, *Explicit plethystic formulas for Macdonald q,t-Kostka coefficients*, Sém. Lothar. Combin. **42** (1999), Art. B42m, 45. The Andrews Festschrift (Maratea, 1998). MR1701592

[19] A. Garsia and A. Mellit, *Five-term relation and Macdonald polynomials*, J. Combin. Theory Ser. A **163** (2019), 182–194, DOI 10.1016/j.jcta.2018.12.003. MR3885219

[20] A. M. Garsia and G. Tesler, *Plethystic formulas for Macdonald q,t-Kostka coefficients*, Adv. Math. **123** (1996), no. 2, 144–222, DOI 10.1006/aima.1996.0071. MR1420484

[21] E. Gorsky, G. Hawkes, A. Schilling, and J. Rainbolt, *Generalized q,t-Catalan numbers*, Algebr. Comb. **3** (2020), no. 4, 855–886, DOI 10.5802/alco.120. MR4145982

[22] J. Haglund, M. Haiman, and N. Loehr, *A combinatorial formula for Macdonald polynomials*, J. Amer. Math. Soc. **18** (2005), no. 3, 735–761, DOI 10.1090/S0894-0347-05-00485-6. MR2138143

[23] J. Haglund, M. Haiman, N. Loehr, J. B. Remmel, and A. Ulyanov, *A combinatorial formula for the character of the diagonal coinvariants*, Duke Math. J. **126** (2005), no. 2, 195–232, DOI 10.1215/S0012-7094-04-12621-1. MR2115257

[24] J. Haglund, J. Morse, and M. Zabrocki, *A Compositional Shuffle Conjecture Specifying Touch Points of the Dyck Path*, Can. J. Math. **64** (2012), no. 4, 822–844, DOI 10.4153/CJM-2011-078-4. MR2957232

[25] J. Haglund, J. B. Remmel, and A. T. Wilson, *The delta conjecture*, Trans. Amer. Math. Soc. **370** (2018), no. 6, 4029–4057, DOI 10.1090/tran/7096. MR3811519

[26] M. Haiman, *Hilbert schemes, polygraphs and the Macdonald positivity conjecture*, J. Amer. Math. Soc. **14** (2001), no. 4, 941–1006, DOI 10.1090/S0894-0347-01-00373-3. MR1839919

[27] M. Haiman, *Vanishing theorems and character formulas for the Hilbert scheme of points in the plane*, Invent. Math. **149** (2002), no. 2, 371–407, DOI 10.1007/s002220200219. MR1918676

[28] A. Hicks and M. Romero, *Delta operators at $q = 1$ and polyominoes*, Sém. Lothar. Combin. **80B** (2018), Art. 77, 12. MR3940652

[29] A. Iraci, B. Rhoades, and M. Romero, *A proof of the fermionic theta coinvariant conjecture*, Discrete Math. **346** (2023), no. 7, Paper No. 113474, 11, DOI 10.1016/j.disc.2023.113474. MR4579856

[30] A. Iraci and M. Romero, *Delta and theta operator expansions*, Forum Math. Sigma **12** (2024), Paper No. e30, 37, DOI 10.1017/fms.2024.14. MR4713626

[31] I. G. Macdonald, *Symmetric functions and Hall polynomials*, 2nd ed., Oxford Mathematical Monographs, The Clarendon Press, Oxford University Press, New York, 1995. With contributions by A. Zelevinsky; Oxford Science Publications. MR1354144

[32] A. Negut, *The shuffle algebra revisited*, Int. Math. Res. Not. IMRN **22** (2014), 6242–6275, DOI 10.1093/imrn/rnt156. MR3283004

[33] M. Romero, *The delta conjecture at $q = 1$*, Trans. Amer. Math. Soc. **369** (2017), no. 10, 7509–7530, DOI 10.1090/tran/7140. MR3683116

[34] M. Romero, *Delta eigenoperators and e-positivities in the theory of Macdonald polynomials*, Ph.D. Thesis, 2019.

[35] M. Romero, *A proof of the theta operator conjecture*, J. Combin. Theory Ser. A **185** (2022), Paper No. 105535, 7, DOI 10.1016/j.jcta.2021.105535. MR4310025

Universität Wien Fakultät für Mathematik
Email address: `marino.romero@univie.ac.at`

Contemporary Mathematics
Volume **815**, 2025
https://doi.org/10.1090/conm/815/16318

A Littlewood-Richardson rule for SSV polynomials

Jason Saied

This paper is dedicated to my PhD advisor, Siddhartha Sahi.

ABSTRACT. Sahi, Stokman, and Venkateswaran introduced a family of quasi-polynomials called SSV polynomials, generalizing Macdonald polynomials and metaplectic Iwahori Whittaker functions. We prove a Littlewood-Richardson rule for the product of a Macdonald polynomial and an SSV polynomial, directly generalizing Yip's results for Macdonald polynomials. This work appeared previously in the author's PhD thesis, building on the author's alcove walk formula for SSV polynomials.

1. Introduction

In [**SSV**,**SSV2**], Sahi, Stokman, and Venkateswaran introduced a new family of quasi-polynomials, a generalization of both Macdonald polynomials and metaplectic Iwahori Whittaker functions. These quasi-polynomials, called SSV polynomials in [**Sai**], arise from a new representation of the double affine Hecke algebra, called the metaplectic representation in [**SSV**] and generalized to the *quasi-polynomial representation* in [**SSV2**]. Similarly to Macdonald polynomials, SSV polynomials are the simultaneous eigenfunctions of the Cherednik operators under the quasi-polynomial representation, and therefore may be studied using the same intertwiner theory.

In the case of Macdonald polynomials, Ram and Yip [**RY**] proved an expansion formula for the product of intertwiners and applied it in order to give an explicit formula for Macdonald polynomials in terms of combinatorial objects called alcove walks. Yip [**Yi**] then used the alcove walk formula in order to prove a generalization of the Littlewood-Richardson rule, expanding the product of two Macdonald polynomials with respect to the Macdonald basis, as a sum indexed by (a variant of) alcove walks. (Note that Sahi had previously given a Littlewood-Richardson rule for Macdonald polynomials in [**S**] as a special case of a result for interpolation polynomials. That theorem involves fewer terms and does not use alcove walks.) In [**Sai**], the author applied Ram and Yip's intertwiner formula in order to prove an alcove walk formula for SSV polynomials, generalizing the alcove walk formula of Ram and Yip for Macdonald polynomials. In the present work, we study the

2020 *Mathematics Subject Classification.* Primary 05E10, 20C08; Secondary 11F70.
Key words and phrases. Macdonald polynomials, Whittaker functions, Alcove walks, Combinatorial formulas, Hecke algebras, Metaplectic groups.
The author was supported in part by NSF Grant #DMS-2001537.

natural extension of Yip's Littlewood-Richardson rule to the SSV case, expanding the product of an SSV polynomial and a Macdonald polynomial in the basis of SSV polynomials. For convenience, we consider only the type A case studied in [**Sai**, **Sai2**]. However, these results may be generalized: the quasi-polynomial representation given in [**SSV2**] works for any root system, so one may use the appropriate intertwiners and adapt the techniques of [**Yi**] and the present work to obtain similar results for other types. Beyond the additional care needed to distinguish roots and coroots and introduce new parameters corresponding to the different root lengths, we expect no serious difficulties.

We briefly introduce some notation. Fix a positive integer $r \geq 2$. Let $E = \mathbb{R}^r$ and $P = \mathbb{Z}^r$, and consider the space of quasi-polynomials,

$$\mathbb{F}[E] = \mathrm{span}_{\mathbb{F}}\{x^\mu : \mu \in E\},$$

where $\mathbb{F}$ is a field of characteristic 0 with many parameters that will be discussed in Section 2.3. For arbitrary $\mu \in E$ and dominant $\lambda \in E$, we have the corresponding *SSV (quasi-)polynomial* $E_\mu \in \mathbb{F}[E]$ and *symmetric SSV (quasi-)polynomial* $P_\mu \in \mathbb{F}[E]$, originally defined in the works [**SSV**, **SSV2**]. Note that P_μ is not in general symmetric under the usual action of the symmetric group, but is instead symmetric with respect to a variant of the Weyl group action of Chinta and Gunnells [**CG1**, **CG2**]: see [**Sai**] Proposition 4.5 for details. When $\mu, \lambda \in P$, E_μ (respectively P_λ) is the corresponding nonsymmetric (respectively symmetric) Macdonald polynomial. Further, suitable specializations of SSV polynomials recover values of Iwahori Whittaker functions for principal series representations of metaplectic n-fold covers of $GL_r(K)$, where K is a non-Archimedean local field. These metaplectic n-fold covers and their Whittaker functions are discussed in [**BBBG**, **CO**, **KP**, **Mc**, **PP**], among other works. In particular, as discussed in Section 2.3 below, SSV polynomials involve two families of parameters in $\mathbb{F}$, here referred to as q (analogous to the parameter q in Macdonald polynomials) and G_s, $s \in \mathbb{R}$ (generalizing the parameters typically called k or t in Macdonald polynomials, with $G_0 = k$). By specializing the G_s parameters to certain Gauss sums in K and specializing $q \to 0, \infty$, one recovers certain values of metaplectic Iwahori Whittaker functions: see [**Sai**] Proposition 5.5. (The $q \to \infty$ case was first proven in [**SSV2**, Section 10].) The alcove walk formula for SSV polynomials gives new and very simple expressions for both of these q-specializations, proven in [**Sai**] Theorems 5.1 and 5.2. Thus the study of SSV polynomials can be applied to the study of both Macdonald polynomials and metaplectic Iwahori Whittaker functions.

The main results, a series of Littlewood-Richardson-type rules for the product of a Macdonald polynomial and an SSV polynomial, are summarized in the following theorem. The detailed results and their proofs are given in Section 4. The essential content is that one may express the product of a (possibly nonsymmetric) Macdonald polynomial and a (possibly nonsymmetric) SSV polynomial as a sum of nonsymmetric SSV polynomials with coefficients that can be computed combinatorially in terms of alcove walks. Further, if both polynomials are symmetric, then the product may be expressed in terms of only symmetric SSV polynomials.

THEOREM 1.1. *Let z be an affine Weyl group element, and let $\mu, \eta \in P$ and $\nu, \lambda \in E$, with η and λ dominant. We have*

$$X^z E_\nu = \sum_p \mathcal{A}_p E_{e(p)^{-1}m(\nu)} \tag{1.1}$$

$$E_\mu E_\nu = \sum_h \sum_p \mathcal{B}_{h,p} E_{e(p)^{-1}m(\nu)} \tag{1.2}$$

$$E_\mu P_\lambda = \sum_v \sum_h \sum_p \mathcal{C}_{v,h,p} E_{e(p)^{-1}m(\lambda)} \tag{1.3}$$

$$P_\eta P_\lambda = \sum_v \sum_h \sum_p \mathcal{D}_{v,h,p} P_{(e(p)^{-1}m(\lambda))_+}, \tag{1.4}$$

where $\mathcal{A}_p, \mathcal{B}_{h,p}, \mathcal{C}_{v,h,p}$, and $\mathcal{D}_{v,h,p} \in \mathbb{F}$ are scalars given by explicit formulas in Section 4, h and p are summed over particular sets of alcove walks, v is summed over the stabilizer of λ in the symmetric group $W_0 = S_r$, $e(p)$ is the affine Weyl group element corresponding to the endpoint of the alcove walk p, and all other notation is defined in Section 2.

Theorem 1.1 is proven analogously to the results of [**Yi**]. First we use a result of [**Yi**] to expand the product of a monomial multiplication operator X^μ with an intertwiner $\tau_w^\vee$, stated as Theorem 4.1 below. We then apply this expansion formula to an appropriate monomial to obtain a formula for the product of a monomial and an SSV polynomial, (1.1). Equation (1.2) then follows by expanding the Macdonald polynomial E_μ in the monomial basis using the alcove walk formula of [**RY**]. To get (1.3), we first expand P_λ as a sum of nonsymmetric SSV polynomials (using the new Proposition 3.16 below) and expand the Macdonald polynomial E_μ as a sum of monomials using the alcove walk formula of [**RY**], then apply (1.1). Finally, we apply a symmetrizing operator to (1.3) in order to obtain (1.4). Each step after the first requires careful calculation but is analogous to the corresponding step in [**Yi**].

Recall that, as discussed above, q-specializations of SSV polynomials yield metaplectic Iwahori Whittaker functions, special functions related to the representation theory of metaplectic covers of reductive groups. Thus, q-specializations of Theorem 1.1 directly imply expansion formulas for the product of metaplectic Iwahori Whittaker functions (or, more precisely, the product of a non-metaplectic Iwahori Whittaker function and a metaplectic one). The significance of this formula to the representation theory of metaplectic covers is still open.

Note that the formulas of Theorem 1.1 have a different form than those in [**Yi**]: namely, we present (1.3) and (1.4) as sums over pairs of alcove walks, while Yip's corresponding results are given in terms of one sum over "colored" alcove walks. This is not a significant conceptual difference: in our setting, the second alcove walk is essentially obtained by introducing new folds in the first; this is precisely what is captured by the notion of colored alcove walk in [**Yi**]. As we discuss in Remark 4.8, however, there is a subtle technical difference that causes the formulas in the present work to have fewer terms than those of [**Yi**] (in their common setting, the Macdonald polynomial case).

Throughout this work, we will use results from [**Sai**], in which the author gave an alcove walk formula for SSV polynomials. Technically, that work discussed only SSV polynomials of the form E_μ, where $\mu \in \mathbb{Q}^r \subseteq E$. The same results hold in the general case $\mu \in E$, and the more general versions are given in [**Sai2**]. Further, [**Sai**] (following [**SSV**]) used the variant notation $E_\mu^{(n)}$ for $\mu \in P$ and $n \in \mathbb{Z}_+$, where $E_\mu^{(n)} = E_{\mu/n}$ up to a change of variables. The conversion between these two perspectives is discussed in detail in [**Sai2**, Section 6.3].

A Littlewood-Richardson rule for Koornwinder polynomials was recently given by Yamaguchi [**Y**], as an application of Orr and Shimozono's analogue of the alcove walk formula for Koornwinder polynomials [**OS**]. Some arguments and notation pertaining to the Littlewood-Richardson rule will parallel that of Yamaguchi.

The structure of the work is as follows. In Section 2, we recall basic definitions and notation, including that for double affine Hecke algebras (DAHAs), alcove walks, and the alcove walk formula of [**RY**]. In Section 3, we recall the quasi-polynomial representation of [**SSV, SSV2**] and the SSV polynomials, and prove some basic results that will be necessary for the Littlewood-Richardson formulas. Section 3.1 focuses on the nonsymmetric SSV polynomials and their intertwiners, and Section 3.2 focuses on the symmetric SSV polynomials and the relevant symmetrizers. Of particular note in Section 3.2 is Proposition 3.16, an explicit formula for the expansion of a symmetric SSV polynomial in the (nonsymmetric) SSV polynomial basis. In Section 4, we prove Theorem 1.1, giving the Littlewood-Richardson rules for SSV polynomials. In Remark 4.9, we then briefly discuss the difficulties with obtaining Littlewood-Richardson rules for the product of two arbitrary SSV polynomials. We conclude with a brief pedagogical section studying a particular example and carrying out detailed calculations for some of the terms in (1.4).

The results in this work also appear in the author's PhD thesis [**Sai2**].

2. Background and notation

In this section, we establish the background and notation required for the rest of the work. Section 2.1 recalls basic notions related to finite and affine root systems and Weyl groups in Type A. Section 2.2 establishes notation for alcove walks, mostly following that of [**Yi**]. Section 2.3 discusses the fields over which SSV polynomials are defined. Section 2.4 introduces the double affine Hecke algebra, Section 2.5 discusses the intertwiners, and Section 2.6 discusses a necessary piece of notation used in [**RY, Yi**]. Finally, Section 2.7 states the alcove walk formula of [**RY**] for Macdonald polynomials.

2.1. Root systems and Weyl groups. We begin by setting the basic notation for root systems and Weyl groups in type A. Fix $r \geq 2$. Let E be the Euclidean space $\mathbb{R}^r$ with inner product $(\cdot,\cdot)$, norm $\|\cdot\|$ and standard orthonormal basis $\{\varepsilon_i : 1 \leq i \leq r\}$. Let $\Phi = \{\varepsilon_i - \varepsilon_j : 1 \leq i \neq j \leq r\}$ be the root system of type A_{r-1} with positive roots $\Phi_+ = \{\varepsilon_i - \varepsilon_j : 1 \leq i < j \leq r\}$ and simple roots $\Delta = \{\alpha_1, \dots, \alpha_{r-1}\}$, where $\alpha_i = \varepsilon_i - \varepsilon_{i+1}$. The highest root is $\theta = \varepsilon_1 - \varepsilon_r$. We have the root lattice $Q = \operatorname{span}_{\mathbb{Z}} \Phi$ and the GL_r weight lattice $P = \operatorname{span}_{\mathbb{Z}}\{\varepsilon_1, \dots, \varepsilon_r\}$, which we identify with $\mathbb{Z}^r$. For $v \in E$, we will use any of the following notations interchangeably:

$$v = \sum_{i=1}^{r} v_i \varepsilon_i = (v_1, \dots, v_r).$$

$\lambda \in E$ is called *dominant* if, for all $1 \leq i < r$, $\lambda_i \geq \lambda_{i+1}$. Let E_+ (respectively P_+) be the set of dominant elements of E (respectively P).

Let $W_0 = S_r$, the Weyl group of type A_{r-1}, with generators $s_i = (i\, i+1)$. For $v \in E$, let $v_+ \in E_+$ be the unique dominant element in the W_0-orbit of v.

For any root $\alpha \in \Phi$, we have the reflection s_α, a transformation of E that preserves P, given by

$$s_\alpha v = v - (v, \alpha)\alpha. \tag{2.1}$$

We identify W_0 with the group generated by $\{s_\alpha : \alpha \in \Phi\}$ via the map $s_i \mapsto s_{\alpha_i}$.

Let $E_a = \mathbb{R}^{r+1}$. We identify the subspace $\{(v_1, \ldots, v_r, 0) : v_1, \ldots, v_r \in \mathbb{R}\}$ with E and let $\delta = (0, \ldots, 0, 1)$, so that $E_a = E \oplus \mathbb{R}\delta$. We consider the subgroup $P_a = P \oplus \mathbb{Z}\delta \cong \mathbb{Z}^r \oplus \mathbb{Z}\delta$. Extend $(\cdot, \cdot)$ to a symmetric bilinear form on E_a by defining $(\delta, v + s\delta) = 0$ for $v \in P$ and $s \in \mathbb{R}$.

We have the set of (real) affine roots $\widetilde{\Phi} = \{\alpha + s\delta : \alpha \in \Phi, s \in \mathbb{Z}\} \subseteq P_a$, with *positive affine roots* $\widetilde{\Phi}_+ = \Phi_+ \cup \{\alpha + s\delta : s \in \mathbb{Z}, s > 0, \alpha \in \Phi\}$. Writing $\alpha_0 = \delta - \theta \in \widetilde{\Phi}$, every element of $\widetilde{\Phi}_+$ is a nonnegative integral linear combination of $\alpha_0, \ldots, \alpha_{r-1}$.

We define a pairing between P_a and E by

$$\langle \mu + s\delta, v \rangle = (\mu, v) + s,$$

where $\mu \in P$, $s \in \mathbb{Z}$, and $v \in E$. This pairing gives the association between affine reflections and (real) affine roots: for $\widehat{\alpha} = \alpha + s\delta \in \widetilde{\Phi}$, we have the affine transformation $s_{\widehat{\alpha}} : E \to E$ given by

$$s_{\widehat{\alpha}} v = v - \langle \widehat{\alpha}, v \rangle \alpha$$

for $v \in E$.

We have the extended affine Weyl group

$$W = P \rtimes W_0 = \{\tau(\mu) w : \mu \in P, w \in W_0\},$$

with $w\tau(\mu)w^{-1} = \tau(w\mu)$. We consider the elements $s_0 = \tau(\theta)s_\theta$ and

$$\omega = s_1 s_2 \cdots s_{r-1} \tau(\varepsilon_r) = \tau(\varepsilon_1) s_1 s_2 \cdots s_{r-1},$$

satisfying

$$\omega s_i = s_{i+1}\omega,$$

where the subscripts are taken modulo r. W is generated by $s_0, \ldots, s_{r-1}, \omega$, and every $w \in W$ has a *reduced expression* of the form $w = \omega^s s_{i_1} \cdots s_{i_\ell}$, where $s \in \mathbb{Z}$ and ℓ is minimal. We define the *length* of w to be $\ell(w) = \ell$. (Thus ω^s has length 0.) We consider the subgroup

$$W_{\text{Cox}} = Q \rtimes W_0 \subseteq W,$$

a Coxeter group with generators $s_0, \ldots, s_{r-1}$, where these notions of reduced expression and length are the standard ones in the theory of Coxeter groups.

We now recall the linear action of W on E_a and the affine action of W on E, both of which will be important throughout this work. First, for $1 \le i \le r-1$, $\mu \in P$, and $\widehat{v} = v + s\delta \in E_a$, we have the linear action given by

$$s_i * \widehat{v} = \widehat{v} - (\widehat{v}, \alpha_i)\alpha_i = (s_i v) + s\delta$$

and

$$\tau(\mu) * \widehat{v} = \widehat{v} - (\widehat{v}, \mu)\delta = v + (s - (v, \mu))\delta,$$

which restricts to an action on P_a. For the affine action, $s_i \in W_0$ acts on $v \in E$ by the reflection s_{α_i} as in (2.1), and $\tau(\mu)v = v + \mu$ for $\mu \in P$. It is easy to see directly that the affine action is faithful, and we identify W with the corresponding group of affine transformations of E. For all $\widehat{\alpha} \in \widetilde{\Phi}$, we have $s_{\widehat{\alpha}} \in W_{\mathrm{Cox}} \subseteq W$, and we also have $s_0 = s_{\alpha_0}$.

Explicitly, under the affine action, we have

$$s_0(v_1, \ldots, v_r) = (v_r + 1, v_2, v_3, \ldots, v_{r-1}, v_1 - 1)$$

and

$$\omega(v_1, \ldots, v_r) = (v_r + 1, v_1, v_2, \ldots, v_{r-1}).$$

The relationship between the two actions of W is given as follows: for $w \in W$, $\widehat{\mu} \in P_a$, and $v \in E$,

$$\langle w * \widehat{\mu}, wv \rangle = \langle \mu, v \rangle.$$

REMARK 2.1. Notice that for $w \in W$, we denote the linear action on $v \in E_a$ by $w * v$ and the affine action on $\widehat{v} \in E$ by $w\widehat{v}$. For $w \in W_0$, where the two actions are essentially the same, we will often drop the $*$ from our notation.

REMARK 2.2. Both the affine and linear actions of W arise from the action of W on the Cartan subalgebra of the corresponding untwisted affine Lie algebra. Details may be found in Sections 6.5 and 6.6 of [**K**].

For $\mu \in E$, let $W_\mu \subseteq W_0$ be the stabilizer of μ under the action of the finite Weyl group, and let W^μ be a set of minimal length coset representatives of W_0/W_μ. Let w_μ and v_μ be the longest elements of W_μ and W^μ respectively. Further define $m(\mu) \in \mathcal{A}_W$ to be the unique W-conjugate of μ in $\mathcal{A}_W$, and let $w(\mu) \in W$ be the unique shortest element such that $w(\mu)m(\mu) = \mu$.

EXAMPLE 2.3. Note that all elements of P are W-conjugate to $\mathbf{0} = (0, 0, \ldots, 0)$. Then for $\mu \in P$, $m(\mu) = \mathbf{0}$, and $w(\mu)$ is the shortest element such that $w(\mu)\mathbf{0} = \mu$. In the Macdonald case, then, one only needs to consider $m(\mu) = \mathbf{0}$. More generally, we will consider $\mu \in E$, and we may instead have $m(\mu)$ an arbitrary element of $\mathcal{A}_W$. For example, if $\mu = (3/2, 0)$, then $m(\mu) = (1/2, 0)$ and $w(\mu) = \omega s_1$.

2.2. Alcove walks. We now recall basic notions related to alcove walks. The notation given here will mostly follow that of [**RY, Yi**]. Note that the alcoves used here are different from those in [**Sai**], since we use all of W rather than restricting to W_{Cox}: see Remark 2.4 for a more detailed explanation.

We choose the *fundamental alcove*

$$\begin{aligned}\mathcal{A}_W &= \{(v_1, \ldots, v_r) \in E : v_1 \geq v_2 \geq \cdots \geq v_r, 0 \leq v_i < 1 \text{ for } 1 \leq i \leq r\} \\ &= \{v \in E : \langle \alpha_j, v \rangle \geq 0 \text{ for } 0 \leq j \leq r-1, 0 \leq v_i < 1 \text{ for } 1 \leq i \leq r\}.\end{aligned}$$

This is a choice of fundamental domain for the action of W on E. The *alcoves* are the images of $\mathcal{A}_W$ under the action of W. The alcoves are in bijection with elements of W, and we naturally identify $w \in W$ with $w\mathcal{A}_W$. Thus we may view the action of W on the alcoves induced by the affine action as arising from left multiplication in W.

For $s \in \mathbb{Z}$, let the *sth sheet* E_s be

$$E_s = \bigcup_{w \in W_{\mathrm{Cox}}} \omega^s w \mathcal{A}_W.$$

For $\widehat{\alpha} \in \widetilde{\Phi}$, define the hyperplane

$$\mathfrak{h}^{\widehat{\alpha}} = \{v \in E : \langle \widehat{\alpha}, v \rangle = 0\}.$$

The connected components of

$$E_s \setminus \left(\bigcup_{\widehat{\alpha} \in \widetilde{\Phi}} \mathfrak{h}^{\widehat{\alpha}} \right)$$

are the interiors of the alcoves in E_s. We call the hyperplanes bounding an alcove its *walls*. In particular, for $s \in \mathbb{Z}$, the alcove $\omega^s \mathcal{A}_W$ has walls $\mathfrak{h}^{\alpha_0}, \ldots, \mathfrak{h}^{\alpha_{r-1}}$. The action of W_{Cox} permutes the alcoves of E_s, with $s_{\widehat{\alpha}}$ corresponding to the reflection over the wall $\mathfrak{h}^{\widehat{\alpha}}$.

The choice of positive finite roots determines the following orientation on the set of hyperplanes. Let $\widehat{\alpha} \in \widetilde{\Phi}$ with

$$\widehat{\alpha} = \alpha + s\delta, \alpha \in \Phi_+.$$

For $v \in E$, we say that v is on the positive side of $\mathfrak{h}^{\widehat{\alpha}}$ if $\langle \widehat{\alpha}, v \rangle > 0$ and is on the negative side of $\mathfrak{h}^{\widehat{\alpha}}$ if $\langle \widehat{\alpha}, v \rangle < 0$. An alcove is said to be on the positive (respectively negative) side of a hyperplane if the points in its interior are on the positive (respectively negative) side of that hyperplane. For example, letting $s \in \mathbb{Z}$, $\omega^s \mathcal{A}_W$ is on the positive side of $\mathfrak{h}^{\alpha_i}$ for $1 \leq i \leq r-1$ but is on the *negative* side of $\mathfrak{h}^{\alpha_0} = \mathfrak{h}^{\theta-\delta}$. This is equivalent to the orientation used in [**RY**, **Yi**, **Y**, **Sai**].

We now define alcove walks. For $w \in W$, fix a reduced expression $w = \omega^s s_{i_1} \cdots s_{i_\ell}$, and let $\vec{w} = (s, i_1, \ldots, i_\ell)$. An *alcove walk* of type $\vec{w}$ beginning at $u \in W$ is a sequence

$$p = (A_0, A_1, \ldots, A_\ell) \tag{2.2}$$

of alcoves, where $A_0 = u\omega^s$ and, for each $1 \leq j \leq \ell$, either $A_j = A_{j-1}$ or $A_j = A_{j-1}s_{i_j}$ (where the notation $A_{j-1}s_{i_j}$ refers to the product of the corresponding elements of W). The hyperplane separating the alcoves A_{j-1} and $A_{j-1}s_{i_j}$ is called the jth *active hyperplane*, and the positive affine root indexing that hyperplane is denoted by $p_j^\vee \in \widetilde{\Phi}_+$. We say that p has a *fold* at step j if $A_j = A_{j-1}$ and a *crossing* otherwise: this fold or crossing is *positive* if A_j is on the positive side of the jth active hyperplane and is negative otherwise. Let $f_+(p)$ (respectively $f_-(p)$) be the set of all j such that p has a positive (respectively negative) fold at step j, and let $f(p) = f_+(p) \cup f_-(p)$. We say p is *unfolded* if it has no folds. We define the *length* of p to be the number of hyperplanes of the form $\mathfrak{h}^{\widehat{\alpha}}$ between A_0 and A_ℓ, or equivalently the number of crossings in p. If p is unfolded, then the length of p is ℓ. The jth step of p is called *ascending* if $\ell(A_{j-1}s_{i_j}) > \ell(A_{j-1})$ and *descending* otherwise. Let $\xi_{\text{des}}(p)$ be the set of indices at which p has a descending crossing, and let $f_{\text{asc}}(p)$ be the set of indices at which h has an ascending fold. We illustrate some of these concepts in Figure 1.

For $u \in W$ and $\vec{w}$ as above, let $\Gamma(u, \vec{w})$ be the set of all alcove walks of type $\vec{w}$ beginning at u. Note that this is the set of alcove walks denoted by $\Gamma(\vec{w}, u)$ in [**Yi**]: we choose the present notation to more closely match the notation of [**RY**, **Sai**].

Given an alcove walk p as in (2.2), let $e(p) = A_\ell$. (This was referred to as $\text{end}(p)$ in [**RY**, **Sai**].) We have the decomposition

$$e(p) = \tau(\text{wt}(p))\phi(p),$$

where $\text{wt}(p) \in P$ and $\phi(p) \in W_0$. Suppose further that p has a crossing at step j. We say that an alcove walk $p' = (A'_0, \ldots, A'_\ell)$ of the same type as p is obtained by *folding p at step j* if $A'_i = A_i$ for $i < j$, $A'_j = A_{j-1}$, and $A'_i = s_{p_j^\vee} A_i$ for

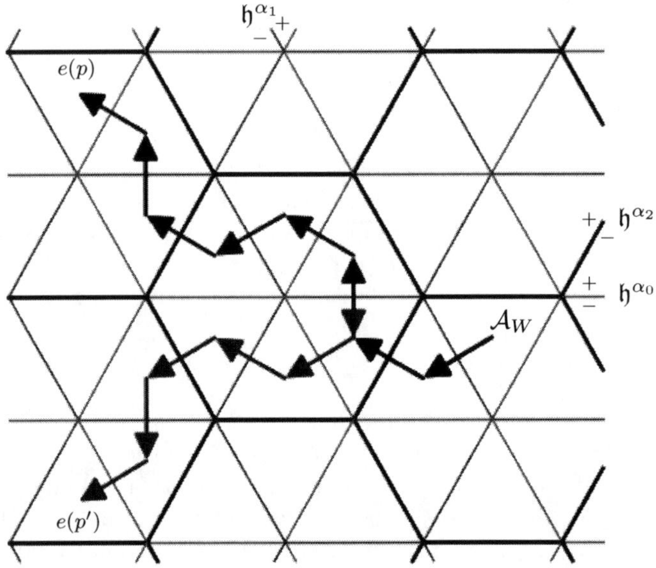

FIGURE 1. We illustrate alcoves and alcove walks in the case $r = 3$, restricting our attention to $\mathrm{span}_{\mathbb{R}}\{\alpha_1, \alpha_2\}$ so that the picture is two-dimensional. (The image is reproduced from [**Sai**].) Let $w = \omega^0 s_1 s_0 s_1 s_2 s_1 s_0 s_1 s_2$. We see two alcove walks of type $\vec{w} = (0, 1, 0, 1, 2, 1, 0, 1, 2)$ starting at the fundamental alcove $\mathcal{A}_W$: we label them p (black) and p' (blue once it deviates from p). The alcove walk p is unfolded, and p' is obtained by folding p at step 3; this is a negative ascending fold. The walk p has length 8, and p' has length 7. We also note that replacing w with any $\omega^s w$ would result in a similar picture, with $\mathcal{A}_W$ replaced by $\mathcal{A}_W \omega^s$.

$i > j$. Geometrically, to obtain p' from p, we reflect the tail of p in the jth active hyperplane $\mathfrak{h}^{p_j^\vee}$. Note that

$$e(p') = s_{p_j^\vee} e(p).$$

For $\lambda \in E$ and $u \in W$, let $\Gamma_\lambda(u, \vec{w})$ be the set of alcove walks $p \in \Gamma(u, \vec{w})$ such that $e(p)^{-1} \in W$ is the shortest element in its left W_λ-coset. This restriction will first arise in Lemma 3.8 and later appear in the Littlewood-Richardson rules of Section 4.

REMARK 2.4. Let $u \in W$, $w = \omega^s s_{i_1} \cdots s_{i_\ell} \in W$, and $w' = s_{i_1} \cdots s_{i_\ell} \in W_{\mathrm{Cox}}$, with $\vec{w} = (s, i_1, \ldots, i_\ell)$ and $\vec{w'} = (0, i_1, \ldots, i_\ell)$. We consider alcove walks of type $\vec{w}$. Note that we do not fold at the length 0 elements ω^s, so in fact $\Gamma(u, \vec{w}) = \Gamma(u\omega^s, \vec{w'})$. Thus we may think of these alcove walks as corresponding to w' and starting at the alcove $u\omega^s$ on the appropriate sheet. Further, as discussed in detail in [**Sai**] Remark 2.3, the alcove walks of type $\vec{w'}$ are in bijection with subsequences of $(i_1, \ldots, i_\ell)$, where i_j is included in the subsequence if and only if the alcove walk has a crossing at step j. This may be equivalently phrased in terms of subwords of a reduced expression for w'. Thus, although the *alcoves* considered in [**Sai, RY, Yi**] are technically different sets from those used here, the theory of alcove walks is the same. The alcove walks in [**Sai**] may be viewed as alcove walks restricted to the 0th sheet in the present setting (although, once again, the alcoves themselves are different subsets of E). In fact, for the fundamental alcove $\overline{\mathcal{A}}$ used in [**Sai**], we have

$$\overline{\mathcal{A}} = \bigcup_{s\in\mathbb{Z}} \omega^s \mathcal{A}_W.$$

Similarly, [**RY, Yi**] describe the sheets abstractly as a discrete set of copies of $E' = \{v \in E : \sum_i v_i = 0\}$, and their version of the sth sheet corresponds in our setting to $\omega^s E' = \{v \in E : \sum_i v_i = s\}$. This focus on integral sheets is the right choice in [**RY, Yi**], because Macdonald polynomials are indexed by $\mu \in P$, and therefore it is sufficient to consider only E' and its ω-translates. However, SSV polynomials are indexed by arbitrary $\mu \in E$, and therefore we are required to consider these larger alcoves and sheets that contain all the non-integral points.

Following [**Yi, Y**], we now introduce an additional piece of notation that will be of use in Section 4. For $w \in W_0$ with reduced expression $w = s_{i_1} \cdots s_{i_\ell}$, define

$$\mathcal{L}(w) = \{\alpha_{i_1}, s_{i_1}\alpha_{i_2}, \ldots, s_{i_1} \cdots s_{i_{\ell-1}}\alpha_{i_\ell}\}.$$

$\mathcal{L}(w)$ is the set of positive roots indexing the hyperplanes separating the alcoves corresponding to 1 and w. For $u, w \in W_0$, we define

$$\mathcal{L}(u, w) = (\mathcal{L}(u) \cup \mathcal{L}(w)) \setminus (\mathcal{L}(u) \cap \mathcal{L}(w)). \tag{2.3}$$

$\mathcal{L}(u, w)$ is the set of positive roots indexing the hyperplanes separating the alcoves corresponding to u and w.

REMARK 2.5. For $\widehat{\alpha} \in \widetilde{\Phi}$, we have $\mathfrak{h}^{\widehat{\alpha}} = \mathfrak{h}^{-\widehat{\alpha}}$. Note that we have used two different conventions regarding which affine root should be used to index this hyperplane. When discussing the jth active hyperplane in an alcove walk, and when defining $\mathcal{L}(u, w)$, we made the natural choice to index the hyperplanes using positive affine roots, elements of $\widetilde{\Phi}_+$. When determining the orientation of a hyperplane, however, we index it using $\widehat{\alpha} = \alpha + s\delta$ with $\alpha \in \Phi_+$, in which case $\widehat{\alpha}$ is not necessarily a positive affine root. This latter choice is required for the orientation to be "periodic," in the sense that the orientations on $\mathfrak{h}^{\widehat{\alpha}}$ and $\mathfrak{h}^{\widehat{\alpha}+m\delta}$ are parallel for all $m \in \mathbb{Z}$.

2.3. Field of definition. We now briefly comment on the fields over which the SSV polynomials are defined. This is discussed in far greater generality in [**SSV2**, Section 3.1].

Fix an invertible parameter k, which will be the parameter involved in the quadratic relations of the Hecke algebras defined in Section 2.4. Also fix a set of parameters $G = \{G_s, s \in \mathbb{R}\}$, satisfying the following conditions:

(1) $G_0 = k$.
(2) $G_s G_{-s} = 1$.
(3) $G_s = G_{\lfloor s \rfloor}$, where $\lfloor s \rfloor$ is the floor of the real number s.

REMARK 2.6. These conditions imply that G is determined by $\{k\} \cup \{G_s : s \in (0, \frac{1}{2}]\}$, and that $G_{1/2} = \pm 1$. A variant of these parameters depending on a positive integer n was introduced in [**SSV**] Definition 3.5, where the parameters were called g_j, with the dictionary being $G_{j/n} = -kg_{-j}$ for $j \in \mathbb{Z}$. In that work, the scalar $G_{1/2} \in \{\pm 1\}$ was called ε. We will continue to call it $G_{1/2}$ in order to keep the notation uniform, but will keep in mind that it is not an entirely free parameter. We also note that the conditions on the parameters G_s are motivated by the comparison with metaplectic Iwahori Whittaker functions, in which the G_s parameters are specialized to certain Gauss sums naturally satisfying (1)–(3). This specialization is discussed in [**Sai**] Proposition 5.5.

We now consider another family of parameters $S = \{q^s : s \in \mathbb{R}\}$, with $q^a q^b = q^{a+b}$ and $q^0 = 1$. We write $q = q^1$. We introduce the fields

$$\mathbb{K} = \mathbb{C}(k, q) \text{ and } \mathbb{F} = \mathbb{C}(G, S).$$

The field $\mathbb{K}$ is the one studied in the standard Macdonald theory. The passage to the quasi-polynomial representation and the corresponding SSV polynomials requires the much larger field $\mathbb{F}$. Note, however, that any particular SSV polynomial will only involve finitely many of the parameters in $G \cup S$.

2.4. The double affine Hecke algebra. We now recall the double affine Hecke algebra (DAHA). The DAHA defined here matches the version defined in Section 2.3 of [**RY**] in the GL_n case, with $t = k^2$.

DEFINITION 2.7. The *double affine Hecke algebra* $\mathbb{H}$ is the unital associative algebra over $\mathbb{C}(k)$ generated by q, ω, ω^{-1}, T_0, T_1, ..., T_{r-1}, and $X^{\widehat{\mu}}$ for $\widehat{\mu} \in P_a$, with the following relations, where $0 \le i, j \le r-1$ and the indices of the T_i and T_j are taken modulo r:

(1) (Braid relations.) If $r \geq 3$, $T_i T_{i+1} T_i = T_{i+1} T_i T_{i+1}$. For $j \not\equiv i \pm 1 \bmod r$, $T_i T_j = T_j T_i$.
(2) (Hecke relations.) $(T_i - k)(T_i + k^{-1}) = 0$.
(3) $\omega\omega^{-1} = 1 = \omega^{-1}\omega$ and $\omega T_i = T_{i+1}\omega$.
(4) For $\widehat{\mu}, \widehat{\lambda} \in P_a$, $X^{\widehat{\mu}} X^{\widehat{\lambda}} = X^{\widehat{\mu}+\widehat{\lambda}}$.
(5) q is central.
(6) $X^\delta = q$.
(7) (Cross relations.) For $\widehat{\mu} = \mu + a\delta \in P_a$,

$$T_i X^{\widehat{\mu}} - X^{s_i * \widehat{\mu}} T_i = (k - k^{-1}) \left(\frac{X^{\widehat{\mu}} - X^{s_i * \widehat{\mu}}}{1 - X^{\alpha_i}} \right) \text{ and } \omega X^{\widehat{\mu}} = X^{\omega * \widehat{\mu}} \omega.$$

As stated above, the indices of the T_i are taken modulo r, so for $i = r-1$, relation (3) reads $\omega T_{r-1} = T_0 \omega$.

For $w \in W_{\mathrm{Cox}}$ with reduced expression $w = s_{i_1} \cdots s_{i_\ell}$, define

$$T_w = T_{i_1} \cdots T_{i_\ell}.$$

By a standard argument, since the T_i satisfy the braid relations (1), T_w is independent of the choice of reduced expression.

For $1 \le i \le r-1$, we will use the notation $T_i^\vee = T_i \in \mathbb{H}$. Also define

$$T_0^\vee = (X^\theta T_{s_\theta})^{-1} \in \mathbb{H}.$$

Define the following elements of $\mathbb{H}$, for $1 \le i \le r$:

$$Y_i = Y^{\varepsilon_i} = T_{i-1}^{-1} \cdots T_1^{-1} \omega T_{r-1} \cdots T_i. \tag{2.4}$$

These elements are invertible and commute with one another. (The commutativity is proven in much greater generality in [**Lu**] Lemma 3.4(a).) We then define, for $\widehat{\mu} = \sum \mu_i \varepsilon_i + s\delta \in P_a$,

$$Y^{\widehat{\mu}} = q^{-s} Y_1^{\mu_1} \cdots Y_r^{\mu_r}.$$

These elements satisfy the Bernstein-Zelevinsky cross relations

$$T_i^\vee Y^{\widehat{\mu}} - Y^{s_i * \widehat{\mu}} T_i^\vee = (k - k^{-1}) \left(\frac{Y^{\widehat{\mu}} - Y^{s_i * \widehat{\mu}}}{1 - Y^{-\alpha_i}} \right) \tag{2.5}$$

for $0 \leq i \leq r-1$ and $\widehat{\mu} \in P_a$. Equation (2.5) is best understood as a corollary of Definition 2.7(7) and the following proposition, due to Cherednik (see [**C1**, **C2**]). Such involutions are discussed in greater generality in [**IS**].

PROPOSITION 2.8. *There exists a unique involution* ϵ *on* $\mathbb{H}$ *such that, for* $\widehat{\mu} \in P_a$ *and* $0 \leq i \leq r-1$, $X^{\widehat{\mu}} \mapsto Y^{\widehat{\mu}}$, $T_i \mapsto (T_i^\vee)^{-1}$, *and* $k \mapsto k^{-1}$.

The DAHA $\mathbb{H}$ has a natural representation on the space of Laurent polynomials $\mathbb{K}[P]$ called the *polynomial representation*, and the simultaneous eigenfunctions $\{E_\mu : \mu \in P\}$ of the operators $\{Y^{\widehat{\nu}} : \widehat{\nu} \in P_a\}$ are the *Macdonald polynomials* (see, for example, [**M**]). We refrain from giving details on the polynomial representation and Macdonald polynomials, as they will arise as special cases of the notions in Section 3.1.

2.5. Intertwiners. We now introduce the intertwiners, which allow one to construct SSV polynomials recursively. These are the same intertwiners as in the standard Macdonald theory, for example [**C2**].

For $0 \leq i \leq r-1$, define the *intertwiners* $\tau_i^\vee$ by

$$\tau_i^\vee = T_i^\vee + \frac{k^{-1} - k}{1 - Y^{-\alpha_i}} = S_i (1 - Y^{-\alpha_i})^{-1}.$$

These intertwiners precisely match those defined in (2.34) of [**RY**] and used throughout that work and [**Yi**]. The intertwiners are not elements of $\mathbb{H}$, but rather of a localization of $\mathbb{H}$: see [**Sai**, Section 3] for details. We have

$$\tau_i^\vee Y^{\widehat{\nu}} = Y^{s_i * \widehat{\nu}} \tau_i^\vee.$$

Further define the *omega intertwiner* $\tau_\omega^\vee \in \mathbb{H}$ by

$$\tau_\omega^\vee = X^{\varepsilon_1} T_1 \cdots T_{r-1} = \epsilon(\omega),$$

where the latter equality follows from (2.4) and Proposition 2.8. By Definition 2.7(3) and Proposition 2.8, we have

$$\tau_\omega^\vee (T_i^\vee)^{\pm 1} = (T_{i+1}^\vee)^{\pm 1} \tau_\omega^\vee, \tag{2.6}$$

where as usual we interpret the subscripts of the $T_j^\vee$ modulo r.

For $w \in W$ with reduced expression $w = \omega^s s_{i_1} \cdots s_{i_\ell}$, define

$$\tau_w^\vee = (\tau_\omega^\vee)^s \, \tau_{i_1}^\vee \cdots \tau_{i_\ell}^\vee.$$

It is well-known that $\tau_w^\vee$ is independent of the choice of reduced expression (see, for instance, Section 2 of [**C2**]).

2.6. Generalized monomial notation. We briefly introduce one further piece of notation, generalizing the notation X^μ. This notation was used in [**RY**] and was modified slightly in [**Yi**], as discussed below.

For $z \in W$ with expression $z = \omega^s s_{i_1} \cdots s_{i_\ell}$ (not necessarily reduced), define

$$X^z = (\tau_\omega^\vee)^s (T_{i_1}^\vee)^{\epsilon_1} \cdots (T_{i_\ell}^\vee)^{\epsilon_\ell}, \tag{2.7}$$

where ϵ_j is 1 if the jth step of the corresponding unfolded alcove walk beginning at 1 is a negative crossing and -1 if it is a positive crossing (see (2.26) in [**RY**]). This is independent of the choice of expression for z: see Theorem 1.1.1 of [**G**]. We have the following comparison with the X^μ notation: see [**Yi**] (3.1.4). This is discussed in detail in [**Sai2**].

LEMMA 2.9. *For $\mu \in P$ and $u \in W_0$, we have*

$$X^{\tau(\mu)u} = X^\mu T_u$$

We now discuss an alternative expression for X^z that is used in [**Yi**] and will be helpful in Section 4. Given $z \in W$, write its reduced expression $z = s_{i_\ell} \cdots s_{i_1}\omega^{-s}$. Note that this is opposite to our earlier convention: this choice is made so that $z^{-1} = \omega^s s_{i_1} \cdots s_{i_\ell}$ has the more standard-looking reduced expression, as we will be considering alcove walks of type

$$\vec{z}^{-1} = \vec{z^{-1}} = (s, i_1, \dots, i_\ell)$$

(see Theorem 4.1). Since $z = \omega^{-s} s_{i_\ell+s} \cdots s_{i_1+s}$, by (2.7) and (2.6) we have

$$X^z = (\tau_\omega^\vee)^{-s}(T_{i_\ell+s}^\vee)^{\epsilon_\ell} \cdots (T_{i_1+s}^\vee)^{\epsilon_1} = (T_{i_\ell}^\vee)^{\epsilon_\ell} \cdots (T_{i_1}^\vee)^{\epsilon_1}(\tau_\omega^\vee)^{-s}$$

(with all indices of the s_j and $T_j^\vee$ taken modulo r). Due to the change of indices, we rephrase the condition on the ϵ_j, following [**Yi**] (2.3.3). Consider the unfolded alcove walk b of type $\vec{z}^{-1} = (s, i_1, \dots, i_\ell)$ beginning at the alcove z (which ends at the alcove 1). Then ϵ_j is 1 if the jth step of b is a positive crossing and -1 if it is a negative crossing.

Now let $w \in W$ and $h \in \Gamma(w, \vec{z}^{-1})$. Following [**Yi**] (3.1.6), define $\psi(h)$ to be the set of indices j at which h has an ascending fold with $\epsilon_j = -1$ or a descending fold with $\epsilon_j = 1$, where the ϵ_j are defined as in the previous paragraph.

2.7. Formulas of Ram and Yip. We now recall the alcove walk formulas of [**RY**]. Given $w \in W$ and a reduced expression $w = \omega^s s_{i_1} \cdots s_{i_\ell}$, define the positive roots $\beta_1, \dots, \beta_\ell$ by

$$\beta_j = s_{i_\ell} s_{i_{\ell-1}} \cdots s_{i_{j+1}} \alpha_{i_j} \in \widetilde{\Phi}_+. \tag{2.8}$$

Then we have the following:

THEOREM 2.10 ([**RY**], Theorem 2.2). *Let $w \in W$. Write $w = \omega^s s_{i_1} \cdots s_{i_\ell}$, a reduced expression. For $1 \le j \le \ell$, let β_j be as defined in (2.8). For $z \in W$,*

$$X^z \tau_w^\vee = \sum_{p \in \Gamma(z, \vec{w})} X^{\mathrm{wt}(p)} T_{\phi(p)} \left(\prod_{j \in f^+(p)} \frac{k^{-1} - k}{1 - Y^{-\beta_j}} \right) \left(\prod_{j \in f^-(p)} \frac{(k^{-1} - k)Y^{-\beta_j}}{1 - Y^{-\beta_j}} \right).$$

In particular, taking $z = u \in W_0$, this gives an expansion of $T_u \tau_w^\vee$.

As a consequence, we obtain Ram and Yip's alcove walk formula for nonsymmetric Macdonald polynomials. Here, we use the notation $f \sim g$ to mean that f and g are nonzero scalar multiples of one another.

THEOREM 2.11 ([**RY**], Theorem 3.1). *Let $\mu \in P$. Write $w(\mu) = \omega^s s_{i_1} \cdots s_{i_\ell}$ and define β_j as in (2.8). Then*

$$E_\mu \sim \tau_{w(\mu)}^\vee 1 = \sum_{p \in \Gamma(1, \vec{w}(\mu))} x^{\mathrm{wt}(p)} k^{\ell(\phi(p))} c_p,$$

where

$$c_p = \left(\prod_{j \in f^+(p)} \frac{k^{-1} - k}{1 - \gamma(-\beta_j; 0)} \right) \left(\prod_{j \in f^-(p)} \frac{(k^{-1} - k)\gamma(-\beta_j; 0)}{1 - \gamma(-\beta_j; 0)} \right) \in \mathbb{K} \tag{2.9}$$

and $\gamma(-\beta_j; 0)$ *is a scalar given in* (3.2).

Recalling that $X^{e(p)} = X^{\mathrm{wt}(p)} T_{\phi(p)}$, we may also express the previous result in the following form, which will be useful in Section 4:

$$E_\mu \sim \tau^\vee_{w(\mu)} 1 = \sum_{p \in \Gamma(1, \vec{w}(\mu))} c_p X^{e(p)} 1. \tag{2.10}$$

3. SSV polynomials

3.1. (Nonsymmetric) SSV polynomials and intertwiners. We now briefly recall the *quasi-polynomial representation* and the SSV (quasi-)polynomials, due to [**SSV**, **SSV2**], and some related notation. We then give some basic results for these quasi-polynomials.

First, recalling the notation $E = \mathbb{R}^r$, we define the space of *quasi-polynomials* to be

$$\mathbb{F}[E] = \mathrm{span}_{\mathbb{F}}\{x^\mu : \mu \in E\}.$$

For any $s \in \mathbb{R}$, we define $x^{s\delta} = q^s$, so that $x^{\widehat{\mu}}$ is well-defined for any $\widehat{\mu} \in E_a$.

We now present the quasi-polynomial representation. This is a special case of [**SSV2**] Theorem 8.19. Restricted to $\mathbb{F}\left[\frac{1}{n}P\right]$ for any positive integer n, this is equivalent to Theorem 5.4 of [**SSV**], which was restated using terminology more similar to our present notation in [**Sai**] Theorem 2.16.

THEOREM 3.1. *[[**SSV2**], Theorem 8.19] The following formulas define a representation π of $\mathbb{H}$ on $\mathbb{F}[E]$:*

$$\begin{aligned}
\pi(T_i)x^\lambda &= (k - k^{-1}) \frac{1 - x^{-\lfloor (\lambda, \alpha_i) \rfloor \alpha_i}}{1 - x^{\alpha_i}} x^\lambda + G_{(\lambda, \alpha_i)} x^{s_i * \lambda} \\
\pi(X^\mu)x^\lambda &= x^{\mu + \lambda}, \\
\pi(\omega)x^\lambda &= x^{\omega * \lambda} = q^{-\lambda_r} x^{s_1 \cdots s_{r-1} \lambda}, \\
\pi(q)x^\lambda &= q x^\lambda
\end{aligned}$$

where $0 \le i \le r-1$, $\lambda \in E$, *and* $\mu \in P$.

The restriction of π to $\mathbb{F}[P]$ is the polynomial representation (over $\mathbb{F}$). In what follows, we will often omit the π notation, simply writing $\pi(h)f(x) = hf(x)$, where $h \in \mathbb{H}$ and $f(x) \in \mathbb{F}[P]$.

Define $\sigma : \mathbb{Z} \to \mathbb{F}$ by

$$\sigma(a) = \begin{cases} k^{-1} & : a \in \mathbb{Z}_{>0} \\ G_a & : a \in \mathbb{R} \setminus \mathbb{Z}_{>0} \end{cases}$$

for $a \in \mathbb{Z}$. For $\widehat{\mu} = \mu + s\delta \in P_a$ and $\lambda \in E$, define the scalars

$$\gamma(\widehat{\mu}; \lambda) = q^{-\langle \widehat{\mu}, \lambda \rangle} \prod_{\alpha \in \Phi_+} \sigma((\lambda, \alpha))^{(\mu, \alpha)}. \tag{3.1}$$

Letting $\rho = \frac{1}{2}\sum_{\alpha\in\Phi_+}\alpha$, when $\lambda = 0$ this simplifies to

$$\gamma(\widehat{\mu};0) = q^{-s}k^{(\mu,2\rho)}, \tag{3.2}$$

as appearing in (2.9).

The following theorem, constructing the SSV polynomials, was originally stated in [**SSV**] Theorem 5.7 and is a special case of [**SSV2**] Theorem 6.2.

THEOREM 3.2 ([**SSV2**] Theorem 6.2). *There exists a unique family of quasi-polynomials $\{E_\mu \in \mathbb{F}[P] : \mu \in E\}$ such that for each $\mu \in E$, the coefficient of x^μ in E_μ is 1 and, for all $\widehat{\nu} \in P_a$,*

$$\pi(Y^{\widehat{\nu}})E_\mu = \gamma(\widehat{\nu};\mu)E_\mu. \tag{3.3}$$

Call the family $\{E_\mu : \mu \in P\}$ the (nonsymmetric) *SSV polynomials.* (Note that these are actually *quasi-polynomials.*) We will sometimes write

$$E_\mu = E_\mu(x;q,k)$$

to emphasize the dependence on the parameters q and k. This notation does not conflict with the standard notation for Macdonald polynomials: in fact, for $\nu \in P$, the SSV polynomial E_μ is precisely equal to the corresponding Macdonald polynomial. For further examples, we refer the reader to the appendices of [**Sai**], where we give explicit alcove walk formulas for $E_{\mu/n}$ for several small values of μ and n.

Another prominent case is the following, proven in [**Sai**] Proposition 3.2 (see also [**SSV2**] Lemma 6.5):

PROPOSITION 3.3. *Let $\lambda \in \omega^s\mathcal{A}_W$, where $s \in \mathbb{Z}$. Then $E_\lambda = x^\lambda$.*

More generally, SSV polynomials satisfy the following triangularity result. The triangularity is originally due to Sahi, Stokman, and Venkateswaran and was proven in [**SSV2**] Theorem 6.2, and the fact that the scalars c_ν are nonzero is due to [**Sai**] Corollary 3.14.

COROLLARY 3.4. *For $\mu \in E$,*

$$E_\mu = x^\mu + \sum_{\nu<\mu} c_\nu x^\nu$$

for some scalars $c_\nu \in \mathbb{F}$, where $<$ is a partial ordering on E induced from the Bruhat order on W_{Cox}. In particular, the family $\{E_\mu : \mu \in E\}$ is a basis for $\mathbb{F}[E]$ over $\mathbb{F}$. Further, the scalars c_ν for $\nu < \mu$ are all nonzero, so $\operatorname{supp}(E_\mu) = \{\nu \in E : \nu \leq \mu\}$.

REMARK 3.5. We briefly discuss the partial order used above, considered in more detail in [**Sai**, Section 3] and [**SSV2**, Section 5.4]. (For notation more similar to the present work, see [**Sai2**, Section 4.1].) For $\mu \in E$, let $m'(\mu)$ be the unique W_{Cox}-conjugate of μ in $\overline{\mathcal{A}} = \bigcup_{s\in\mathbb{Z}}\omega^s\mathcal{A}_W$, and let $w'(\mu)$ be the unique shortest element such that $w'(\mu)m'(\mu) = \mu$. (Note these are essentially the same as the notions $m(\mu), w(\mu)$ introduced in Section 2.1, but considering only conjugacy up to elements of W_{Cox} and accordingly expanding the fundamental alcove.) Then for $\nu \in E$, $\nu \leq \mu$ if $\nu = vm'(\mu)$, where $v \in W_{\text{Cox}}$ satisfies $v \leq w'(\mu)$ with respect to the Bruhat ordering on W_{Cox}. As a consequence, since elements of W_{Cox} have finite length, we see that $\operatorname{supp}(E_\mu)$ is always a finite set.

The SSV polynomials have the following relationship with the intertwiners. Equation (3.5) follows from [**Sai**] Proposition 3.5 and direct calculation, and (3.4) is similar (see [**Sai2**] Proposition 3.9 for a proof sketch).

PROPOSITION 3.6. *Let $\mu \in E$. Then*

$$\pi(\tau_\omega^\vee)E_\mu \sim E_{\omega\mu} \tag{3.4}$$

and

$$\pi(\tau_i^\vee)E_\mu \sim \begin{cases} E_{s_i\mu} & s_i\mu \neq \mu \\ 0 & s_i\mu = \mu, \end{cases} \tag{3.5}$$

where $f \sim g$ means that f and g are nonzero scalar multiples of one another.

Note that, although $\tau_i^\vee \notin \mathbb{H}$, the representation π extends to the appropriate localization of $\mathbb{H}$, and $\pi(\tau_w^\vee)$ is well-defined for arbitrary $w \in W$. (See [**Sai**] Proposition 3.4 for details.)

We have the following corollary of (3.5):

COROLLARY 3.7. *Let $w \in W$ and $\mu \in E$. If w is the shortest element in its left W_μ-coset, then*

$$\pi(\tau_w^\vee)E_\mu \sim E_{w\mu}.$$

Otherwise,

$$\pi(\tau_w^\vee)E_\mu = 0.$$

As a consequence, recalling the notation $\Gamma_\lambda(w, \vec{z})$ from Section 2.2, we have

LEMMA 3.8. *If $\lambda \in E$, $z, w \in W$, and $h \in \Gamma(w, \vec{z}) \setminus \Gamma_\lambda(w, \vec{z})$ (in other words, $e(h)^{-1}$ is not the shortest element of its left W_λ-coset), then*

$$\tau_{e(h)^{-1}}^\vee E_\lambda = 0.$$

The next result follows from Proposition 3.3 and Corollary 3.7.

PROPOSITION 3.9. *Let $\mu \in E$.*

$$\pi(\tau_{w(\mu)}^\vee)x^{m(\mu)} = \pi(\tau_{w(\mu)}^\vee)E_{m(\mu)} \sim E_\mu,$$

where we recall the notation $m(\mu) \in \mathcal{A}_W$ and $w(\mu) \in W$ from Section 2.1.

This motivates the following normalization of the SSV polynomials. For $\mu \in E$, define

$$\widetilde{E}_\mu = \pi(\tau_{w(\mu)}^\vee)x^{m(\mu)} \sim E_\mu. \tag{3.6}$$

We now record the generalization of a standard result for Macdonald polynomials (see [**Sai2**] Lemma 4.7):

LEMMA 3.10. *Let $\mu \in E$ and $u \in W_\mu$. Then $T_u E_\mu = k^{\ell(u)}E_\mu$.*

PROOF. Suppose that, for $\mu \in E$ and $i \neq 0$, $s_i\mu = \mu$. We prove that $T_iE_\mu = kE_\mu$. The result then follows by induction on $\ell(u)$. Since the SSV polynomials are a basis for the space $F[E]$ of quasi-polynomials, by Corollary 3.4, we may write $T_iE_\mu = \sum_\nu d_\nu E_\nu$ for some scalars $d_\nu \in \mathbb{F}$. We will show that $d_\nu = 0$ for $\nu \neq \mu$. Once this is proven, the fact that $d_\mu = k$ is clear from the explicit formulas of Theorem 3.1. By (2.5), for all $\widehat{\lambda} \in P_a$,

$$\left(T_iY^{\widehat{\lambda}} - Y^{s_i * \widehat{\lambda}}T_i\right)E_\mu = (k - k^{-1})\left(\frac{Y^{\widehat{\lambda}} - Y^{s_i*\widehat{\lambda}}}{1 - Y^{-\alpha_i}}\right)E_\mu.$$

Applying (3.3), we have

$$\sum_{\nu}\left(\gamma(\widehat{\lambda};\mu)d_\nu - \gamma(s_i*\widehat{\lambda};\nu)d_\nu\right)E_\nu = (k-k^{-1})\left(\frac{\gamma(\widehat{\lambda};\mu)-\gamma(s_i*\widehat{\lambda};\mu)}{1-\gamma(-\alpha_i;\mu)}\right)E_\mu.$$

In particular, for $\nu \neq \mu$ and all $\lambda \in P_a$,

$$\gamma(\widehat{\lambda};\mu)d_\nu = \gamma(s_i*\widehat{\lambda};\nu)d_\nu.$$

Suppose for the sake of contradiction that there exists $\nu \neq \mu$ such that $d_\nu \neq 0$. Then $\gamma(\widehat{\lambda};\mu) = \gamma(s_i*\widehat{\lambda};\nu)$. Considering $\widehat{\lambda} = \varepsilon_1, \ldots, \varepsilon_r$, the power of q in (3.1) implies that this is possible only if $\nu = s_i\mu = \mu$, which contradicts our assumption. □

3.2. Symmetric SSV polynomials and symmetrizers. We now recall the symmetrized version of the SSV polynomials, due to [**SSV2**]. The basic theory in this setting is given in detail in [**Sai, Sai2**]. We prove some fundamental results for these quasi-polynomials, generalizing results of [**Yi**] that will be needed in Section 4. In particular, we give an explicit expansion of $\widetilde{P}_\lambda$ as a linear combination of nonsymmetric SSV polynomials $\widetilde{E}_\mu$ and record the extra scalar factor that arises when one applies the symmetrizer U to E_μ, $\mu \notin E_+$.

Define the Hecke symmetrizer

$$U_\lambda = k^{-\ell(w_\lambda)}\sum_{w\in W_\lambda} k^{\ell(w)}T_w.$$

We will use the notation $U = U_0$. For convenience, we will also denote $\pi(U) = U$. For $\mu \in E$, define the *symmetric SSV polynomial*

$$\widetilde{P}_\mu = \frac{1}{\overline{W}_{\mu_+}(k)}U\widetilde{E}_\mu, \tag{3.7}$$

where, for $\lambda \in E_+$,

$$\overline{W}_\lambda(k) = k^{-\ell(w_\lambda)}\sum_{u\in W_\lambda} k^{2\ell(u)}. \tag{3.8}$$

For $\mu \in E$, we also define P_μ to be the normalization of $\widetilde{P}_\mu$ such that x^{μ_+} has coefficient 1. This is well-defined by [**Sai**] Corollaries 4.7 and 3.14. However, when studying alcove walks, it is often more natural to work directly with $\widetilde{P}_\mu$.

By [**Sai**] Proposition 4.4, $\widetilde{P}_{w\mu} \sim \widetilde{P}_\mu$ for $w \in W_0$, so it suffices to consider the case where μ is dominant.

The following lemma is well-known and can be found in [**M**] (5.5.9).

LEMMA 3.11. *For $1 \leq i \leq r-1$,*

$$(T_i - k)U = U(T_i - k) = 0.$$

We call a quasi-polynomial $f(x) \in \mathbb{F}[P]$ *Hecke-symmetric* if $T_i f(x) = kf(x)$ for all $1 \leq i \leq r-1$. By Lemma 3.11, it is easy to see that $f(x)$ is Hecke-symmetric if and only if $f(x) = Ug(x)$ for some quasi-polynomial $g(x)$. Note that Hecke-symmetry and standard symmetry are equivalent for polynomials, but they are *not* equivalent for arbitrary quasi-polynomials.

By Lemma 3.11, we have

PROPOSITION 3.12. *For $\mu \in E$, $\widetilde{P}_\mu$ is Hecke-symmetric.*

We now make a straightforward but important observation (see [**Sai2**] Lemma 3.23), allowing one to simplify the product of a monomial and a quasi-polynomial to a single quasi-polynomial. Here the representation π is explicitly included in the notation for clarity.

LEMMA 3.13. *For $z \in W$ and $f(x)$ a Hecke-symmetric quasi-polynomial,*

$$(\pi(X^z)1)f(x) = \pi(X^z)f(x).$$

In particular, this holds for $f(x) = Ug(x)$, where $g(x)$ is an arbitrary quasi-polynomial. If z is a translation, then the assumption that $f(x)$ is Hecke-symmetric is not required.

PROOF. Write $z = \tau(\mathrm{wt}(z))\phi(z)$. We have

$$(\pi(X^z)1)f(x) = (\pi(X^{\tau(\mathrm{wt}(z))}T_{\phi(z)})1)f(x) = k^{\ell(\phi(z))}x^{\mathrm{wt}(z)}f(x).$$

Since $f(x)$ is Hecke-symmetric, we have $\pi(T_{\phi(z)})f(x) = k^{\ell(\phi(z))}f(x)$, and therefore the above becomes

$$x^{\mathrm{wt}(z)}\pi(T_{\phi(z)})f(x) = \pi(X^z)f(x).$$

If $z = \tau(\mathrm{wt}(z))$ is a translation, then there is no $T_{\phi(z)}$ factor and $f(x)$ is not required to be Hecke-symmetric. $\square$

The following result is a consequence of the proof of Proposition 3.6(a) of [**Yi**], as discussed in [**Y**]. Note that the only dependence on $\lambda \in E_+$ is in the stabilizer $W_\lambda \subseteq W_0$. These are in fact the same Coxeter subgroups that arise in the integral (Macdonald) case, and this result is no more general than the analogous statement in that setting. (See also Corollary 3.2.2 of [**Y**] for the analogue in the Koornwinder case.)

PROPOSITION 3.14. *For $\lambda \in E_+$,*

$$U_\lambda = \sum_{u \in W_\lambda} \tau_u^\vee b_{(u^{-1}, w_\lambda^{-1})}(Y),$$

where

$$b_{(w_1,w_2)}(Y^{-\alpha}) = \prod_{\alpha \in \mathcal{L}(w_1,w_2)} b(Y^{-\alpha}),$$

$\mathcal{L}(w_1, w_2)$ for $w_1, w_2 \in W_0$ is defined in (2.3), and

$$b(Y^{-\alpha}) = k\frac{1 - k^{-2}Y^{-\alpha}}{1 - Y^{-\alpha}}.$$

We also have the following lemma, which, for $\lambda \in P_+$, is used in the proof of Proposition 3.6(b) in [**Yi**] (see also the proof of Lemma 3.2.3 in [**Y**]). As with the previous result, the case $\lambda \in E_+$ is no more general than the result studied in those works. We omit the proof of this lemma, which is a consequence of Proposition 3.14 and a calculation in the finite root system Φ ultimately following from the observation that, if $u, v \in W_0$ with $\ell(uv) = \ell(u) + \ell(v)$, then $\mathcal{L}(uv) = \mathcal{L}(u) \cup u\mathcal{L}(v)$.

LEMMA 3.15. *For $\lambda \in E_+$,*

$$U = \left(\sum_{v \in W^\lambda} \tau_v^\vee b_{(v^{-1}, v_\lambda^{-1})}(Y)\right) U_\lambda.$$

Following the proof of Proposition 3.6(b) in [**Yi**], we have the following result, explicitly giving the expansion of a symmetrized SSV polynomial in terms of nonsymmetric SSV polynomials:

PROPOSITION 3.16. *Let* $\lambda \in E_+$.

$$\widetilde{P}_\lambda = \sum_{v \in W^\lambda} \rho_\lambda(v) \widetilde{E}_{v\lambda},$$

where

$$\rho_\lambda(v) = \prod_{\alpha \in \mathcal{L}(v^{-1}, v_\lambda^{-1})} \rho_\lambda(\alpha)$$

and

$$\rho_\lambda(\alpha) = k \frac{1 - k^{-2}\gamma(-w(\lambda)^{-1}\alpha; m(\lambda))}{1 - \gamma(-w(\lambda)^{-1}\alpha; m(\lambda))} = k \frac{1 - k^{-2}\gamma(-\alpha; \lambda)}{1 - \gamma(-\alpha; \lambda)}.$$

PROOF. By (3.7) and Lemma 3.15,

$$\begin{aligned}
\overline{W}_\lambda(k)\widetilde{P}_\lambda &= U\widetilde{E}_\lambda \\
&= \left(\sum_{v \in W^\lambda} \tau_v^\vee b_{(v^{-1}, v_\lambda^{-1})}(Y) \right) U_\lambda \widetilde{E}_\lambda \\
&= \left(\sum_{v \in W^\lambda} \tau_v^\vee b_{(v^{-1}, v_\lambda^{-1})}(Y) \right) \left(k^{-\ell(w_\lambda)} \sum_{w \in W_\lambda} k^{\ell(w)} T_w \right) \widetilde{E}_\lambda.
\end{aligned}$$

By Lemma 3.10 and (3.8), this reduces to

$$\begin{aligned}
&\left(\sum_{v \in W^\lambda} \tau_v^\vee b_{(v^{-1}, v_\lambda^{-1})}(Y) \right) \left(k^{-\ell(w_\lambda)} \sum_{w \in W_\lambda} k^{2\ell(w)} \right) \widetilde{E}_\lambda \\
=&\overline{W}_\lambda(k) \left(\sum_{v \in W^\lambda} \tau_v^\vee b_{(v^{-1}, v_\lambda^{-1})}(Y) \right) \widetilde{E}_\lambda.
\end{aligned}$$

Applying (3.3) and (3.6), we obtain the result. □

We now observe the following analogue of Proposition 3.7 of [**Yi**]. The proof of (3.9) is the same as Yip's proof, replacing 1 with $x^{m(\lambda)}$ and using (3.3) rather than the eigenvalues in the integral setting. Equation (3.10) follows from (3.6), (3.9), and the fact that the intertwiners are independent of the choice of reduced expression.

LEMMA 3.17. *For* $v \in W_0$ *and* $\lambda \in E_+$,

$$U\tau_v^\vee \tau_{w(\lambda)}^\vee x^{m(\lambda)} = \left(\prod_{\alpha \in \mathcal{L}(v^{-1})} \rho_\lambda(\alpha) \right) U\tau_{w(\lambda)}^\vee x^{m(\lambda)} \tag{3.9}$$

and

$$U\widetilde{E}_{v\lambda} = \left(\prod_{\alpha \in \mathcal{L}(v^{-1})} \rho_\lambda(\alpha) \right) \widetilde{P}_\lambda. \tag{3.10}$$

4. Littlewood-Richardson rules

We now prove the Littlewood-Richardson rules for SSV polynomials. These formulas give the expansion of the product of a Macdonald polynomial and an SSV polynomial as a sum of SSV polynomials, indexed by alcove walks. Note that these formulas are given in terms of the normalizations $\widetilde{E}_\mu$ and $\widetilde{P}_\lambda$. They can easily be rewritten in terms of the monic versions E_μ and P_λ, as the scalar factors converting between the two conventions can be written down explicitly using [**Sai**] Theorems 3.11 and 4.6.

We begin by stating Yip's expansion formula for the product of a monomial and an intertwiner [**Yi**], which, in addition to Ram and Yip's alcove walk formula for Macdonald polynomials, is the key ingredient in the Littlewood-Richardson rules of [**Yi**] and the present paper. We then prove the Littlewood-Richardson rules for SSV polynomials, following the methods of [**Yi**]. We conclude with a remark explaining the aesthetic differences between these formulas and the formulas of [**Yi, Y**].

We now state the formula of Yip for the product of a monomial and an intertwiner.

THEOREM 4.1 ([**Yi**], Theorem 3.3). *Let $z, w \in W$, and write $z = s_{i_\ell} \dots s_{i_1} \omega^{-s}$, a reduced expression.*

$$X^z \tau_w^\vee = \sum_{h \in \Gamma(w^{-1}, \vec{z}^{-1})} (-1)^{|f_{\mathrm{asc}}(h)|} \tau_{e(h)^{-1}}^\vee g_h(Y) n_h(Y),$$

where

$$g_h(Y) = \left(\prod_{j \in f(h)} \frac{k^{-1} - k}{1 - Y^{-h_j^\vee}} \right) \left(\prod_{j \in \psi(h)} Y^{-h_j^\vee} \right)$$

$$n_h(Y) = \prod_{j \in \xi_{\mathrm{des}}(h)} \frac{1 - k^{-2} Y^{-h_j^\vee}}{1 - Y^{-h_j^\vee}} \frac{1 - k^2 Y^{-h_j^\vee}}{1 - Y^{-h_j^\vee}}.$$

Here $f(h)$, f_{asc}, $\xi_{\mathrm{des}}(h)$, and $h_j^\vee$ are as defined in Section 2.2 *(where the alcove walk was called p and the positive roots indexing the active hyperplanes were called $p_j^\vee$ rather than $h_j^\vee$), and $\psi(h)$ is as defined in Section* 2.6.

In [**Yi**], Yip applied this result and (2.10) to obtain expansion formulas for the products of Macdonald polynomials $E_\mu E_\nu$, $E_\mu P_\lambda$, and $P_\eta P_\lambda$, where $\mu, \nu \in P$, and $\lambda, \eta \in P_+$. Yamaguchi [**Y**] recently generalized these results to the Koornwinder setting. We now give the expansion formulas in the SSV polynomial setting. At each step, our arguments closely parallel those of [**Yi**] and [**Y**]. As stated above, the main ingredients are Theorem 4.1 and (2.10).

We begin by applying Theorem 4.1 to obtain an expansion for $X^z \widetilde{E}_\lambda$ in terms of SSV polynomials.

COROLLARY 4.2. *Let $z \in W$, $\lambda \in E$.*

$$X^z \widetilde{E}_\lambda = \sum_{p \in \Gamma_{m(\lambda)}(w(\lambda)^{-1}, \vec{z}^{-1})} (-1)^{|f_{\mathrm{asc}}(p)|} g_p(m(\lambda)) n_p(m(\lambda)) \widetilde{E}_{e(p)^{-1} m(\lambda)},$$

where

$$g_p(m(\lambda)) = \left(\prod_{j \in f(p)} \frac{k^{-1} - k}{1 - \gamma(-h_j^\vee; m(\lambda))} \right) \left(\prod_{j \in \psi(p)} \gamma(-h_j^\vee; m(\lambda)) \right)$$

and

$$n_p(m(\lambda)) = \prod_{j\in\psi_{des}(p)} \frac{1-k^{-1}\gamma(-h_j^\vee; m(\lambda))}{1-\gamma(-h_j^\vee; m(\lambda))}\frac{1-k\gamma(-h_j^\vee; m(\lambda))}{1-\gamma(-h_j^\vee; m(\lambda))}.$$

PROOF. First, we have

$$X^z\widetilde{E}_\lambda = X^z\tau^\vee_{w(\lambda)}x^{m(\lambda)}$$

by Proposition 3.9. Expand $X^z\tau^\vee_{w(\lambda)}$ according to Theorem 4.1 to obtain

$$X^z\widetilde{E}_\lambda = \sum_{p\in\Gamma(w(\lambda)^{-1},\vec{z}^{-1})} (-1)^{|f_{\mathrm{asc}}(p)|}\tau^\vee_{e(p)^{-1}}g_p(Y)n_p(Y)x^{m(\lambda)}.$$

By (3.3), $g_p(Y)n_p(Y)$ acts by the scalar $g_p(m(\lambda))n_p(m(\lambda))$ on $x^{m(\lambda)}$, allowing us to apply the intertwiners directly to $x^{m(\lambda)}$. By Lemma 3.8, all terms vanish except those with $p\in\Gamma_{m(\lambda)}(w(\lambda)^{-1},\vec{z}^{-1})$ (those where $e(p)^{-1}$ is the shortest element in its left $W_{m(\lambda)}$-coset). Finally, we apply (3.6) to rewrite $\tau^\vee_{e(p)^{-1}}x^{m(\lambda)} = \widetilde{E}_{e(p)^{-1}m(\lambda)}$, giving the result. □

In the special case $z=\tau(\mu)$, this gives a formula for the product $x^\mu\widetilde{E}_\lambda$, where $\mu\in P$ and $\lambda\in E$.

REMARK 4.3. Recall from Example 2.3 that if $\lambda\in P$, then $m(\lambda)=\mathbf{0}$. In this case, $W_{m(\lambda)} = W_0$, the entire finite Weyl group. Then the requirement that $e(p)^{-1}$ be the shortest element in its left $W_{m(\lambda)}$-coset translates to a requirement that the alcove walk p remains inside the closed dominant chamber. See the proof of Corollary 3.5 in [**Yi**], where the $\lambda\in P$ case is studied. For general $\lambda\in E$, such simplification is not possible: for example, if $m(\lambda)=(1/2,0)$, then $W_{m(\lambda)}$ is the trivial group and the condition that $p\in\Gamma_{m(\lambda)}(w(\lambda)^{-1},\vec{z}^{-1})$ leads to no further restrictions on the alcove walk.

We note that the coefficients $g_p(m(\lambda)), n_p(m(\lambda))$ arising above are no different from those arising in [**Yi**], except the eigenvalues (3.2) are replaced with the more general eigenvalues (3.1).

We immediately obtain the expansion formula for the product of a nonsymmetric Macdonald polynomial and a nonsymmetric SSV polynomial:

THEOREM 4.4. *Let $\mu\in P$ and $\lambda\in E$. We have*

$$\widetilde{E}_\mu\widetilde{E}_\lambda = \sum_{h\in\Gamma(1,\vec{w}(\mu))} c_h k^{\ell(\phi(h))} \sum_{p\in\Gamma_{m(\lambda)}(w(\lambda)^{-1},\vec{\mathrm{wt}}(h)^{-1})} (-1)^{|f_{\mathrm{asc}}(p)|}g_p(m(\lambda))\cdot$$
$$\cdot\, n_p(m(\lambda))\widetilde{E}_{e(p)^{-1}m(\lambda)},$$

recalling the notation c_h from (2.9).

PROOF. We first expand $\widetilde{E}_\mu$ according to the alcove walk formula (2.10):

$$\widetilde{E}_\mu\widetilde{E}_\lambda = \sum_{h\in\Gamma(1,\vec{w}(\mu))} c_h k^{\ell(\phi(h))}x^{\mathrm{wt}(h)}\widetilde{E}_\lambda.$$

We then expand $x^{\mathrm{wt}(h)}\widetilde{E}_\lambda = X^{\tau(\mathrm{wt}(h))}\widetilde{E}_\lambda$ (these are equal by Lemma 3.13) using Corollary 4.2, obtaining the result. □

We now expand the product $\widetilde{E}_\mu\widetilde{P}_\lambda$, where $\mu\in P$ and $\lambda\in E_+$.

THEOREM 4.5. *Let $\mu \in P$ and $\lambda \in E_+$. We have*

$$\widetilde{E}_\mu \widetilde{P}_\lambda = \sum_{v\in W^\lambda} \rho_\lambda(v) \sum_{h\in\Gamma(1,\vec{w}(\mu))} c_h \cdot \sum_{p\in\Gamma_{m(\lambda)}((vw(\lambda))^{-1},\vec{e(h)}^{-1})} (-1)^{|f_{\mathrm{asc}}(p)|} g_p(m(\lambda)) n_p(m(\lambda)) \widetilde{E}_{e(p)^{-1}m(\lambda)},$$

where $\rho_\lambda(v)$ is as in Proposition 3.16.

PROOF. By the alcove walk formula (2.10) and Proposition 3.16, we have

$$\widetilde{E}_\mu \widetilde{P}_\lambda = \left(\sum_{h\in\Gamma(1,\vec{w}(\mu))} c_h X^{e(h)} 1 \right) \left(\sum_{v\in W^\lambda} \rho_\lambda(v) \widetilde{E}_{v\lambda} \right).$$

By Lemma 3.13, since $\widetilde{P}_\lambda$ is Hecke-symmetric,

$$\widetilde{E}_\mu \widetilde{P}_\lambda = \left(\sum_{h\in\Gamma(1,\vec{w}(\mu))} c_h X^{e(h)} 1 \right) \sum_{v\in W^\lambda} \rho_\lambda(v) \widetilde{E}_{v\lambda} \tag{4.1}$$

$$= \sum_{v\in W^\lambda} \rho_\lambda(v) \sum_{h\in\Gamma(1,\vec{w}(\mu))} c_h X^{e(h)} \widetilde{E}_{v\lambda}. \tag{4.2}$$

We then apply Corollary 4.2 to obtain the result. □

REMARK 4.6. Note that in Theorem 4.4, the alcove walk p is determined by $\mathrm{wt}(h)$ rather than $e(h)$. One may also write down a version of Theorem 4.5 in which p depends on $\mathrm{wt}(h)$. However, the version above leads to an alcove walk formula that is easier to compute: one first calculates a reduced expression for $w(\mu)$, then the alcove walks h are obtained by omitting factors in that reduced expression, and the alcove walks p are obtained by reducing the expression for $e(h)^{-1}$ and omitting still more factors. (In particular, one does not need to calculate a reduced expression for each $\mathrm{wt}(h)$, only $e(h)$.) Unfortunately, one cannot rewrite Theorem 4.4 in such a form, because $\widetilde{E}_\lambda$ is not Hecke-symmetric and therefore the analogue of (4.2) does not hold. The same issue arises in [**Yi**]: see Theorem 4.2 and Remark 4.5 in that work.

We now apply the symmetrizer U to Theorem 4.5 and simplify using (3.10).

THEOREM 4.7. *Let $\mu \in P_+$ and $\lambda \in E_+$. Then*

$$\overline{W}_\mu(k) \widetilde{P}_\mu \widetilde{P}_\lambda = \sum_{v\in W^\lambda} \rho_\lambda(v) \sum_{h\in\Gamma(1,\vec{w}(\mu))} c_h \sum_{p\in\Gamma_{m(\lambda)}((vw(\lambda))^{-1},\vec{e(h)}^{-1})} (-1)^{|f_{\mathrm{asc}}(p)|} g_p(m(\lambda)) n_p(m(\lambda)) \cdot \left(\prod_{\alpha\in\mathcal{L}(u_\lambda(p)^{-1})} \rho_{(e(p)^{-1}m(\lambda))_+}(\alpha) \right) \widetilde{P}_{(e(p)^{-1}m(\lambda))_+},$$

where $u_\lambda(p) \in W_0$ is the shortest element so that

$$u_\lambda(p)(e(p)^{-1}m(\lambda))_+ = e(p)^{-1}m(\lambda).$$

PROOF. Consider the expansion for the product $\widetilde{E}_\mu \widetilde{P}_\lambda$ given in Theorem 4.5 and apply the symmetrizer U to both sides. By (3.7) and the fact that $\widetilde{P}_\lambda$ is Hecke-symmetric, the left-hand side is equal to $\overline{W}_\mu(k)\widetilde{P}_\mu \widetilde{P}_\lambda$. The right-hand side is equal to

$$\sum_{v\in W^\lambda} \rho_\lambda(v) \sum_{h\in\Gamma(1,\vec{w}(\mu))} c_h \sum_{p\in\Gamma_{m(\lambda)}((vw(\lambda))^{-1},e(\vec{h})^{-1})} (-1)^{|f_{\rm asc}(p)|} g_p(m(\lambda))n_p(m(\lambda))\cdot$$
$$\cdot U\widetilde{E}_{e(p)^{-1}m(\lambda)}.$$

Applying (3.10) gives the result. □

REMARK 4.8. Yip and Yamaguchi [**Yi**, **Y**] write the analogues of Theorems 4.5 and 4.7 using slightly different terminology, that of colored alcove walks. Note that since $p \in \Gamma_{m(\lambda)}((vw(\lambda))^{-1}, e(\vec{h})^{-1})$, it is obtained by reversing the alcove walk h, translating its beginning to $(vw(\lambda))^{-1}$, and introducing more folds with certain restrictions. Thus we may view h and p as both arising from a single "colored" alcove walk with two types of folds, black (corresponding to the folds of h) and gray (corresponding to the folds of p). The results of [**Yi**, **Y**] are given as sums over such colored alcove walks. However, these sums introduce more (cancelling) terms than in the results above, because the set of colored alcove walks of the appropriate type is not in bijection with the set of pairs (h,p) as above. We see that p has type $e(\vec{h})^{-1}$ and is therefore determined only by the *endpoint* of h. In particular, there are $2^{\ell(e(h))}$ possible alcove walks p before taking into account the $\Gamma_{m(\lambda)}$ restriction. On the other hand, the colored alcove walk perspective allows one to introduce gray folds at *any* of the unfolded steps of h, giving as many as $2^{\text{number of steps of } h}$ possible alcove walks p before taking into account the $\Gamma_{m(\lambda)}$ restriction. There may be a significant difference between these two numbers: for example, h could be an alcove walk with many steps but with $e(h) = 1$, in which case there would be a unique possible p above (a trivial alcove walk with 0 steps). (This is seen in the example in Section 5 below, in which h_1 is an alcove walk with 2 unfolded steps and $e(h) = s_0s_0 = 1$.) To see that the extra terms introduced in [**Yi**, **Y**] cancel, one must verify that Theorem 4.1 (Theorem 3.3 in [**Yi**], analogous to Proposition 3.1.3 in [**Y**]) holds even if the expression given for z is not reduced. We choose not to introduce this additional complication, instead writing the formulas in this more efficient, if perhaps less aesthetically pleasing, form.

REMARK 4.9. We recall that the above Littlewood-Richardson rules give expansions for the product of an SSV polynomial with a Macdonald polynomial. A natural question, of course, is how one can expand the product of two general SSV polynomials. This seems to be significantly more difficult, with the difficulty arising immediately in attempting to generalize Corollary 4.2 to products $x^\mu E_\lambda$ where $\mu, \lambda \in E$. In the version above, where we restrict to $\mu \in P$, we consider the corresponding element X^μ of the DAHA, apply Theorem 4.1 to obtain a nice expansion of $X^\mu \tau_{w(\lambda)}$ in terms of DAHA elements, then simply apply that expansion to the monomial $x^{m(\lambda)}$. If $\mu \notin P$, then there is no corresponding element X^μ in the DAHA, the relations of Definition 2.7 are not satisfied for μ, and there is no clear analogue of Theorem 4.1. Thus to generalize the Littlewood-Richardson rule to a product of arbitrary SSV polynomials, we expect that one needs to use techniques fundamentally different from those of [**Yi**, **Y**] and the present work.

We note that Lenart, Naito, Nomoto, and Sagaki [**LNNS**] have recently given a new combinatorial formula for Macdonald polynomials in terms of a generalization of Lakshmibai-Seshadri paths. They conjecture that their formula naturally leads to a simpler Littlewood-Richardson rule than that of [**Yi**]. It would be of great interest to investigate whether their alternative techniques also extend to Littlewood-Richardson rules for products of Macdonald and SSV polynomials, or even products of two arbitrary SSV polynomials.

5. Example

We conclude with a brief example illustrating the calculation of some of the quantities involved in Theorem 4.7 above. We will take $r = 2$, $\mu = (4, 0)$, and $\lambda = (\zeta, 0)$, where $0 < \zeta < 1$. We calculate some terms in the expansion of $\overline{W}_\mu(k)\widetilde{P}_\mu\widetilde{P}_\lambda$. We begin by observing:

$$\omega^4 s_0 s_1 s_0(0,0) = \mu, \text{ so } m(\mu) = (0,0), \quad w(\mu) = \omega^4 s_0 s_1 s_0,$$

and

$$\lambda \in \mathcal{A}_W, \text{ so } m(\lambda) = \lambda = (\zeta, 0), \quad w(\lambda) = 1.$$

Then the elements $h \in \Gamma(1, \vec{w}(\mu))$ correspond to subwords of $\omega^4 s_0 s_1 s_0$ with some of the s_i factors omitted. In particular, the possible values of $\omega^{-4}e(h)$ are as follows (where we multiply by ω^{-4} to avoid writing ω^4 before every term): $s_0 s_1 s_0$, corresponding to the unfolded walk; $s_1 s_0$, $s_0 s_1$, and $1 = s_0 s_0$, corresponding to the walks with 1 fold; s_0 and s_1, corresponding to the walks with 2 folds (where s_0 appears twice, corresponding to folding at steps $1, 2$ or $2, 3$ respectively); and 1, appearing this time as a walk with 3 folds.

Since λ has distinct entries, W_λ is the trivial group, and therefore $W^\lambda = \{1, s_1\} = W_0$. Similarly, the requirement that $e(p)^{-1}$ be the shortest element in its $W_{m(\lambda)}$-coset eliminates no terms, since $W_{m(\lambda)} = W_\lambda$ is the trivial group.

For the purposes of illustration, we consider the terms corresonding to the two alcove walks h with $\omega^{-4}e(h) = 1$, namely those corresponding to the subwords $\omega^4 s_0 s_0$ (with one fold) and ω^4 (with 3 folds). Call these h_1, h_2 respectively. Since p depends only on $e(h)$, the corresponding terms in Theorem 4.7 are the same other than the initial c_h factor. Since $e(h) = \omega^4$, we have $p \in \Gamma(v^{-1}, \omega^{-4}) = \Gamma((\omega^4 v)^{-1}, 1)$; in other words, the walks p have no steps, so

$$(-1)^{|f_{\text{asc}}(p)|} g_p(m(\lambda)) n_p(m(\lambda)) = 1$$

and $e(p)^{-1} = \omega^4 v$. Write $v = s_1^k$ $(k = 0, 1)$. Then

$$e(p)^{-1} m(\lambda) = \omega^4 s_1^k(\zeta, 0) = s_1^k \omega^4(\zeta, 0) = s_1^k(2+\zeta, 2) \in \{(2+\zeta, 2), (2, 2+\zeta)\}.$$

The $k = 0$ $(v = 1)$ case is dominant, with $u_\lambda(p) = 1$; the $k = 1$ $(v = s_1)$ case is antidominant, with $u_\lambda(p) = s_1$ the shortest Weyl group element such that $u_\lambda(p)(2, 2+\zeta) = (2+\zeta, 2)$ is dominant. Then the sum of the h_1, h_2 terms is

$$(c_{h_1} + c_{h_2})\left(\rho_\lambda(1) + \rho_\lambda(s_1)\rho_{(2+\zeta,2)}(\alpha_1)\right)\widetilde{P}_{(2+\zeta,2)}.$$

We now calculate the ρ factors using the formulas in Proposition 3.16. Since $W^\lambda = W_0$, we have $v_\lambda = s_1$, the longest element in W^λ. The scalar $\rho_\lambda(v)$ is a product over the hyperplanes separating the alcoves of v^{-1} and $v_\lambda^{-1} = s_1$; thus $\rho_\lambda(s_1) = 1$ and $\rho_\lambda(1) = \rho_\lambda(\alpha_1)$. Since $\langle \alpha_1, (2+\zeta, 2)\rangle = \langle \alpha_1, \lambda\rangle$, we see that

$$\gamma(-\alpha_1; \lambda) = \gamma(-\alpha_1; (2+\zeta, \zeta))$$

and thus $\rho_\lambda(\alpha_1) = \rho_{(2+\zeta,2)}(\alpha_1)$. Explicitly calculating

$$\gamma(-\alpha_1;\lambda) = q^\zeta \sigma(\zeta)^{-2} = q^\zeta G_\zeta^{-2},$$

the h_1, h_2 terms reduce to

$$2(c_{h_1} + c_{h_2})\left(k\frac{1-k^{-2}q^\zeta G_\zeta^{-2}}{1-q^\zeta G_\zeta^{-2}}\right)\widetilde{P}_{(2+\zeta,2)}.$$

Finally, we compute c_{h_1}, c_{h_2} using (2.9). For $i = 1,2,3$, we use (2.8) to calculate $\beta_i = (4-i)\delta - \alpha_1$, giving $\gamma(-\beta_i;0) = q^{4-i}k^2$. To use these formulas, we need to be able to label folds as positive or negative. This is particularly easy in the $r=2$ case we are considering here: an alcove walk whose W_{Cox} part starts with s_0 must have type $(s,0,1,0,1,\dots)$, and the first fold is always negative, the second is positive, and so on. (The opposite holds for walks starting with s_1.) Then h_1 has a negative fold at step 2; h_2 has a negative fold at step 1, a positive fold at step 2, and a negative fold at step 3. This gives

$$c_{h_1} = (k^{-1}-k)\frac{q^2k^2}{1-q^2k^2}$$

and

$$c_{h_2} = (k^{-1}-k)^3\frac{q^3k^2}{1-q^3k^2}\cdot\frac{1}{1-q^2k^2}\cdot\frac{qk^2}{1-qk^2}.$$

References

[BBBG] B. Brubaker, V. Buciumas, D. Bump, and H. P. A. Gustafsson, *Metaplectic Iwahori Whittaker functions and supersymmetric lattice models*, arXiv:2012.15778v2 (2021).

[C1] I. Cherednik, *Double affine Hecke algebras, Knizhnik-Zamolodchikov equations, and Macdonald's operators*, Internat. Math. Res. Notices **9** (1992), 171–180, DOI 10.1155/S1073792892000199. MR1185831

[C2] I. Cherednik, *Intertwining operators of double affine Hecke algebras*, Selecta Math. (N.S.) **3** (1997), no. 4, 459–495, DOI 10.1007/s000290050017. MR1613515

[CG1] G. Chinta and P. E. Gunnells, *Weyl group multiple Dirichlet series constructed from quadratic characters*, Invent. Math. **167** (2007), no. 2, 327–353, DOI 10.1007/s00222-006-0014-1. MR2270457

[CG2] G. Chinta and P. E. Gunnells, *Constructing Weyl group multiple Dirichlet series*, J. Amer. Math. Soc. **23** (2010), no. 1, 189–215, DOI 10.1090/S0894-0347-09-00641-9. MR2552251

[CO] G. Chinta and O. Offen, *A metaplectic Casselman-Shalika formula for* GL_r, Amer. J. Math. **135** (2013), no. 2, 403–441, DOI 10.1353/ajm.2013.0013. MR3038716

[G] U. Görtz, *Alcove walks and nearby cycles on affine flag manifolds*, J. Algebraic Combin. **26** (2007), no. 4, 415–430, DOI 10.1007/s10801-007-0063-6. MR2341858

[K] V. G. Kac, *Infinite-dimensional Lie algebras*, 3rd ed., Cambridge University Press, Cambridge, 1990, DOI 10.1017/CBO9780511626234. MR1104219

[KP] D. A. Kazhdan and S. J. Patterson, *Metaplectic forms*, Inst. Hautes Études Sci. Publ. Math. **59** (1984), 35–142. MR743816

[LNNS] C. Lenart, S. Naito, F. Nomoto, and D. Sagaki, *Symmetric and nonsymmetric Macdonald polynomials via a path model with a pseudo-crystal structure*, arXiv:2210.14464 (2022).

[Lu] G. Lusztig, *Affine Hecke algebras and their graded version*, J. Amer. Math. Soc. **2** (1989), no. 3, 599–635, DOI 10.2307/1990945. MR991016

[IS] B. Ion and S. Sahi, *Double affine Hecke algebras and congruence groups*, Mem. Amer. Math. Soc. **268** (2020), no. 1305, xi+90, DOI 10.1090/memo/1305. MR4219824

[M] I. G. Macdonald, *Affine Hecke algebras and orthogonal polynomials*, Cambridge Tracts in Mathematics, vol. 157, Cambridge University Press, Cambridge, 2003, DOI 10.1017/CBO9780511542824. MR1976581

[Mc] P. J. McNamara, *The metaplectic Casselman-Shalika formula*, Trans. Amer. Math. Soc. **368** (2016), no. 4, 2913–2937, DOI 10.1090/tran/6597. MR3449262

[OS] D. Orr and M. Shimozono, *Specializations of nonsymmetric Macdonald-Koornwinder polynomials*, J. Algebraic Combin. **47** (2018), no. 1, 91–127, DOI 10.1007/s10801-017-0770-6. MR3757150

[PP] M. Patnaik and A. Puskás, *On Iwahori-Whittaker functions for metaplectic groups*, Adv. Math. **313** (2017), 875–914, DOI 10.1016/j.aim.2017.04.005. MR3649240

[RY] A. Ram and M. Yip, *A combinatorial formula for Macdonald polynomials*, Adv. Math. **226** (2011), no. 1, 309–331, DOI 10.1016/j.aim.2010.06.022. MR2735761

[S] S. Sahi, *Binomial coefficients and Littlewood-Richardson coefficients for interpolation polynomials and Macdonald polynomials*, Representation theory and mathematical physics, Contemp. Math., vol. 557, Amer. Math. Soc., Providence, RI, 2011, pp. 359–369, DOI 10.1090/conm/557/11039. MR2848933

[SSV] S. Sahi, J. V. Stokman, and V. Venkateswaran, *Metaplectic representations of Hecke algebras, Weyl group actions, and associated polynomials*, Selecta Math. (N.S.) **27** (2021), no. 3, Paper No. 47, 42, DOI 10.1007/s00029-021-00654-1. MR4273644

[SSV2] S. Sahi, J. V. Stokman, and V. Venkateswaran, *Quasi-polynomial representations of double affine Hecke algebras*, arXiv:2204.13729v2 (2022).

[Sai] J. Saied, *A combinatorial formula for Sahi, Stokman, and Venkateswaran's generalization of Macdonald polynomials*. part B, Adv. Math. **404** (2022), no. part B, Paper No. 108440, 51, DOI 10.1016/j.aim.2022.108440. MR4420440

[Sai2] J. Saied, *A Combinatorial Formula and Littlewood-Richardson Rule for SSV Polynomials*, ProQuest LLC, Ann Arbor, MI, 2022. Thesis (Ph.D.)–Rutgers The State University of New Jersey, School of Graduate Studies. MR4463361

[Y] K. Yamaguchi, *A Littlewood-Richardson rule for Koornwinder polynomials*, J. Algebraic Combin. **56** (2022), no. 2, 335–381, DOI 10.1007/s10801-022-01114-5. MR4467958

[Yi] M. Yip, *A Littlewood-Richardson rule for Macdonald polynomials*, Math. Z. **272** (2012), no. 3-4, 1259–1290, DOI 10.1007/s00209-012-0986-z. MR2995168

Department of Mathematics, Rutgers University-New Brunswick, Piscataway, New Jersey 08854

Email address: `jjs435@scarletmail.rutgers.edu`

SELECTED PUBLISHED TITLES IN THIS SERIES

815 **Daniel Orr and Joshua Jeishing Wen, Editors,** Macdonald Theory and Beyond, 2025

814 **Susan Montgomery and Siu-Hung Ng, Editors,** Quantum Groups, Hopf Algebras, and Applications, 2025

813 **Thomas Creutzig, Julia Plavnik, and David Ridout, Editors,** Quantum Symmetries, 2025

812 **Ross Geoghegan, Craig R. Guilbault, and Kim Ruane, Editors,** Topology at Infinity of Discrete Groups, 2025

811 **Krishnendu Gongopadhyay and Shashikant A. Katre, Editors,** Geometry, Groups and Mathematical Philosophy, 2025

810 **Robert J. Berman and Yanir A. Rubinstein, Editors,** Convex and Complex: Perspectives on Positivity in Geometry, 2025

809 **Fabrice Baudoin and Luca Rizzi, Editors,** New Trends in Sub-Riemannian Geometry, 2025

808 **Loring W. Tu, Editor,** Group Actions and Equivariant Cohomology, 2024

807 **Galina Filipuk, Editor,** Recent Progress in Special Functions, 2024

806 **Maximilian Engel, Hildeberto Jardón-Kojakhmetov, and Cinzia Soresina, Editors,** Topics in Multiple Time Scale Dynamics, 2024

805 **Anthony Iarrobino, Pedro Macias Marques, Maria Evelina Rossi, and Jean Vallès, Editors,** Deformation of Artinian Algebras and Jordan Type, 2024

804 **Mahir Bilen Can and Jörg Feldvoss, Editors,** A Glimpse into Geometric Representation Theory, 2024

803 **Peter Gothen, Margarida Melo, and Montserrat Teixidor i Bigas, Editors,** Moduli Spaces and Vector Bundles—New Trends, 2024

802 **Ralph M. Kaufmann, Martin Markl, and Alexander A. Voronov, Editors,** Higher Structures in Topology, Geometry, and Physics, 2024

801 **K. A. Brown, T. J. Hodges, M. Vancliff, and J. J. Zhang, Editors,** Recent Advances in Noncommutative Algebra and Geometry, 2024

800 **Avinoam Mann, Louis H. Rowen, David J. Saltman, Aner Shalev, Lance W. Small, and Uzi Vishne, Editors,** Amitsur Centennial Symposium, 2024

799 **Alberto A. Condori, Elodie Pozzi, William T. Ross, and Alan A. Sola, Editors,** Recent Progress in Function Theory and Operator Theory, 2024

798 **Marat V. Markin, Igor V. Nikolaev, and Carsten Trunk, Editors,** Advances in Functional Analysis and Operator Theory, 2024

797 **Sangita Jha, Mrinal Kanti Roychowdhury, and Saurabh Verma, Editors,** Recent Developments in Fractal Geometry and Dynamical Systems, 2024

796 **John Cremona, John Jones, Jennifer Paulhus, Andrew V. Sutherland, and John Voight, Editors,** LuCaNT: LMFDB, Computation, and Number Theory, 2024

795 **Michael A. Jones, David McCune, and Jennifer M. Wilson, Editors,** Mathematical Analyses of Decisions, Voting and Games, 2024

794 **Jean-Pierre Magnot, Editor,** Recent Advances in Diffeologies and Their Applications, 2024

793 **Abba Gumel, Editor,** Mathematical and Computational Modeling of Phenomena Arising in Population Biology and Nonlinear Oscillations, 2024

792 **Shaoming Guo, Zane Kun Li, and Brian Street, Editors,** Recent Developments in Harmonic Analysis and its Applications, 2024

791 **Mee Seong Im, Bach Nguyen, and Arik Wilbert, Editors,** Algebraic and Topological Aspects of Representation Theory, 2024

For a complete list of titles in this series, visit the
AMS Bookstore at **www.ams.org/bookstore/conmseries/**.